STATISTICS with POWER using SPSS

First Edition

By Steven Davis and Evelyn R. Davis
California State University, Los Angeles

Bassim Hamadeh, CEO and Publisher
Christopher Foster, General Vice President
Michael Simpson, Vice President of Acquisitions
Jamie Giganti, Managing Editor
Kevin Fahey, Marketing Manager
Jess Busch, Senior Graphic Designer
Marissa Applegate, Acquisitions Editor
Luiz Ferreira, Licensing Associate

First published in the United States of America in 2013 by Cognella, Inc.

File licensed by www.depositphotos.com

Printed in the United States of America

ISBN: 978-1-62131-213-0 (perfect bound) / 978-1-62131-214-7 (binder ready)

www.cognella.com 800-200-3908

Contents

CHAPTER 1

Introduction to Types of Data

Objectives

After completing this chapter, a student should be able to:

- ✓ Define statistics
- ✓ Distinguish between a sample space and an event space
- ✓ Know the foundations of probability
- ✓ Classify Data into Nominal, Ordinal and Scale which includes interval and ratio
- ✓ Know the definition of and how to generate random numbers

Statistics

Statistics is the study of data collection, organization, analysis, and interpretation of numerical and categorical information. Modern statistics has its origins in the 17th century where games of chance were very much in vogue. Modern statistics arose out of probability, and probability was first discussed between Pascal and Fermat through a series of correspondence regarding solving different types of gambling problems.[1] The foundation of probability is to determine how many ways an event can occur divided by how many outcomes are possible. Gamblers were very interested in determining probability and the odds for winning games of chance. Problems 5 and 6 in Chapter 5 discuss what might have been the motivating problems behind the beginnings of the exploration of probability theory.

Outcome

An outcome is the result of an experiment such as tossing a coin, rolling a die, or having a child. All possible results from such experiments are called outcomes; the coin can be either heads or tails; the die can be any number between 1 and 6; and in the context of having a child, the outcomes are either a boy or girl. Later on we'll consider other outcomes such as the intelligence of the child, height of the child, and when the child first learns to read.

Sample Space and Event Space

The sample space consists of all possible outcomes in the context of an experiment. An event is the set of outcomes favorable for a desired result. For instance, the sample space of outcomes for the toss of a coin is a set that is equal to heads and tails, i.e., sample space = {heads, tails}. On the other hand, an event could be the set of only those outcomes that show heads, i.e., event = {heads}. In general an event is a subset of the sample space, see **Figure 1**.

Probability

The probability of an event is the number of ways the event can occur divided by the number of possible outcomes. In symbols we write $P(E)=\frac{\#(E)}{\#(O)}$, where $P(E)$ is the probability of the event, $\#(E)$ is the number of ways the event can occur, and $\#(O)$ is the total number of possible outcomes. The event space is a subset of the sample space, see **Figure 1**.

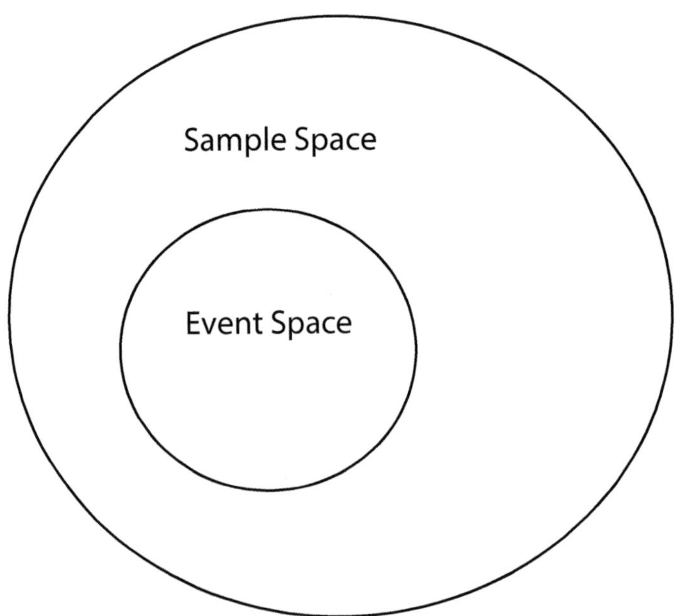

Figure 1

Example1.

What is the probability that a head will appear on a toss of a coin?

Solution: There is only one way for the event "head to appear." The coin must land with heads up. The number of outcomes for the toss of a coin is two: heads or tails. Thus the number of ways a head can appear divided by the number of outcomes for a toss of a coin is 1 divided by 2. The probability for a head is ½ or 50%.

Statistics Today

Probability is the foundation for statistics, but we will not spend time in this text discussing the mathematics of probability. Even though mathematics plays a major role in statistics, our emphasis in this text is to let the computer work the math. However we do expect the reader to have a firm grasp on the mathematics involved, so as not to be amazed by the computer. Therefore we do introduce formulas and demonstrate how to apply the formulas. Using the computer, however, will greatly facilitate our work. When I (Steven) first taught statistics in the early 1980s, I used formulas, combinatorics, and a calculator. Doing statistics was very tedious, time consuming, and prone to errors. Grading tests was very difficult because the students' answers were never the same; I had to go through their calculations to see where they made a mistake and if they really understood the concept. But those painful days are gone. We live in a new decade and a new century.

Today we have personal computers or laptops at our fingertips. This book is intended to be taught in a classroom where each student has access to a personal computer and with the statistical program called IBM® / SPSS®(a product of SPSS, Inc., an IBM Company)[2] downloaded onto the computer. Ideally, the instructor will have a computer attached to an overhead projection device to demonstrate the program. We expect students to be computer-literate by knowing the basics, such as how to use the mouse to move the cursor and being able to open and close files. We will teach statistics using SPSS at every step. The computer will give us a power that was not available 30 years ago, hence the title of this book, *Statistics with Power Using SPSS*. The cover of this text is a picture of Iguazu Falls located on the Igauazu River which forms part of the border between Brazil and Argentina. The purpose of the falls is to show that they represent power, which is a fundamental concept in this book.

In their line of work, modern statisticians use the computer with an appropriate statistical program. We chose to use the SPSS program because it is widely used and constantly being improved. It has a considerable amount of flexibility that makes it very user-friendly and bypasses a lot of computer terminology. The terminology in the program is the same as that used by statisticians in their field and that we'll use in this class; it is therefore not an added subject requirement to this class. The data editor in SPSS is similar to a Microsoft Excel spreadsheet, which is another software application with which students should be familiar. We don't expect today's student to have any difficulty learning statistics and learning the software at the same time because they do go hand in hand. Furthermore we'll itemize each step in the software application when we discuss a new procedure for the first time.

After finishing this book we feel that a student who has completed a suitable degree should be able to land an entry-level position doing basic statistical data analysis. Jobs of this nature exist in marketing, engineering, finance, risk management, pharmaceuticals, biostatistics, and econometrics, to name a few, in both the government and private sectors. The use of statistics is widespread and growing, as evidenced by statistical reports in newspapers and magazines. In fact, for a person to be well informed regarding current events, a fundamental understanding of statistics is necessary. Thus not only research analysts but also all persons will benefit from an introductory-level class in elementary statistics.

Actuaries are well-paid professionals who use the methods of statistics and probability to determine insurance rates. Being an actuary is said to be one of the most satisfying jobs available with regard to income and job satisfaction, although job stress is usually not an issue with actuaries. The classes necessary to enter the actuarial field include calculus, advanced probability, and, of course, statistics. This text provides the basic knowledge of statistics, which will give the student a first step toward completion of classes that can lead to an actuarial career.

Types of Data

There are two types of categories for data, **Qualitative Data** (name only, expresses some type of quality such as male, female, ethnicity, or full- or part-time student) and **Quantitative Data** (numerical value, expresses some number quantity). In the qualitative data category, we have the type of variable called **Nominal** (name only). In the quantitative category, we have the type of variable called **Scale** (a value that can take on any decimal number in an interval). Statisticians often further classify scale variables as <u>interval</u> variable or <u>ratio</u> variable, the difference being in a ratio variable, ratios are meaningful, whereas in an interval variable, ratios need not be meaningful. For example, temperature in Fahrenheit is an interval variable, not a ratio variable, because 10 degrees hotter is the same whether it is 10 degrees to 20 degrees or 90 degrees to 100 degrees, but we cannot say that 20 degrees is twice as hot as 10 degrees. IQ is also another measure that is an interval data value but not a ratio data value. For ratios to be meaningful, a ratio variable needs an absolute zero, which is a point that has no quantity of the units being measured. Zero degrees Fahrenheit still has temperature, although very cold, so it is not a starting point at which there is no temperature. The point at which there is absolutely no temperature is -459.67 degrees Fahrenheit and that is the starting point for the ratio variable called the Rankine scale.

We also have a variable type that can be classified as either qualitative or quantitative, which is the variable **Ordinal** (ranked values: first, second, third, etc.). Ordinal variables have an order but the order may or may not have any significance as to who's first, second, or third. For example, zip codes are ordinal data values because they can be put in order; however, the order has no significance because they represent geographical regions of the United States and geographical regions don't have order. On the other hand, place finishes in a race are ordinal data values and their position is significant. The first-place horse pays more than the second- or the other lower-placed finishers.

Our next example will show how we can input data into an SPSS data editor with the different measures Nominal, Ordinal, and Scale. First download the SPSS program into the computer you are working on if it is not already in the program files. To download (available onthehub), follow the directions given with the disc or the Internet instructions.

Example 2.

The following data represent the social structure of an organization. Input the data as nominal data (Gender), ordinal data (Position), and scale data (Income).

Gender	Position	Income
Female	President	$100,000
Male	Vice President	$50,000
Male	Manager	$40,000
Female	Worker 1	$35,000
Male	Worker 2	$30,000

Solution: Find the SPSS program in your computer and click on SPSS 20.0 (or 21.0) for Windows. It will first show an opening page (may take a little time) and then it should show a window that asks you what you would like to do. **See Figure 1a.**

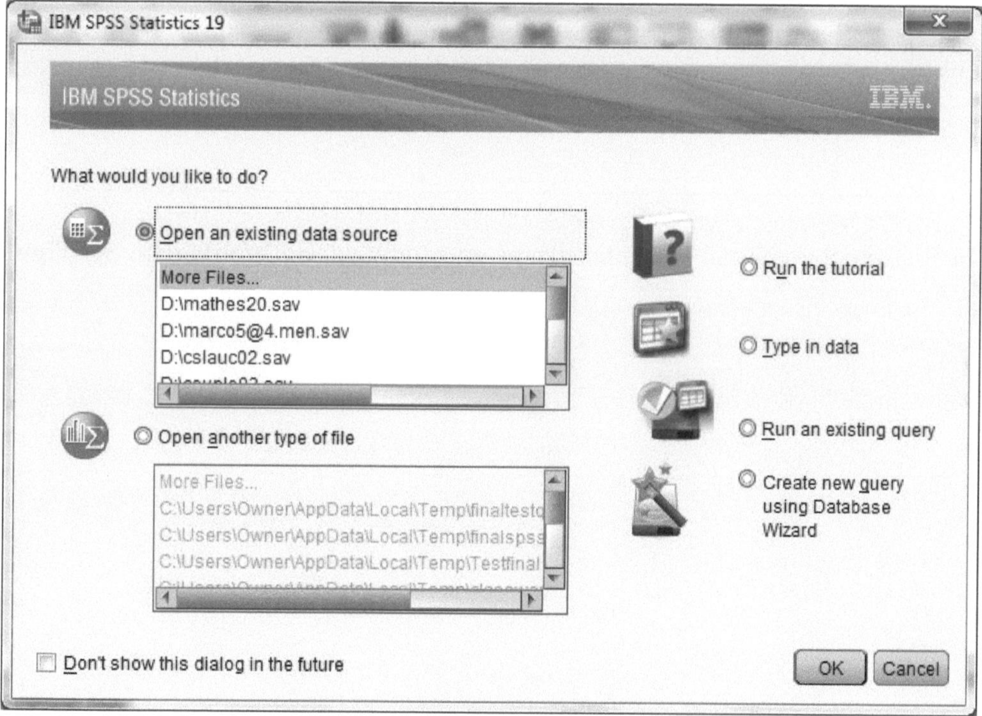

Figure 1a

Either click on Cancel or select Type in data and click on OK. Now you should have a blank data editor screen. It will say Untitled1 [DataSet0] – SPSS Data Editor. **See Figure 1b.**

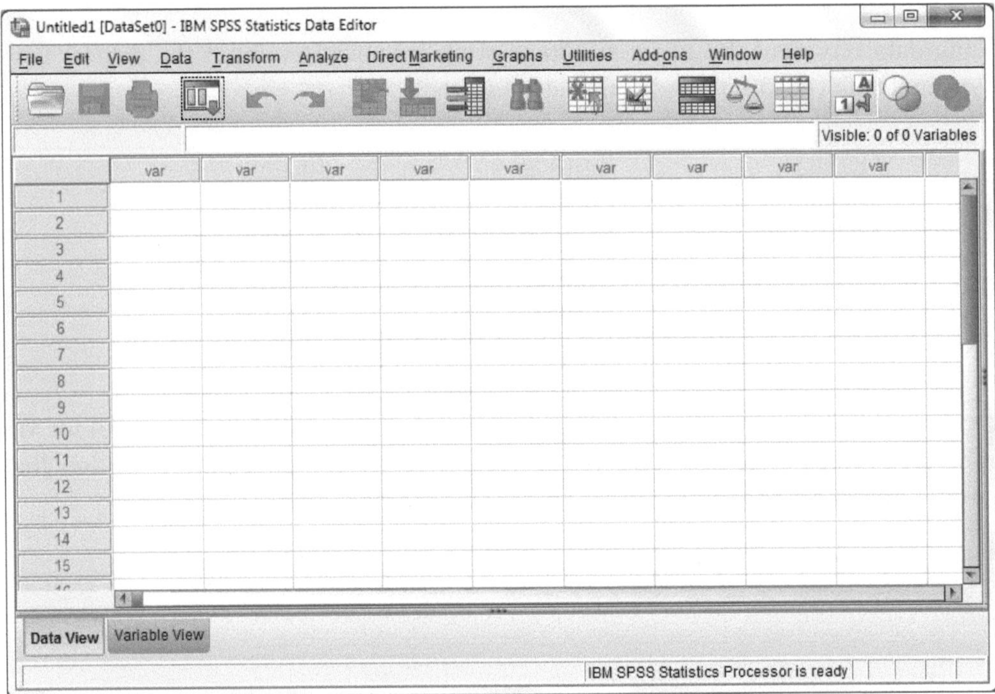

Figure 1b

Go to the first variable column and enter the values *1* for male and *2* for female. **See Figure 1c**.

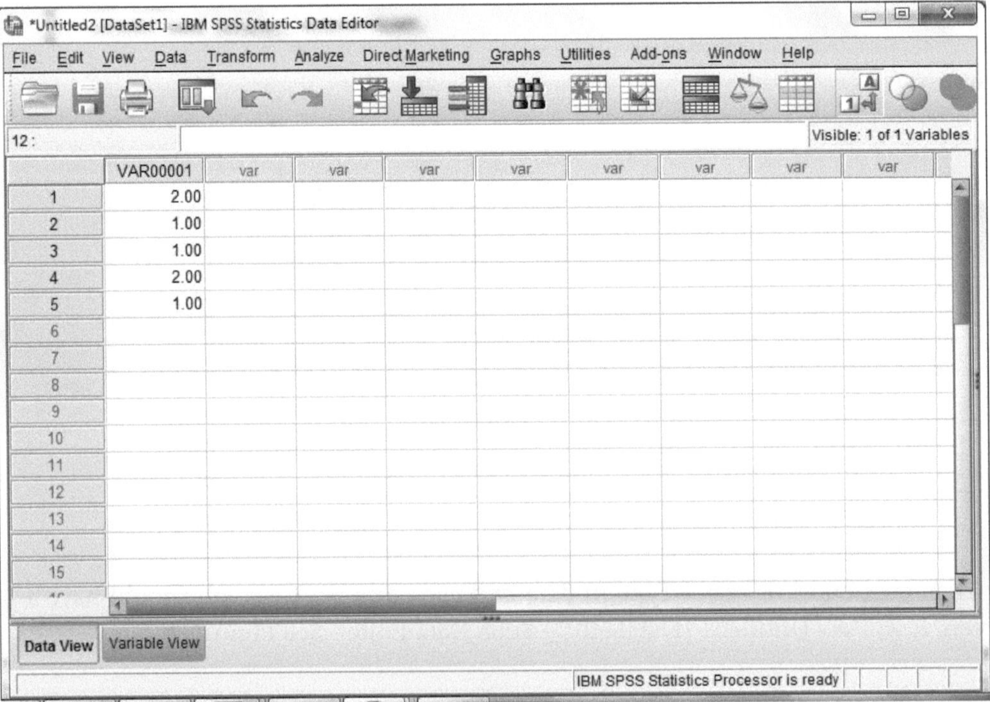

Figure 1c

Now go to the bottom and click on the tab Variable View. You will get a different screen. **See Figure 1d.**

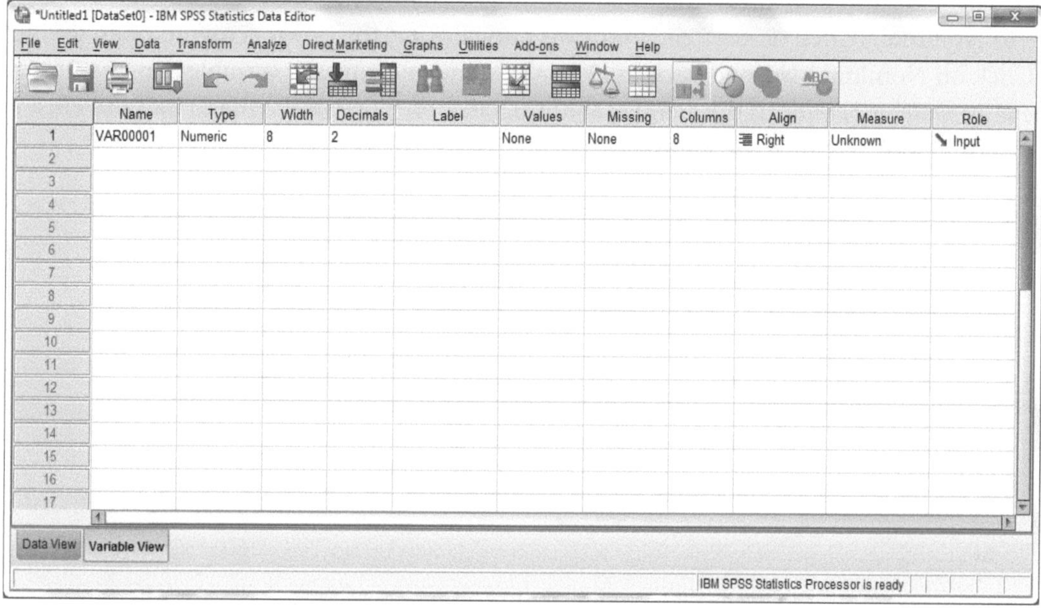

Figure 1d

This is the screen where we can name the type of variable as discussed above. First we will create the variable name Gender. Type in *Gender* where it says VAR00001. The next column asks for the type of variable. This is a nominal variable (also known as a string variable) but we will leave the type as numerical (SPSS assigns a variable as numeric with 2 decimals by default). However, we will change the Measure into Nominal later. For the time being we will leave the Width as 8 and the Decimals as 2. Next we will enter *Example 2 Gender* under Label. A label is a short phrase you can assign a variable so that the variable is easier to identify when dealing with many variables (such as 50 variables). The next column Values is where we will give our numerical values their actual names. Recall 1 was male and 2 was female. Click on the cell labeled None under Values and you will see three dots (ellipsis). Click on those and another window will open. **See Figure 1e.**

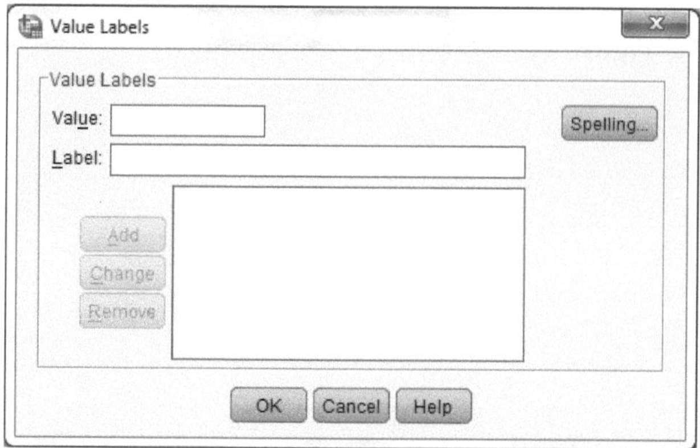

Figure 1e

Put *1* in the Value box and *Male* in the Label box, then click on ADD. Do the same with *2* and *Female*, then click OK. Skip the next three columns—Missing, Columns, Align—and change the last column Measure to Nominal. (Click on the cell Unknown and click on the down arrow to open the pull-down screen. Click on Nominal.) Now go back to Data View and see how the variable appears. If the data do not appear as *Male* or *Female* in the Gender column, click on View in the menu bar and click on Value Labels. It should look like this, **see Figure 1f**.

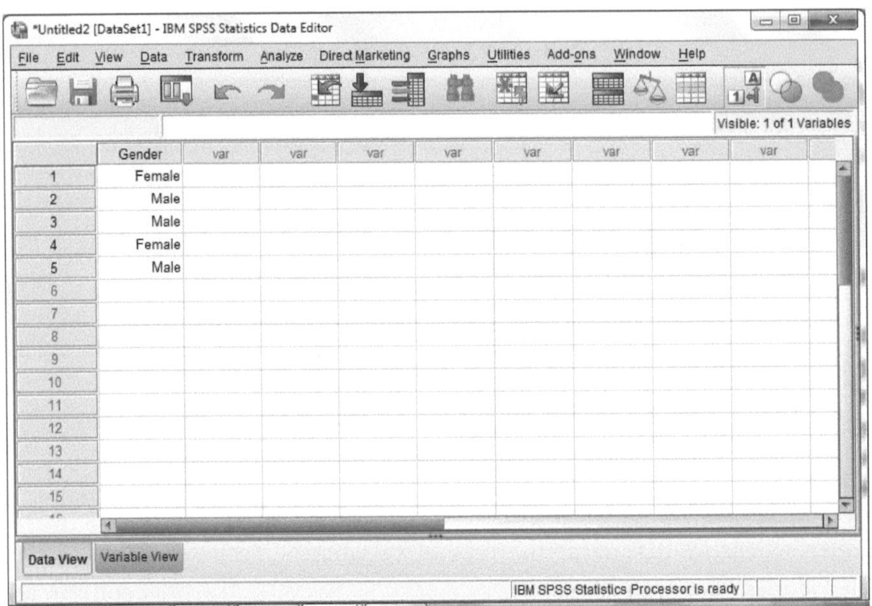

Figure 1f

You have successfully entered nominal data into a Nominal measure variable and called it *Gender*.

Our next column will be the ordinal data. In this example, we will rank the positions President, Vice President, Manager, Worker 1, and Worker 2 as 1, 2, 3, 4, and 5. Type in *1, 2, 3, 4, 5* in the next column in Data View and change the measure to Ordinal (click on Variable View). If you like, you can change the value labels to *President, Vice President, Manager, Worker 1,*and *Worker 2.* Also, you can change the variable name to *Position* and the variable label to *Example 2 Position.* Click on Data View and the Data Editor should now look like **Figure 1g.**

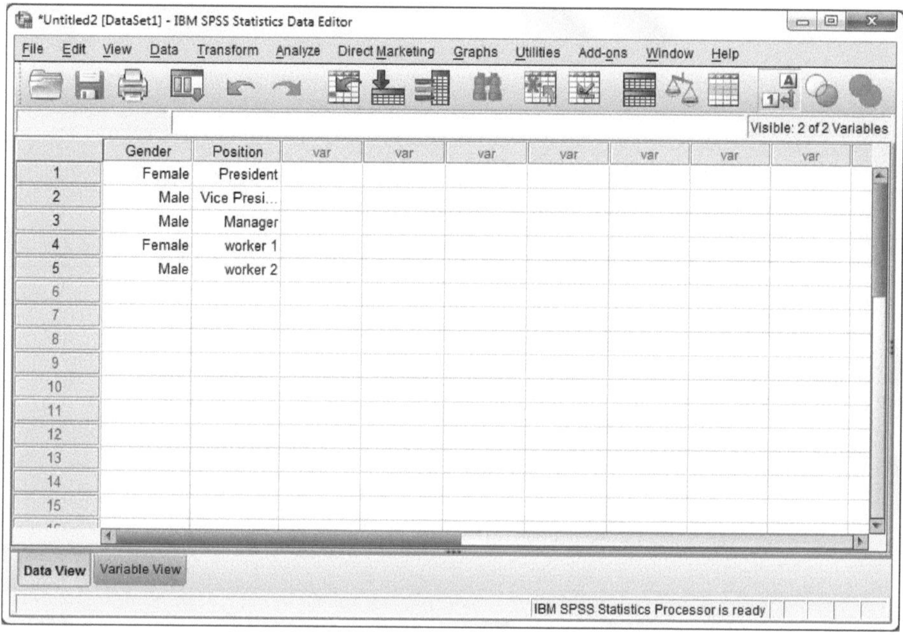

Figure 1g

The next column will be their salaries. Input the values *100000, 50000, 40000, 35000, 30000* and label the variable *Income*. Change the variable label to *Example 2 Income*. Change the Measure to Scale. The data view of your Data Editor should look like **Figure 1h**,

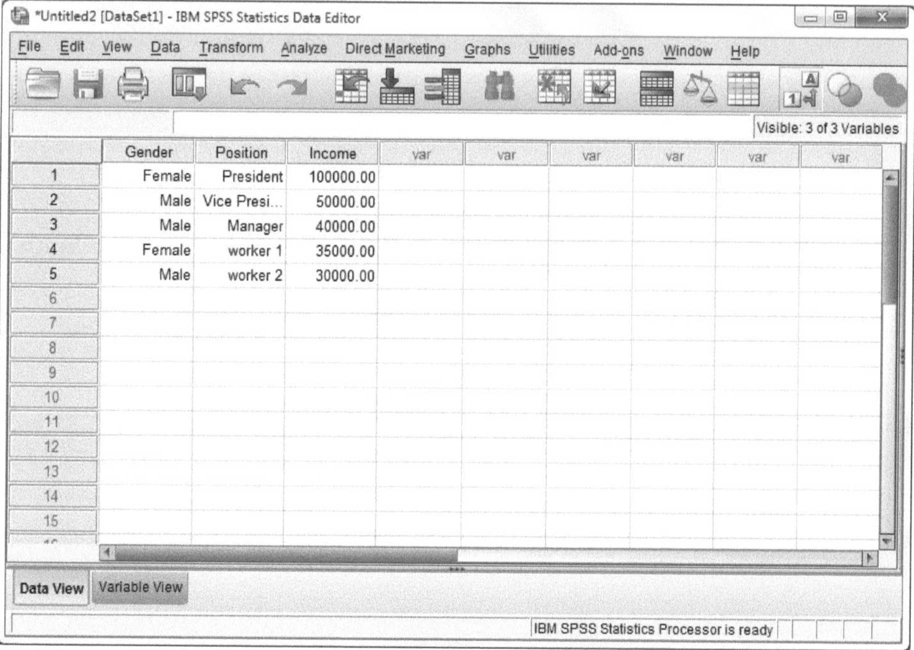

Figure 1h

and the variable view should look like **Figure 1i.**

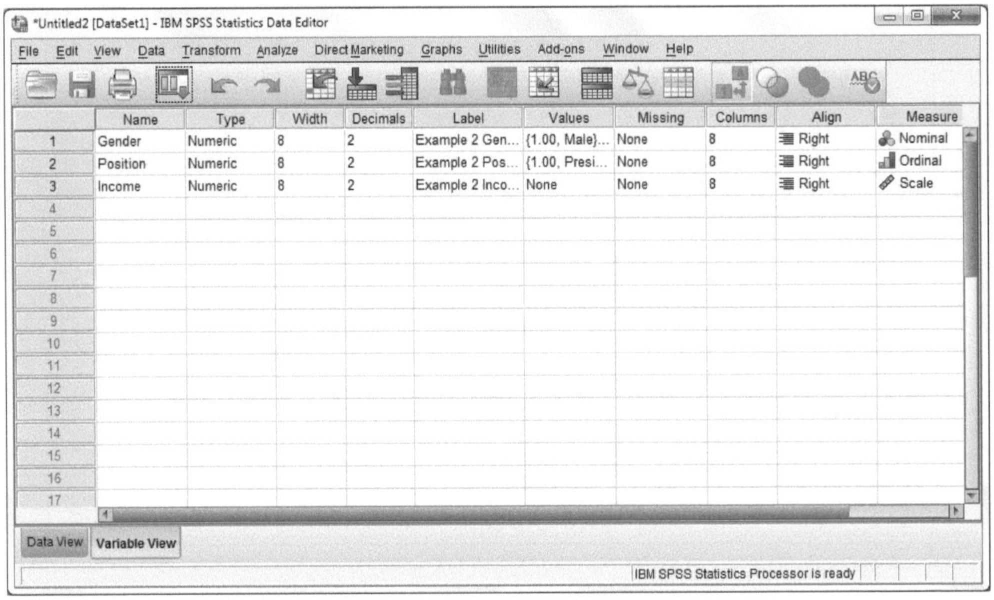

*Untitled2 [DataSet1] - IBM SPSS Statistics Data Editor

File Edit View Data Transform Analyze Direct Marketing Graphs Utilities Add-ons Window Help

	Name	Type	Width	Decimals	Label	Values	Missing	Columns	Align	Measure
1	Gender	Numeric	8	2	Example 2 Gen...	{1.00, Male}...	None	8	Right	Nominal
2	Position	Numeric	8	2	Example 2 Pos...	{1.00, Presi...	None	8	Right	Ordinal
3	Income	Numeric	8	2	Example 2 Inco...	None	None	8	Right	Scale

Data View Variable View

IBM SPSS Statistics Processor is ready

Figure 1i

Now save this data file in a folder labeled *YourNameClass* by clicking on File → Save and type in *StatClassChpt1Example2*. We will use this data file in our next chapter when we talk about mean, median, and mode. If you are doing this example on a school computer, then you need to bring a flash drive to save your work and examples because the administration usually does not like students saving work on their computers. We will assume you know how to create folders and save files.

Categories of Statistics

There are two major categories for Statistics, **Descriptive Statistics** and **Inferential Statistics.**

Descriptive Statistics

Descriptive statistics simply describes the behavior of a sample or a population by organizing, picturing, and summarizing collected data or information. Data organization can be done via a stem-and-leaf display; information can be pictured via a histogram or boxplot; and the mean, median, mode, standard deviation, variance, and range are examples of summarizing data in descriptive statistics. We will work with descriptive statistics in the next three chapters.

Inferential Statistics

The other major category is inferential statistics. We use inferential statistics to infer how a population will behave based on a random sample taken from the population. Inferential statistics involves methods of using information from a sample to draw conclusions regarding the population. The heart of inferential statistics is hypothesis testing and confidence intervals. We will work with inferential statistics in the second half of the text.

Data Values

When we collect data for our statistical research, be it descriptive or inferential, we are interested only in the quantitative or qualitative values. The actual elements the data are gleaned from are not our concern; we are interested only in the data. However, when we interpret the data, we are suggesting how the elements are behaving with respect to our data.

Example 3a.

What data values are we interested in if we need to know the average length of fish in a pond or small lake?

Solution: The data values of interest are the lengths of the fish. The lengths are a scale measure using a unit appropriate for measuring the length of fish (inches or centimeters).

Example 3b.

What data values are we interested in if we want to know if preschool children will be better prepared for primary school if they interact with different playthings while in preschool? What inferences could be made?

Solution: The data value of interest would be a score on a test given to preschool children. The test could measure a child's ability to group the same quantity of items or to identify letters in the English alphabet. It might also test whether or not a child can draw a complete circle or other geometric figures. The motivation behind a test of this sort might be that an instructional toy company wants to make the claim that children interacting with the company's playthings will be better prepared to enter formal education as compared to children interacting with arbitrary playthings. An independent t-test could indicate if there is a statistical difference in the mean scores between the two groups, one group that interacts with the company's toys and another group with the nondescript toys. If there is a statistical significance, the statistician can say only that a significance has occurred. They cannot conclude why any difference exists. That is left for the people who sponsored or implemented the study. Statisticians look only at data and infer how an entire population will behave based on the results of the data analysis; they do not suggest any reason why the data give the results that were obtained. That is not the role of a statistician.

The Role of a Statistician

In the examples and homework throughout this text, we will use events gleaned from the real world. If we just gave you a bunch of numbers and did not say anything about the source of the numbers, then it would just be an exercise in abstract thinking. There would be no connection to reality, which is a study in its own right, but we would like to make this text a little livelier by using real-world applications. Therefore it is okay to suggest possible reasons for differences in statistics when they occur, but remember the role of a statistician is only to work with statistics and present the results. Our job is not to interpret the result in terms of why they are what they are, or suggest why means or statistics are different. Our job is to simply present the results of the data. Thus if we discover crime rates go up in months where ice cream sales go up, then we present that. We do not make any claims that there is a cause-and-effect relationship. To make conclusions about what the data say in regard to the context of the experiment is left for someone with a higher pay grade.

Sample Statistics

When we talk about statistics, represented by Latin letters, we talk about a value obtained from a sample. The statistics of interest are the mean, median, mode, standard deviation, variance, range, correlation coefficient, and few others we will come across as we progress through the text. We use the letter $\overline{x}$ (x bar) for the mean, M for the median, and mode for mode. We use the letter s for standard deviation, s^2 for variance (because the variance is the square of the standard deviation), and range for range. We use r_{xy} for the Pearson Product-Moment Correlation Coefficient. These will all be explained in more detail later in the text, but our point here is to say that these statistics are often used to approximate a population parameter and thus we use a different symbol from the one we use for the population parameter. The population parameter is a number that usually cannot be ascertained unless you sample the entire population. Hence it is thought of as being a value that exists but we can never really put our hands on it.

Example 4.

Considering the entire population of the world as our population of interest, what is the mean height of all persons?

Solution: As you might surmise, to find the mean height of the entire population of the world cannot be done. By the time you finished measuring the heights of half the population, the other half has changed—individuals have grown; births and deaths have occurred—and the heights of the first half have also changed. It is said about three children are born every second somewhere in the world. At that rate we could not even measure the heights of all the new babies to add to our total to find the average height. It is physically impossible to find the mean height of all the people in the entire world at a particular moment. So we call the mean height of the entire world a population parameter that will never be known for certain. But we can use a random sample to infer what the population parameter is.

Population Parameters

Population parameters are represented by Greek letters. We use μ (mu) for the population mean, median and mode are the same as the sample notation, σ (sigma) for the population standard deviation, and σ^2 (sigma squared) for the population variance. There is no special letter for the range of a population. The population Pearson Product-Moment Correlation Coefficient is represented by ρ (rho). If you own a scientific calculator you might have noticed these symbols on the keyboard. If you had no idea what they represented before, it might be a little clearer now. We will learn the differences in sample statistics and population parameters as we progress through the next chapters and you might be able to use your calculator to find these values, but as always, our emphasis will be on using the SPSS program.

Our last concept for this chapter is the concept of a **random sample**. The definition of a random sample:

Random Samples

A sample is random if every set of the same sample size from the population has an equal chance of being selected; that is, when selecting a sample from the population, any element in the population is just as likely to be selected for that sample as any other element. For example, if we need a sample of 30 students from a sample space of 2,000 students, then we have to ensure that each student in the college has the same chance of being selected for our sample. This also implies that every subset of size 30 out of a population size of 2,000 has an equal chance of being selected. How can a random sample be selected using SPSS?

Example 5

5a) Are the students in this particular class a random sample of the students in the college?
5b) Did the Literary Digest pollsters in 1936 pick a random sample to infer the next president of the United States?
5c) Did the Chicago Tribune use proper polling techniques before publishing the winner of the presidential election in 1948?
5d) If every person worldwide is given a universal social security number at birth and we use a computer randomizing program to generate a sample of 2,000 persons to measure their heights, will we have a random sample?

Solution: **A**) No, the students in this particular class do not reflect a random sample of all the students in the college because not all the students in the college had an equal chance of being selected. Not all the students in the college had an equal chance of being in this class.

B) No, in 1936, Franklin Delano Roosevelt, Democrat, had been President for one term. The magazine, *The Literary Digest*, predicted that Alf Landon, Republican, would beat FDR in that year's election by 57 to 43 percent. The *Digest* mailed over 10 million questionnaires to names drawn from lists of magazine subscribers and automobile and telephone owners, and over 2.3 million people responded—a huge sample. This is called a straw poll.

At the same time, a young man named George Gallup sampled only 50,000 people, and using scientific polling techniques, predicted that Roosevelt would win. Gallup's prediction was ridiculed as naive.

After all, the *Digest* had predicted the winner in every election since 1916, and had based its predictions on the largest response to any poll in history. But Roosevelt won with 62% of the vote. The size of the *Digest's* error is staggering. How could they have been so far off?

[Context: The Great Depression had begun in 1929, and Roosevelt had attempted to mitigate its effects with then-controversial policies like Social Security, large government "make-work" jobs programs, and other "socialistic" projects. The underlying economic condition in 1936 was still that of widespread and deep depression.]

The Literary Digest had made two fatal mistakes. Their list of names was biased in favor of those with enough money to buy a magazine subscription, cars, and phones, a much smaller portion of the population in the thirties than it is today. And, more seriously, the *Digest* had depended on voluntary response. FDR was the incumbent, and those who were unhappy with his administration were more likely to respond to the *Digest* survey. When a sample is biased, even a large number of subjects cannot correct for the error.

[The magazine folded, not too much later; some think the wrong prediction was largely responsible. Gallup's scientific polling organization still exists today, highly respected for the quality of their work over 70 years.](2) Also straw polls have been regulated to the status of entertainment or amusement and are not used for predicting the results of any vote.

C) No, aided by erroneous polling, newspapers prematurely called the presidential election for challenger Thomas Dewey, leading to the famous photograph of an elatedly re-elected Harry Truman (see photo below). The problematic polls used then-popular "quota-sampling" techniques. In other words, they sought out various income levels and certain numbers of men, women, blacks, and whites. According to statistician Fritz Scheuren, quota sampling in political polling was abandoned after this debacle in favor of random sampling of the population.(3)Even today if a pollster wants to sample the voting population, they must consider those persons who have no phone contact when drawing their random sample.

Chicago Tribune: November 3, 1948.

D) Yes, SPSS has a random number generator and can be used to generate a random sample of any size. We'll do an example. We won't try to find a random sample of 2,000 persons out of 7 billion; instead we'll aim a little lower and find a random sample of 30 persons out of 2,000 persons. First assign every student in the sample space a number from 1 to 2,000 (or whatever is the total number of students in the sample). Open a new SPSS data editor. (Review above to go through the steps of opening an SPSS Data Editor if needed, or click on File→New→Data). Next enter the numbers 1 through 30 in the first column of the data editor and change the variable to *x*. Also in variable view change the number of decimals by clicking on decimals, lower the number to 0, and change the measure to scale. **See Figure 2a.**

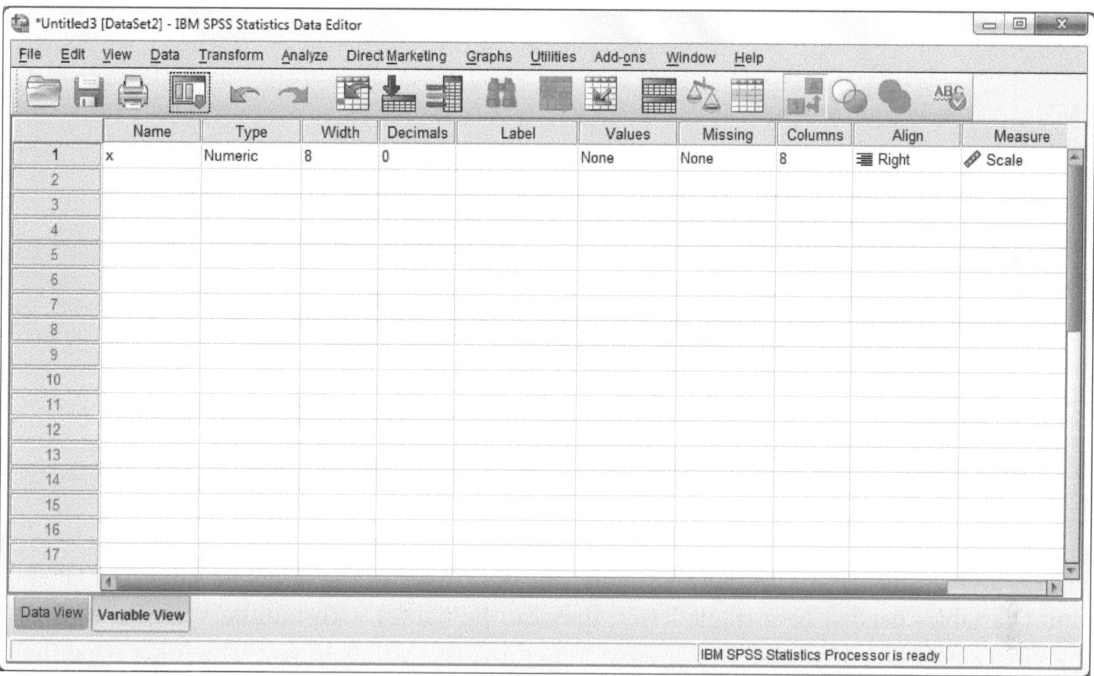

Figure 2a

Your data editor should look like **Figure 2b.**

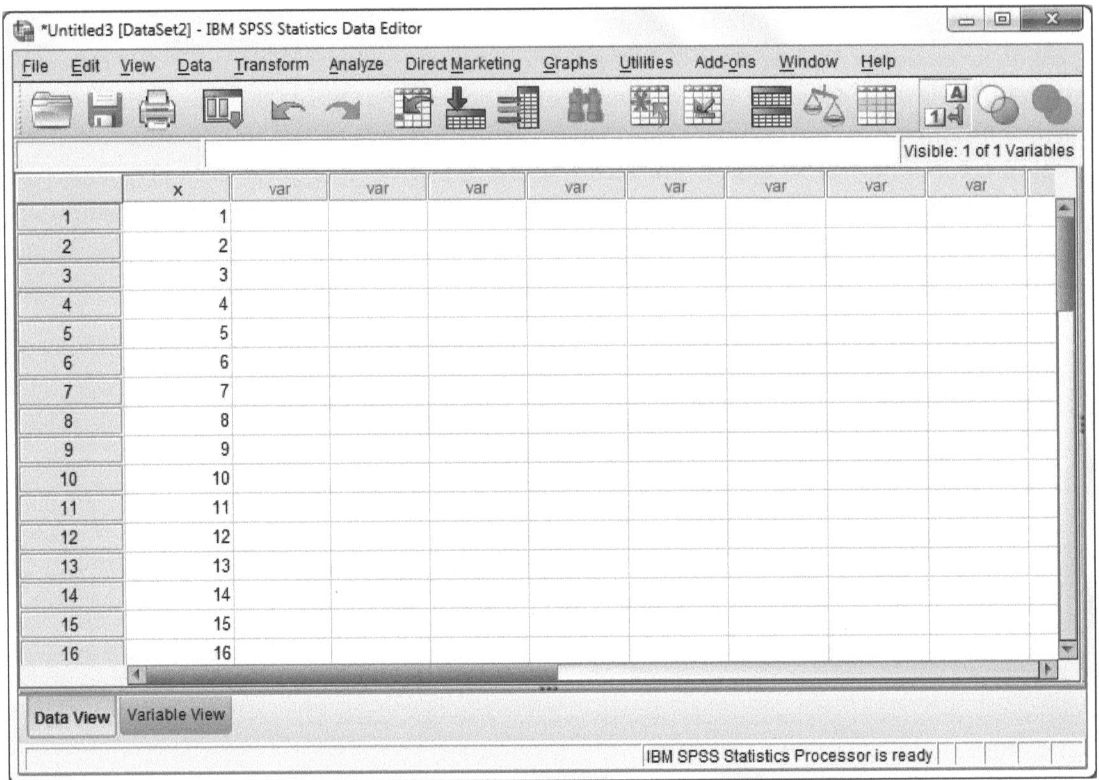

Figure 2b

Now go to Transform → Compute Variable, put the cursor in the target variable and label it *r*, for random. (Variables needn't be a single letter; they can be names with numbers. If you will use many variables in a single file, then label the variable something that will help you remember what they stand for.) Next scroll down the Function group until you find Random Numbers, and click on that. In the Functions and Special Variables section, scroll down until you find RV.Uniform. Double click on that. RV.Uniform should appear in the Numeric Expression. It asks for two parameters, the beginning number and the last number of all the persons in the sample space. Put 1 in place of the first ?and 2000 in place of the second ?. **See Figure 2c.**

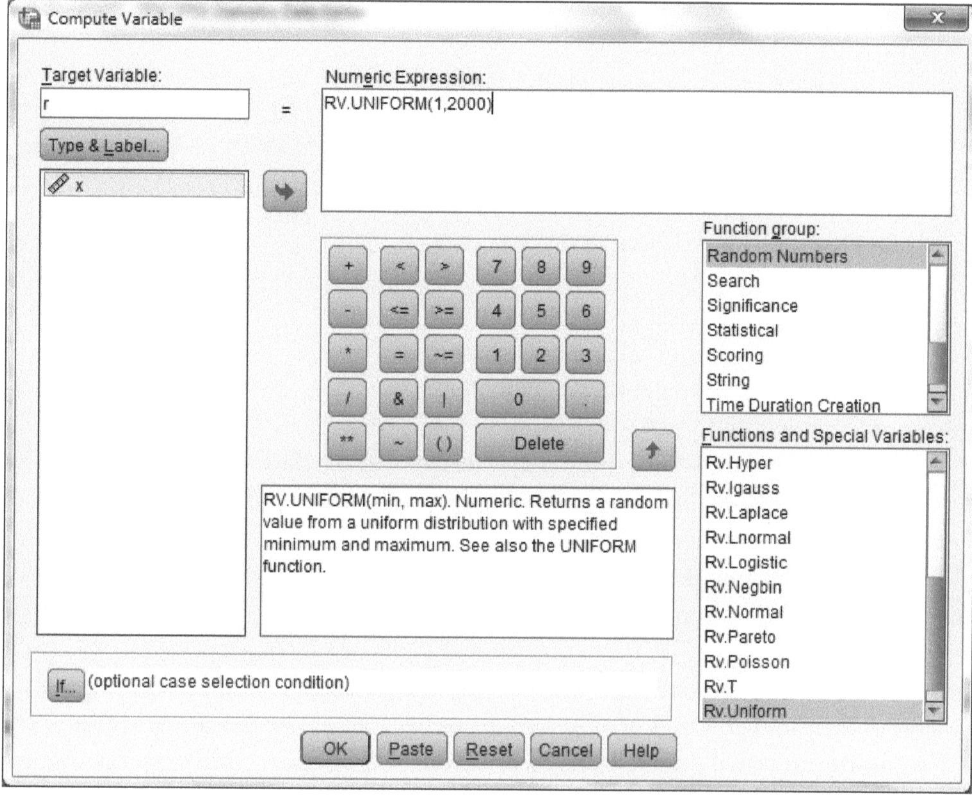

Figure 2c

Hit OK. Go back to variable view and change the decimals to 0 for the variable *r*. You should have 30 random numbers associated with 30 students whom you can now poll to find information regarding whatever it is you want to study, (height, gender, full/part-time student, etc.). However, all the random numbers will be the same every first time we run this program.

SPSS always starts with a random "seed" number of 2 million. To avoid getting the same random numbers every time the program is started, simply change the seed by using Transform → Random Number Generators, click on Set Starting Point, and use either Random or Fixed Value and change the value. Thereafter SPSS will generate different random seeds each time the random-number generator is run until you exit the program. Then it will go back to the default seed of 2,000,000.

By repeating this process we can generate different outcomes on account the random generator will be different each time the procedure is used. Do it several times and you'll see what we mean. Just go back to Transform → Compute and hit ok. When it says, "Change existing variable," click ok.

SPSS will generate decimal numbers between 1 and 2000, exclusive, which means SPSS will not include the numbers 1 and 2000, but then it will round the decimals to the nearest whole number. Another method to get the integers for our sample is to use the TRUNC function (truncate—eliminate the decimal part). In Numeric Expression put Trunc (Rv.Uniform(1,2001)). This will simply chop off the decimal parts and leave the integer part. Because Rv.Uniform does not select the first or last integer, we need to increase our range to 2001 so 2000 will be included.

Let's review what we learned in this chapter.

1. Statistics arose during the 17th century.
2. Probability is the foundation for Statistics.
3. Types of Data
    - A.) Qualitative Data
        - i.) Nominal—Name-only data
        - ii.) Ordinal Data—Categorical data
    - B.) Quantitative Data
        - i.) Ordinal Data—Ranked data
        - ii.) Scale Data—Interval and ratio data
4. Input data into SPSS as nominal, ordinal, or scale.
5. Categories of Statistics
    - A.) Descriptive Statistics—Describes characteristics of the data
    - B.) Inferential Statistics—Infers how a population will behave based on the statistics of a random sample of the population.
6. Sample Statistics and Population Parameters
7. Random samples and random-number generator

A note regarding printing: all data files and output files can be printed in their entirety by clicking on the print icon in the tool bar. Titles of any output can be changed by moving the cursor over the title, left clicking the mouse, and then double clicking to delete or type in any title or name that is relevant. If you wish to print only selected sections of the output, then highlight those sections in the outline pane (the pane on the left) or click on a section in the content pane (pane on the right), hold the shift key, and scroll down to the end of the section you want to print and click on that. The printer will print only those sections; **see Figure 3**. You can also copy and paste those sections into a Word document to turn in as an online assignment.

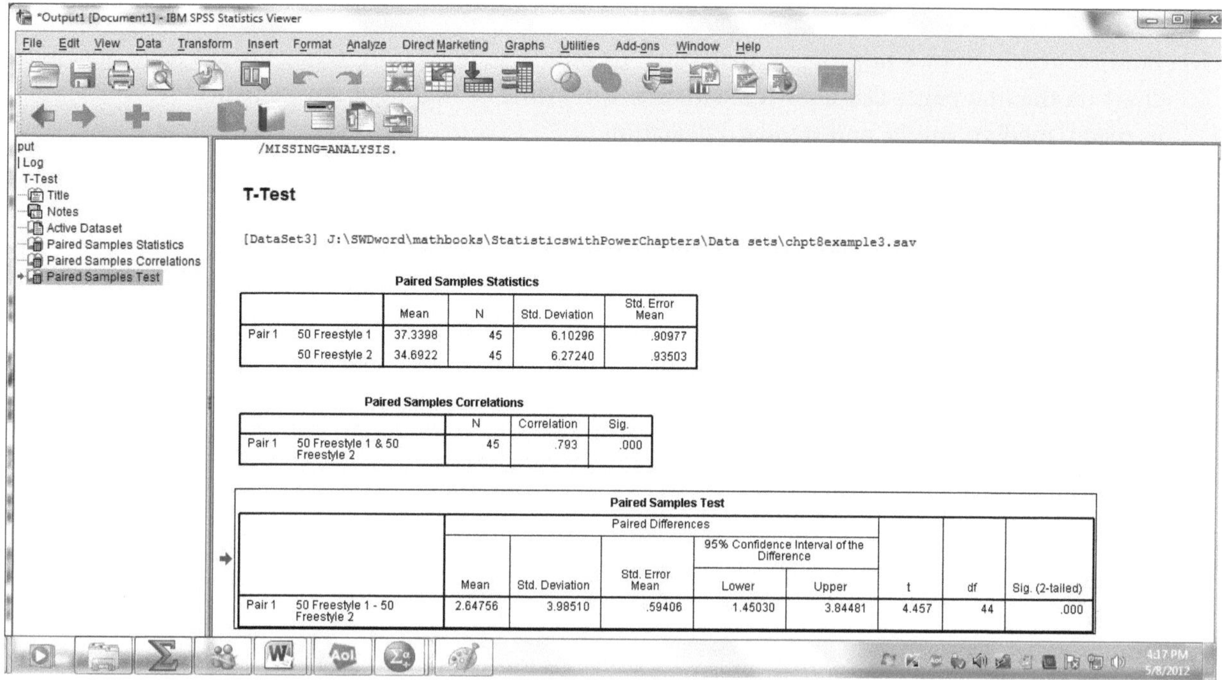

*Output1 [Document1] - IBM SPSS Statistics Viewer

File Edit View Data Transform Insert Format Analyze Direct Marketing Graphs Utilities Add-ons Window Help

put
Log
T-Test
Title
Notes
Active Dataset
Paired Samples Statistics
Paired Samples Correlations
Paired Samples Test

```
/MISSING=ANALYSIS.
```

T-Test

```
[DataSet3] J:\SWDword\mathbooks\StatisticswithPowerChapters\Data sets\chpt8example3.sav
```

Paired Samples Statistics

		Mean	N	Std. Deviation	Std. Error Mean
Pair 1	50 Freestyle 1	37.3398	45	6.10296	.90977
	50 Freestyle 2	34.6922	45	6.27240	.93503

Paired Samples Correlations

		N	Correlation	Sig.
Pair 1	50 Freestyle 1 & 50 Freestyle 2	45	.793	.000

Paired Samples Test

		Paired Differences					t	df	Sig. (2-tailed)
					95% Confidence Interval of the Difference				
		Mean	Std. Deviation	Std. Error Mean	Lower	Upper			
Pair 1	50 Freestyle 1 - 50 Freestyle 2	2.64756	3.98510	.59406	1.45030	3.84481	4.457	44	.000

Figure 3

Class Work 1

1. Classify the data as nominal, ordinal, or scale.

   Ethnicity
   School fees
   Age
   Gender
   Color
   Positions of finishers in a race
   Calendar time
   Likert scale: strongly like, like, no opinion, dislike, strongly dislike
   Temperature
   Class ranking

2. Solve for z: $z = \dfrac{x-100}{15}$, when $x = 136$

Homework 1

1. Find an article from a newspaper, magazine, or online source (*USA Today* always has a statistical chart on the first page) dealing with statistics. Underline or highlight any statistical reference such as mean, median, mode, and standard deviation.
2. Find an example (other than temperature or IQ) of an interval variable that is not a ratio variable and explain why it is not a ratio variable.
3. Use the random-number generator in SPSS to generate a random sample of 15 students out of 2,000 students as the population to test a new weight-loss exercise program. Assume the 2,000 students have all been assigned identification numbers from 1 to 2,000. List the 15 students you found.
4. Baseball has an ordinal variable for both the offensive side and the defensive side. Determine if the rankings on either side are significant in terms of numerical value or just a label.

Offensive: Order of batters; 1, 2, 3, 4, 5, 6, 7, 8, 9

Defensive: Pitcher = 1, Catcher = 2, First Base = 3, Second Base = 4, Third Base = 5, Shortstop = 6, Left Field = 7, Center Field = 8, Right Field = 9

Algebra

5. Solve for x: $6 = \frac{x-2}{3}$

6. Solve for z: $z = \frac{11-4}{7}$

7. Solve for z if $x = 42$. $z = \frac{x-3}{13}$

8. Solve for x in terms of z. $z = \frac{x-3}{13}$

9. Solve for x if $z = -2$. $z = \frac{x-3}{13}$

10. Find the square root: $\sqrt{49t^2}$

11. Solve for t if $x = 17$. $t = \dfrac{x-13.6}{\frac{2.04}{\sqrt{36}}}$

12. Solve for x if $t = 1$. $t = \dfrac{x-13.6}{\frac{2.04}{\sqrt{36}}}$

13. Solve for b_0 if $y = b_1 x + b_0$ and $(x, y) = (2, 3)$, $b_1 = -2$.

Project 1

Fill out the following form and return it to your instructor anonymously. The results will be used as a project later in the class after we learn a little about inferential statistics.

1. What is your gender?

   1.) Male   2.) Female

2. What is your current age?   _____ Years

3. Which of the following best describes your race/ethnicity?

   1.) African American   2.) Asian American   3.)Caucasian   4.)Latino   5.) Other

4. What is your cost for attending the college/university? $__________ per semester/quarter

5. Are you a full-time (12 units or more) or a part-time student?

   1.) Full-time   2.) Part-time

Just for fun: Use the random-number generator to generate lottery numbers for your state or region. It's possible the random-number generator will generate the same number twice (depending on the range and the number of numbers needed for the lottery). If that is the case, just run the generator again or write a program using syntax (beyond this course) with an If-Then statement. What is the likelihood a random-number generator can generate a DNA sequence for a human being?

References

[1]Wikipedia: Statistics

[2]SPSS was acquired by IBM in October, 2009

(1) Wikipedia: 1936 Literary Digest

(2) Whyfiles.org

CHAPTER 2

Graphs

Objectives

After completing this chapter, a student should be able to:

- ✓ Make a stem and leaf plot
- ✓ Create a Histogram and change the bins
- ✓ Create a line graph
- ✓ Find the frequency distribution for each bin and create a Cumulative Histogram
- ✓ Create an Ogive
- ✓ Create Pie graphs and Bar graphs
- ✓ Understand what is a Normal Distribution and Skewness

Organizing Data

Statisticians are employed to do two jobs: 1.) to collect data and 2.) to organize and interpret data. We will not concern ourselves with the collection of data in this text. (The class survey we did in Chapter 1 is an example of data collection. Other examples can be found in research journals and statistical magazines.) Instead we will be concerned with the organization and interpretation of data in this text.

When a statistician is given a bunch of raw data, the first step is to make some sense out of it. We employ our favorite tool for quickly organizing data—the stem-and-leaf plot.

Example 1.

Given the following data, make a stem-and-leaf plot.

47	12	44	41	17	16	35	38	35	36
10	11	14	14	30	20	32	33	34	32
31	31	15	16	17	16	15	19	18	16
25	25	26	26	27	29	29	28	29	27
20	21	21	21	24	24	23	20	21	20

Solution: These data can represent anything that seems reasonable so we'll just deal with them as numbers and we won't give the numbers any specific context. Our first task is to input the data into an SPSS data editor. Open a new SPSS data editor and input the above data into the first column; it doesn't matter if you go down column by column or across row by row as you copy the data into the data editor, but put all the data into one column. (All data sets are saved on the disc or the online source; this data set is called chpt2example1.sav, but it is instructive for you to have personal experience entering data, so occasionally we will have you enter the data.) Change the variable name for the column to a convenient name; we chose *x*. To make a stem-and-leaf display use Analyze → Descriptive Statistics → Explore; see **Figure 1a** and put *x* (or the variable name you chose) into the dependent box. (Highlight the variable and click on the movement arrow; see **Figure 1b**.)

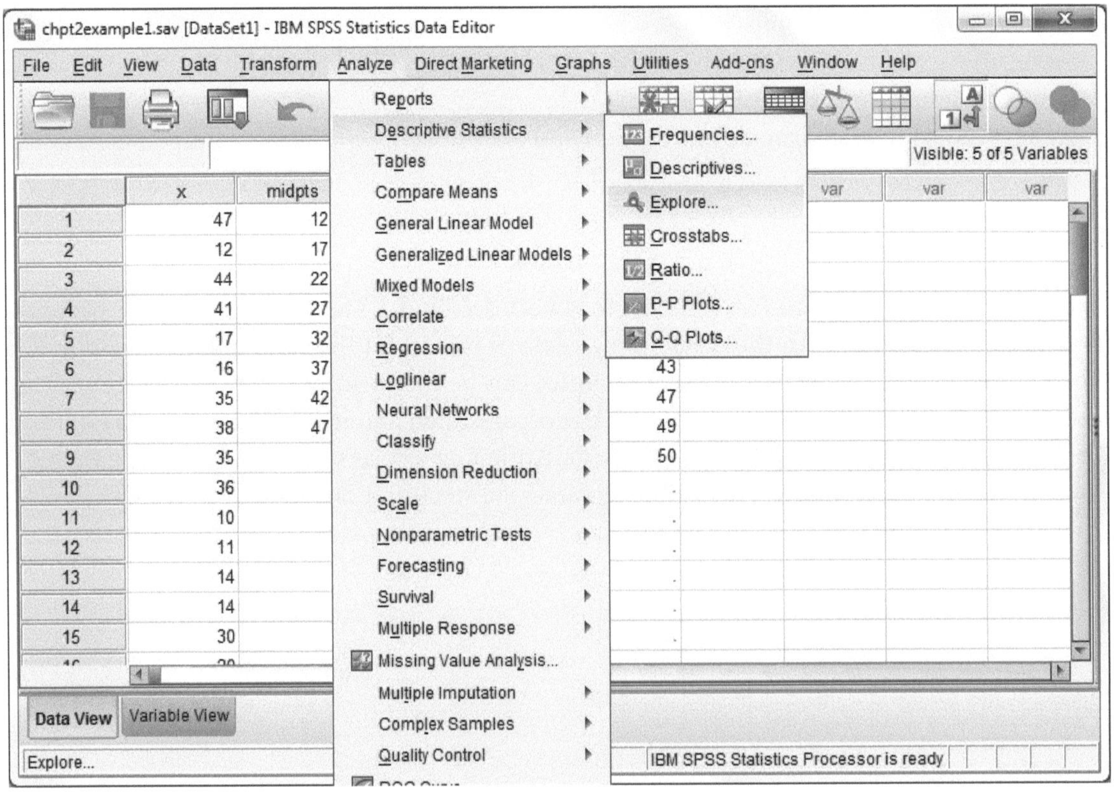

Figure 1a

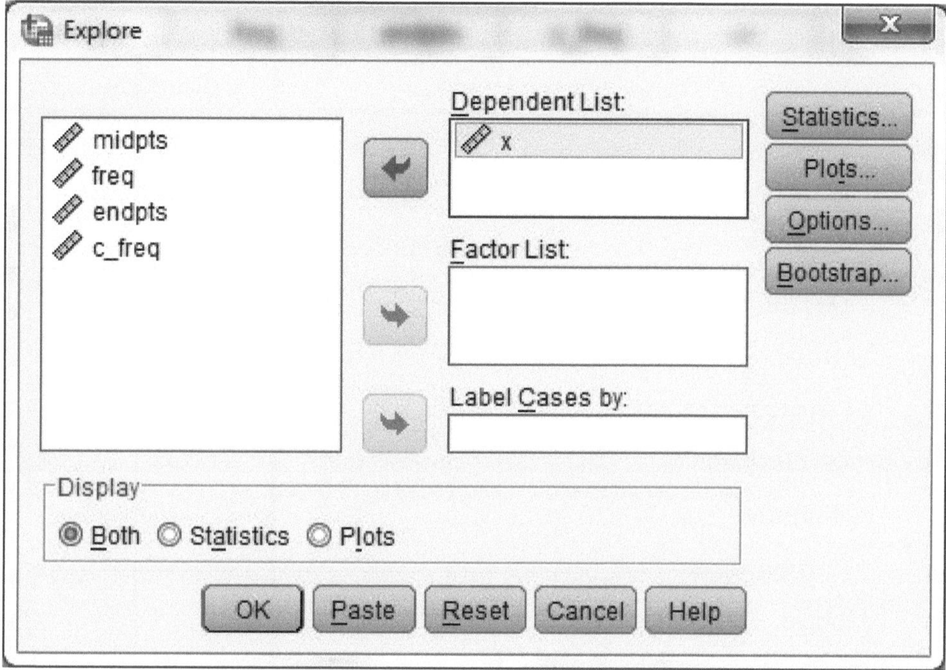

Figure 1b

Click OK. You will have these output boxes, see **figures 1c–1f**.

Case Processing Summary

	Cases					
	Valid		Missing		Total	
	N	Percent	N	Percent	N	Percent
x	50	100.0%	0	.0%	50	100.0%

Figure 1c

The first box, **Figure 1c**, shows the number of data values, if any values are missing, and the total. We should have no data values missing in our example. (In practical applications, many survey forms will have missing values and so consideration must be made for those missing values, which is why we usually use the valid percent instead of the total percent. In this text most of our examples will have no missing values.)

The next box, **Figure 1d**, gives us all the descriptive statistics we'll use.

Descriptives

			Statistic	Std. Error
x	Mean		24.62	1.234
	95% Confidence Interval for Mean	Lower Bound	22.14	
		Upper Bound	27.10	
	5% Trimmed Mean		24.28	
	Median		24.00	
	Variance		76.118	
	Std. Deviation		8.725	
	Minimum		10	
	Maximum		47	
	Range		37	
	Interquartile Range		14	
	Skewness		.508	.337
	Kurtosis		-.237	.662

Figure 1d

These procedures will be used to find descriptive statistics later in the text: Analyze → Descriptive Statistics → Frequencies or Analyze → Descriptive Statistics → Descriptives. The details of this box will be explained in the next two chapters.

Stem-and-Leaf Plot

What we want to look at is the Stem-and-Leaf Plot in **Figure 1e**.

x Stem-and-Leaf Plot

Frequency	Stem & Leaf
5.00	1 . 01244
10.00	1 . 5566667789
11.00	2 . 00001111344
10.00	2 . 5566778999
7.00	3 . 0112234
4.00	3 . 5568
2.00	4 . 14
1.00	4 . 7

Stem width: 10
Each leaf: 1 case(s)

Figure 1e

It has a frequency column and a stem-and-leaf column. At the bottom it says the stem width is 10 and each leaf is 1 case. What this means is SPSS took the tens digit of each data value and made that the stem and took the ones digit and made that the leaf. Furthermore it divided the tens digits into lower and upper values with the lower values being 0 through 4 and the upper values being 5 through 9 for the ones position. So in the first row the frequency is 5, which means that row has 5 data values and the first data value is 10, the next is 11, followed by 12, 14, and 14. In the second row the frequency is 10 and the data values are 15, 15, 16, 16, 16, 16, 17, 17, 18, 19. Not only did SPSS give us a stem-and-leaf display, but it ranked the data values as well (put the data values in order). Can you tell how many data values are in the third row and what the data values are? (Eleven data values, the values are 20, 20, 20, 20, 21, 21, 21, 21, 23, 24, 24.) Also the stem-and-leaf display gives a visual picture of the distribution of the data values. We see the average value should be in the lower twenties and indeed, if we check the Descriptive Statistics box, the mean (average value) is 24.62. The last display in our output is called a boxplot, **Figure 1f**, which will be discussed in the next chapter.

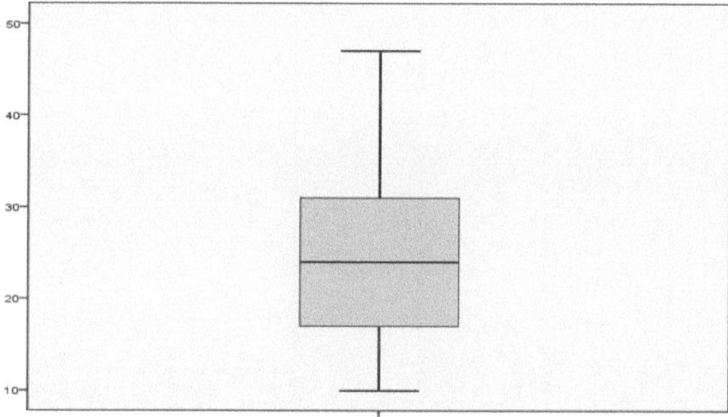

Figure 1f

Discrete Data

Data that is finite in sample size or that can be counted (put into a 1-1 relationship with the Natural Numbers) is discrete. There is always a space between any two data values.

Continuous Data

Data that consist of an interval of the Real Number Line are continuous data. There is no space, no matter how small, between continuous data values. (In practical applications, however, we usually round off data values to a finite number of decimal places so there is always a space between two data values.)

Histograms

A stem-and-leaf plot isn't as easy a picture to grasp as a bar graph is to grasp. You can note that in most published media, statistics are represented by a bar graph. A bar graph is easier for most people to grasp and understand. Most people don't need to know all the numerical data that went into the creation of the statistical graph. A picture is easy enough to understand the results. So we want to make a bar graph; because our data values are continuous, our bars will be connected. A connected bar graph is called a histogram. Go back to Analyze → Descriptive Statistics → Explore and click on Plots. An Explore: Plot window will pop up; put a check mark in the box called Histogram (leave Stem and Leaf checked also). (Can also use Graphs → Legacy Dialog → Histogram, move the variable into the dependent window, and hit OK.) Hit continue and then OK. Scroll down the output window until you see the Histogram:see **Figure 1g**.

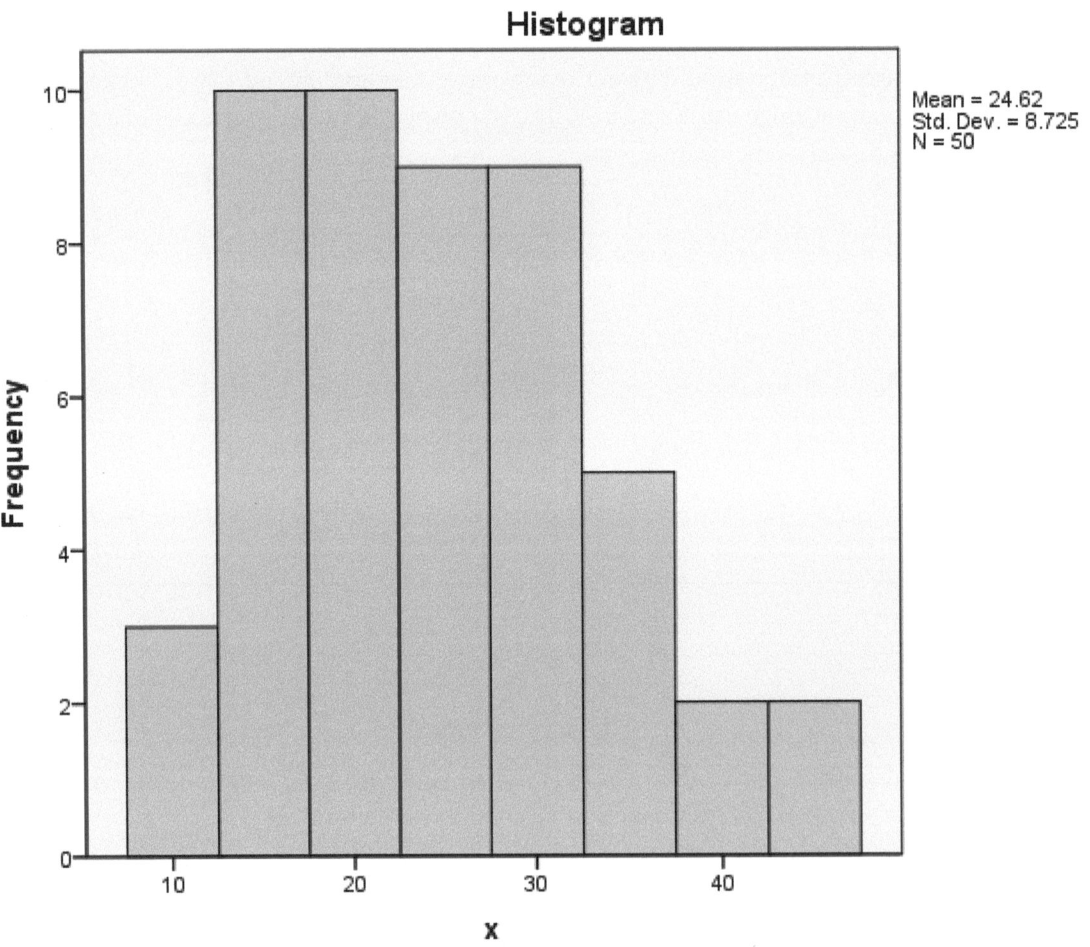

Figure 1g

SPSS makes an automatic histogram depending on criteria inside the program. We can change our histogram to look like our stem-and-leaf plot by moving the cursor over the graph and double click to activate the chart editor. We can also right click on the mouse. A pop-up window will appear where we can check "Edit Content in a separate window" and a chart editor will appear. Click on the tab Options and scroll down to Un-Bin Element. Close the pop-up window that appears and click on Options again. This time scroll to Bin Element and click on that. This window will appear; see **Figure 1h**.

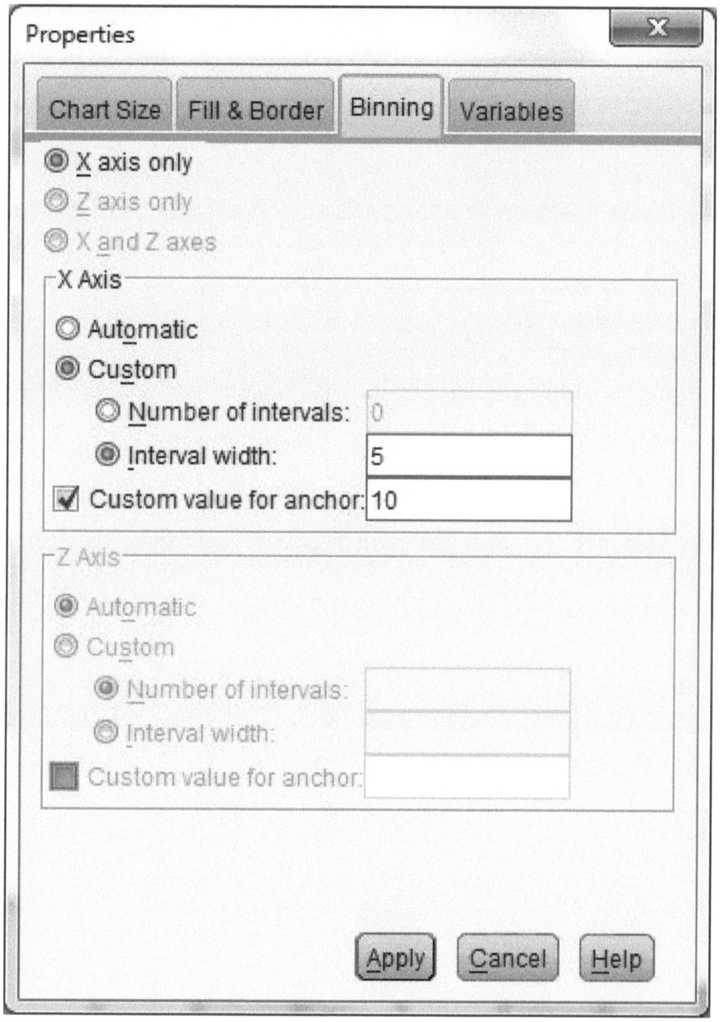

Figure 1h

The Binning tab should be highlighted, if not click on it and check the circle Custom, then check the circle Interval width and make the width 5, the same as our stem-and-leaf plot, or check number of intervals and make it 8. Also click on Custom value for anchor and make it read 10. Click Apply and then Close. Close the Chart Editor and the new Histogram will appear on the output. This Histogram looks like our stem-and-leaf display. The same set of data can give us different histograms, thus not all histograms from the same set of data look exactly alike. This same graph can be done with Graphs → Legacy Dialogs → Histogram and repeat above. See **Figure 1i**

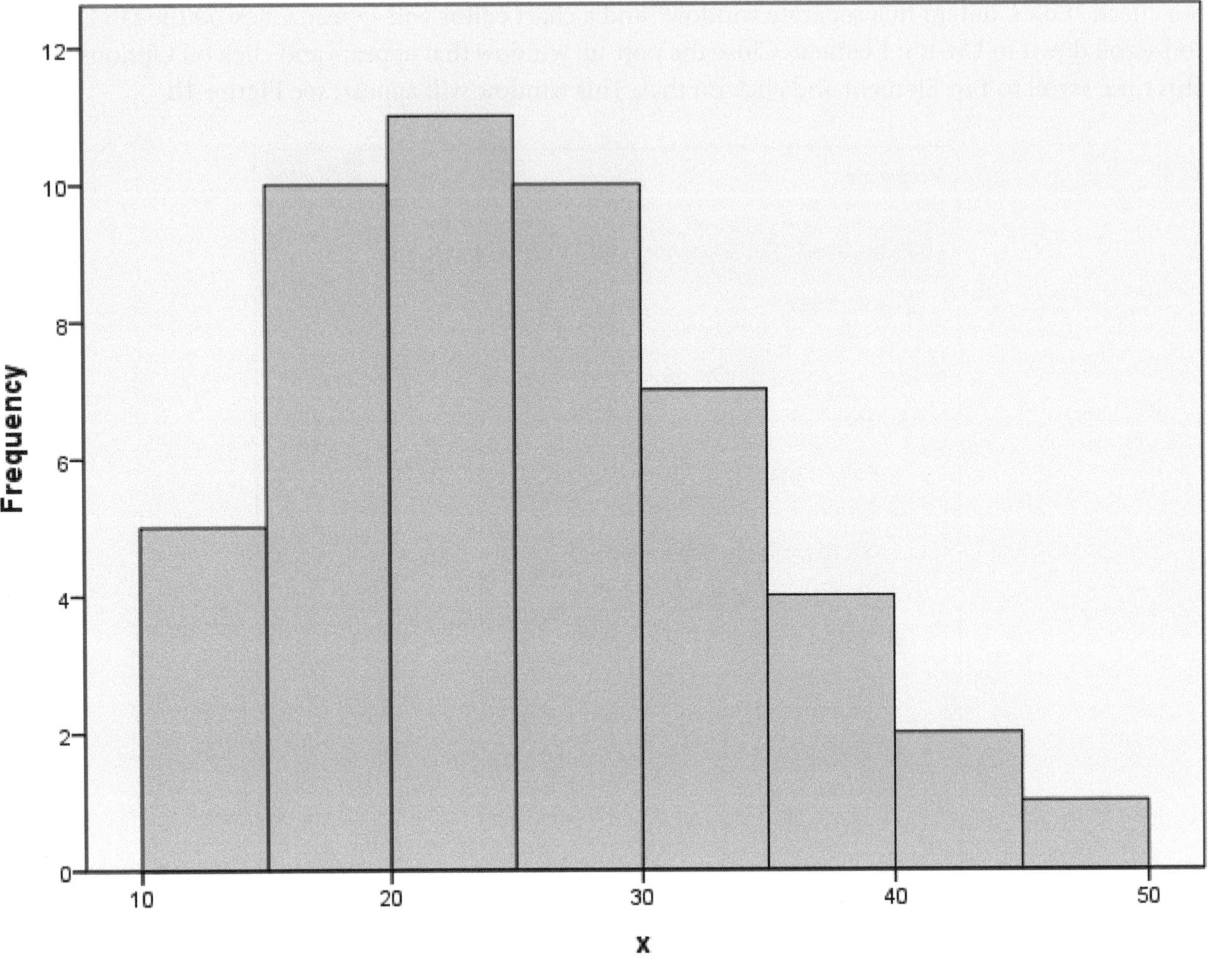

Figure 1i

Bins

When we made histograms in the past we had things called intervals, boundary values, interval limits, and so on. SPSS ignores all that. SPSS calls the intervals bins and the number of bins is determined by an interior calculation. If a data value lands on a bin boundary, then SPSS will always move the data value into the bin to the right, except of course, any data value that lands on the last right-side bin boundary will be placed in the last bin. As we saw we can change our bin width or the number of bins to any value to fit the histogram we feel is most beneficial in displaying the data.

Statistics Can Prove Almost Anything

One note about statistics: By presenting the data with different slants we can almost make it appear the data imply anything we want them to imply. In fact in Project 2 (1) you will have an opportunity to imply dogs can do calculus, which of course can't be true. I (Steven) have a hard enough time trying to get my students to do calculus, and they can read, which I don't think dogs can do. The saying "Lies, damned lies, and Statistics" attributed to Mark Twain has meaning in the sense that statisticians can be deceptive. So even if we want the data to show a certain point of view, we must remember to be objective in our presentation. It does no good to say everything is true.

Mathematics of Voting

Dr. Don Saari, University of Irvine, Mathematics Professor, shows in his works (2) that any outcome in an election can be achieved using different methods of tabulating votes; that is, even a black-horse candidate can be elected president by a valid way of counting the votes. That is, just find the right set of circumstances in which your favorite candidate can win. The best noncontroversial way to elect offices is to use the Borda count—each voter ranks all the candidates in order of preference, 1 being best, and so on. The candidate with the lowest total score wins the election. The point being the method itself of interpreting data needs to be circumspect or the whole method of data analysis can be flawed.

Data Mining

Many programs look through enormous amounts of data to find a common thread for various events. For example, people who are considered terrorists might have a commonality somewhere in all the data collected on them, hence that common thread could be used to identify a potential terrorist. On the other hand if we search hard enough through enormous amounts of data, we can find commonality for most anything. Searching through data and mining it for nuggets that stand out in particular instances is called data mining. Data mining is used by business to determine which type of person is most likely to buy a product, the justice department to determine which type of person is likely to commit a crime, and odds makers to determine the odds of a sporting event given the conditions of the two teams at the time of the game. As might be surmised, data mining isn't always accurate.

Line Graphs

A frequency line graph is obtained from the histogram by making a mark at the top of each bar in the middle and connecting the dots. We can also make a frequency line graph using the data from above: just add an interpolation line in the chart editor and make it a different color so that it will stand out from the histogram; see **Figure 1j** (To make the line, fill, or background a different color, double click on the line, fill, or background and change the color in the pop-up properties box.)

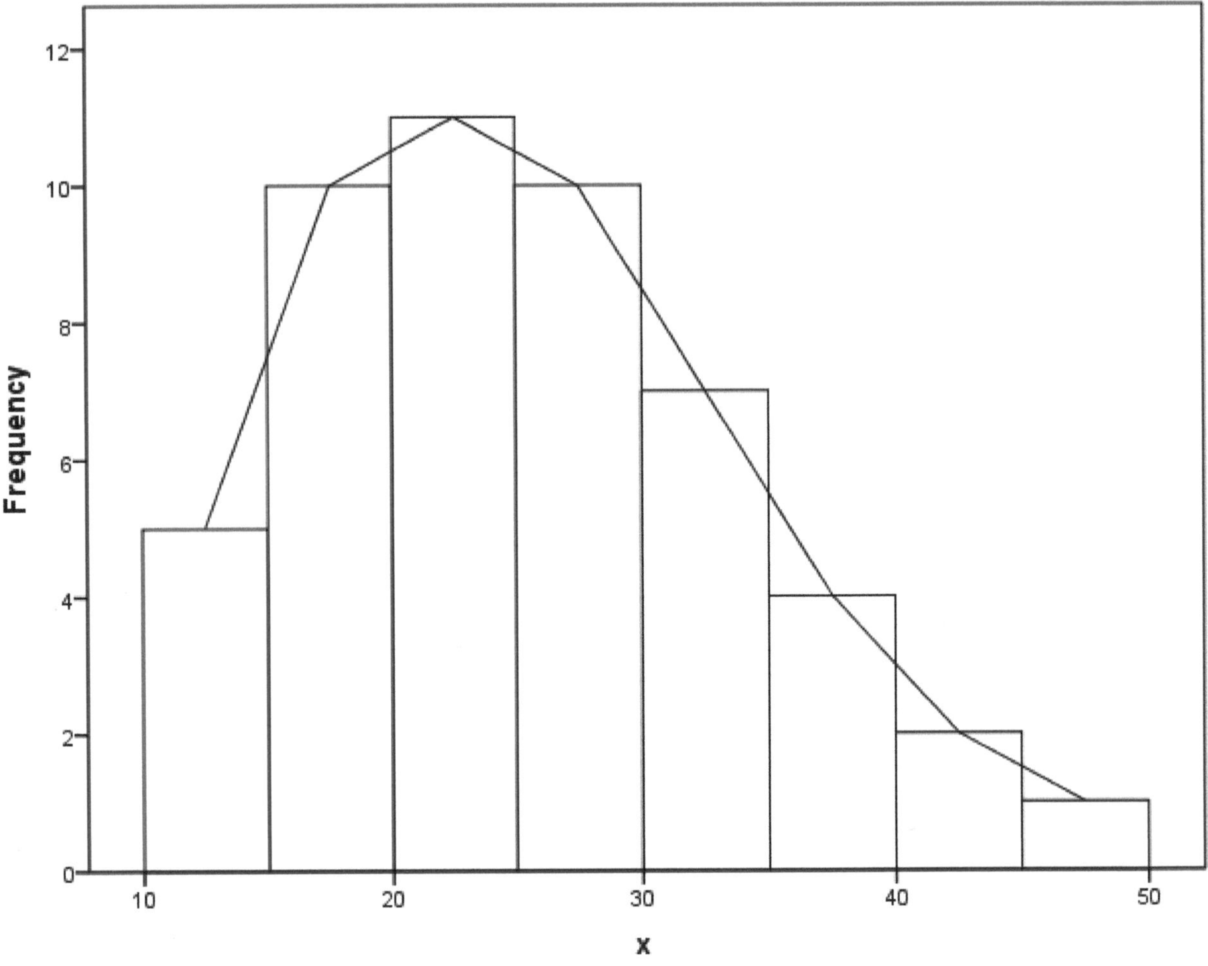

Figure 1j

Or we can rework the data and draw our own line graph. In your chpt2example1.sav data editor put the midpoint of each interval in a new column. Since the interval width is 5 and the first interval starts at 10, then the first midpoint is 12.5 (12.5 = 10 + 5/2), then add 5 to find successive midpoints (or we can find the first midpoint by averaging the two endpoints, (10 + 15)/2 = 12.5). Label the variable midpts. In the next column input the frequency (the height of the bars in the histogram) for each interval. (To find the frequency for each bar, double click on the histogram and in the chart editor move the cursor over the tab Elements and click on the line Show Data labels. The data labels will appear as in **Figure 1k**) The first frequency is 5, the next frequency is 10, etc. Call that variable freq. Reduce the decimals for the variable midpts. to 1 and the decimals for the variable freq to 0. The data editor should look like this, see **Figure 1l.**

Use Graphs ⟶ Legacy Dialogs ⟶ Line; a line chart window should pop up and the default for the line graph will be simple. Click Define, and this new window will appear; see **Figure 1m.**

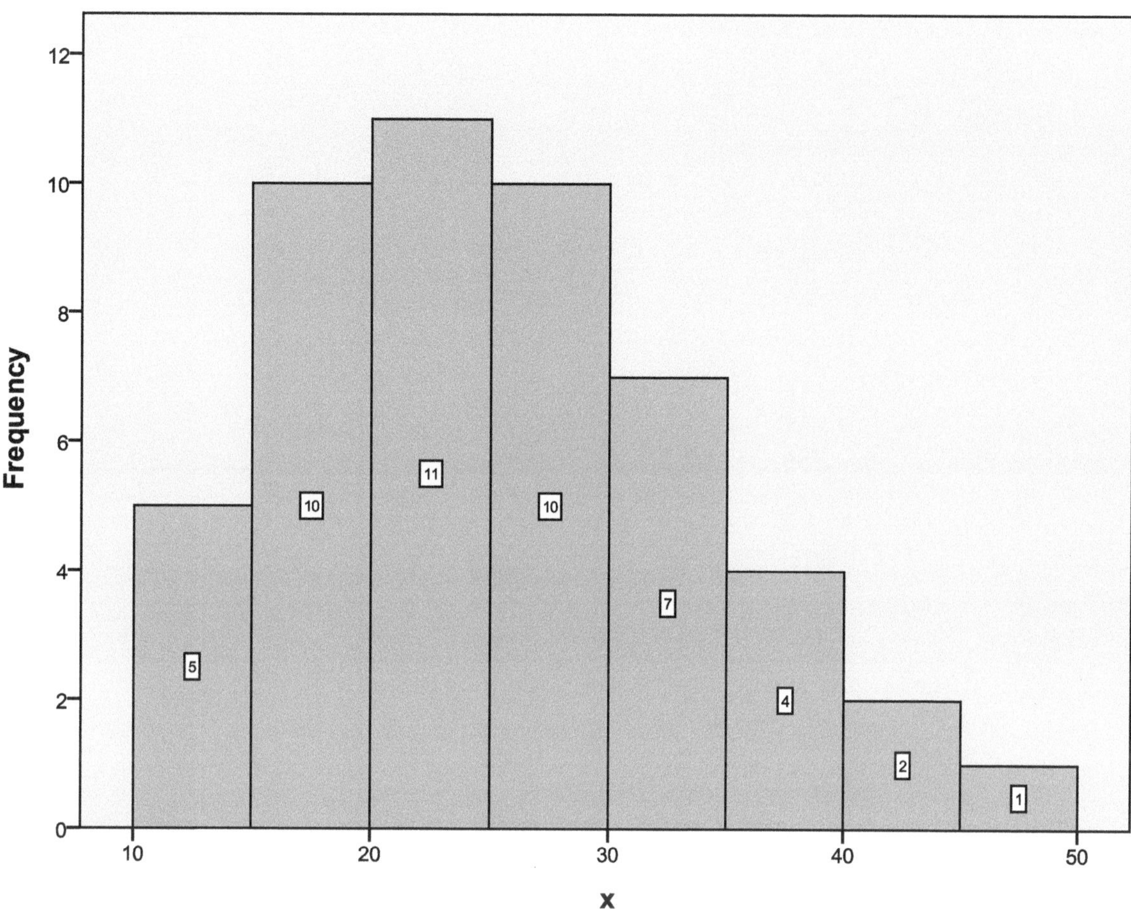

Figure 1k

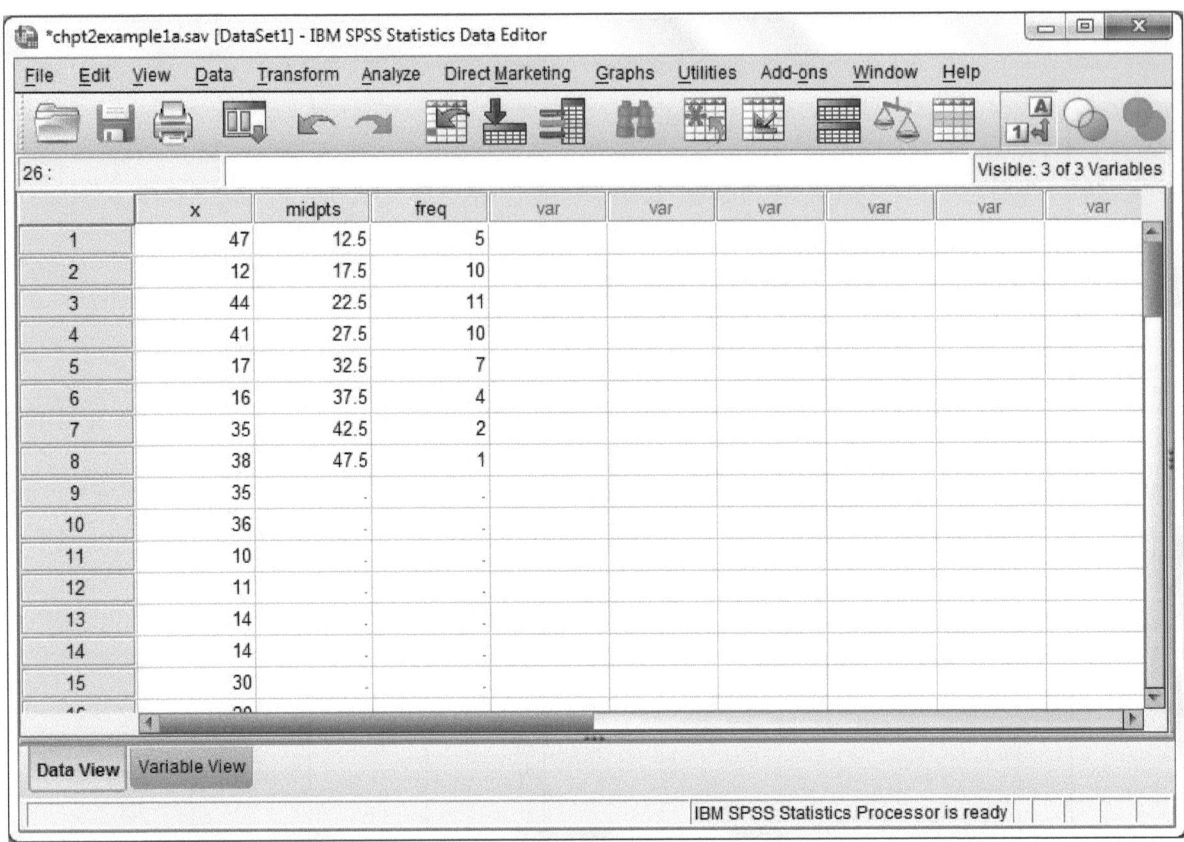
*chpt2example1a.sav [DataSet1] - IBM SPSS Statistics Data Editor
File Edit View Data Transform Analyze Direct Marketing Graphs Utilities Add-ons Window Help
26 :
Visible: 3 of 3 Variables
x midpts freq var var var var var var
1 47 12.5 5
2 12 17.5 10
3 44 22.5 11
4 41 27.5 10
5 17 32.5 7
6 16 37.5 4
7 35 42.5 2
8 38 47.5 1
9 35 . .
10 36 . .
11 10 . .
12 11 . .
13 14 . .
14 14 . .
15 30 . .
Data View
Variable View
IBM SPSS Statistics Processor is ready

Figure 1l

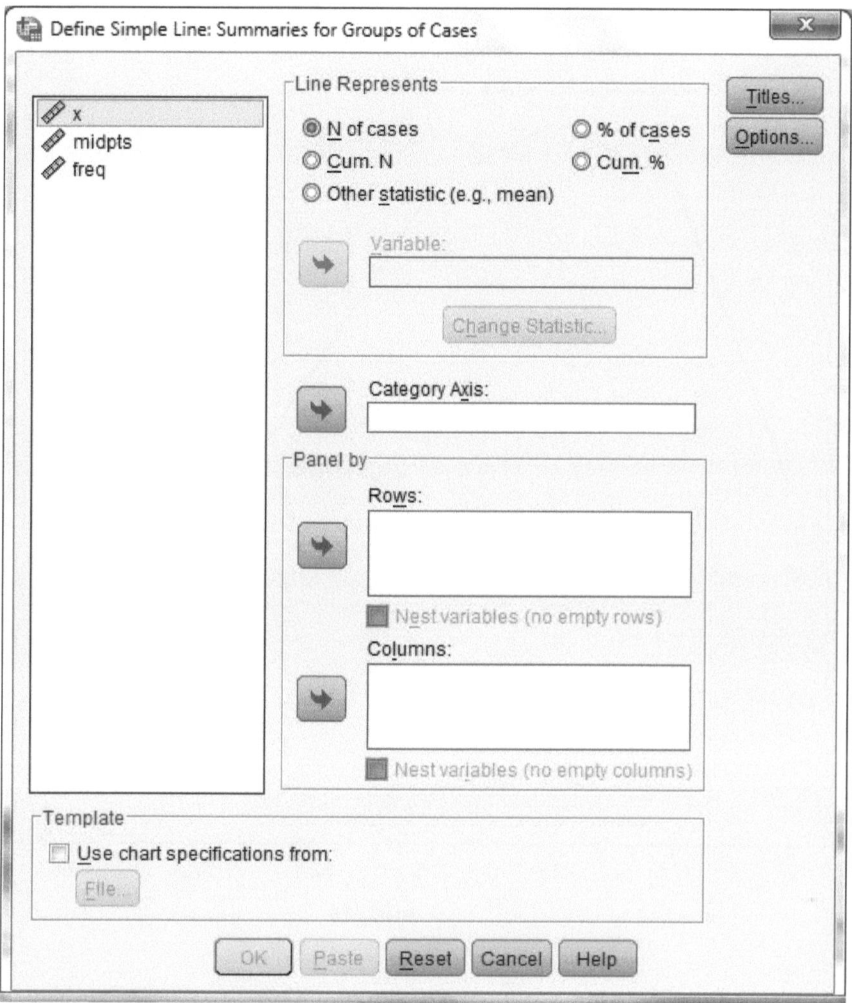

Figure 1m

Click on the circle marked Other statistic, (e.g., mean). Put freq in the Variable box and midpts as the Category Axis (*x*-axis). This will tell SPSS to read the numbers in the freq variables for the vertical-axis values (the *y* values). Hit OK and this graph should appear; **see Figure 1n**. It will look somewhat like our histogram, as it should, because it is the graph obtained by connecting the midpoints on the top of each bar on the histogram.

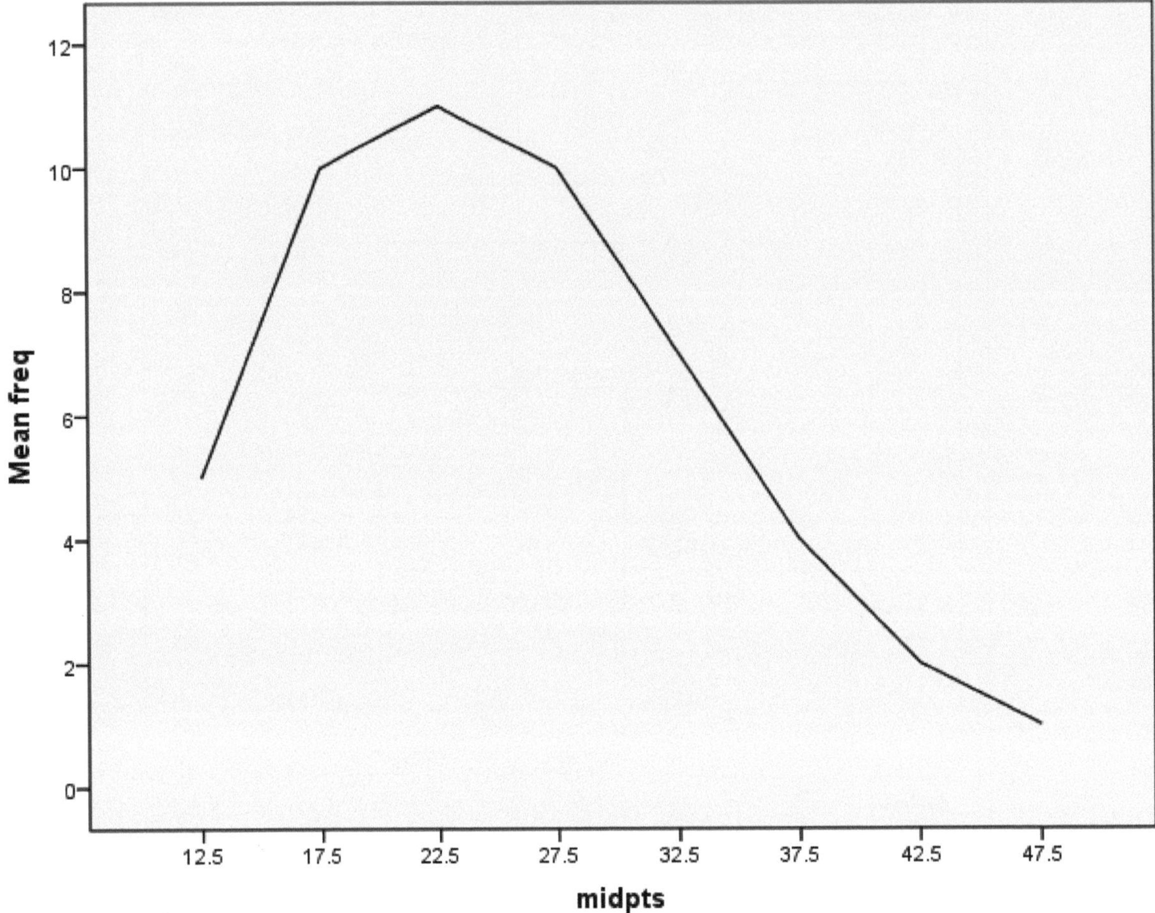

Figure 1n

Cumulative Histograms

The next graph we want to look at is called a cumulative histogram. Cumulative histograms represent the cumulative frequency of the data up to a given point. Before we find the cumulative frequency, we need to find the frequency of each interval on the histogram which was done in drawing the line graph, see **figure 1k**. We can use these data labels to determine the cumulative frequency for our cumulative histogram.

We need to make one column for the endpoints and one column for the frequency as such:

Endpoints	Frequency
10	5
15	10
20	11

25	10
30	7
35	4
40	2
45	1

Now we make another column for the cumulative frequency, which is the sum of all the previous frequencies:

Endpoints	Cumulative Frequency
10	5
15	15
20	26
25	36
30	43
35	47
40	49
45	50

Note that each value in the Cumulative Frequency is the sum of all the values up to that point of the variable Frequency. As you can see, 5 as the first value in Cumulative Frequency is the same as the first value in Frequency. But the next value in Cumulative Frequency is 15, which is the sum of 5 + 10 in Frequency. The next value in Cumulative Frequency is 26, which is the sum of 5 + 10 + 11 in Frequency. 26 + 10 = 36, 36 +7 = 43, 43 + 4 = 47, 47 + 2 = 49 and 49 + 1 = 50. These are the cumulative frequency values: 5, 15, 26, 36, 43, 47, 49, 50.

To make a cumulative histogram, enter the endpoints (endpts) as one variable and the cumulative frequencies (c_freq) as another variable; see **Figure 1o**. Then weight cases using Data → Weight Cases. Click on Weight Cases by and move the cumulative frequency (c_freq) variable into the Frequency Variable window. Click OK. Now create the cumulative histogram using Graphs → Legacy Dialog → Histogram and move the endpoints into the variable window; see **Figure 1p**.

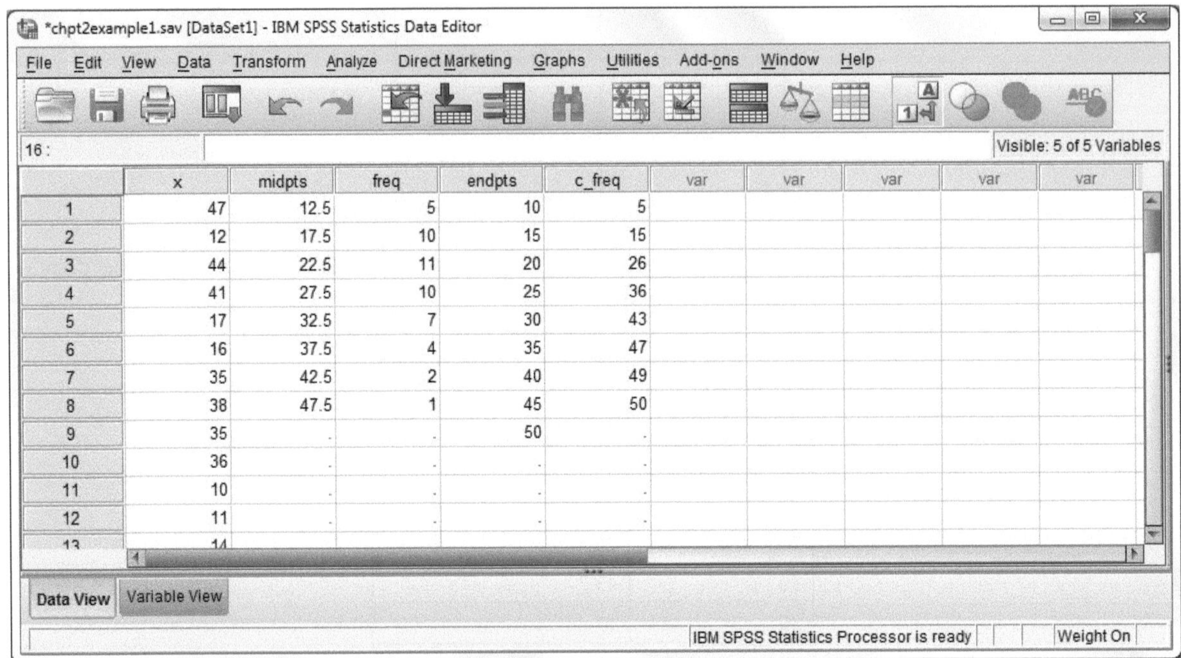

Figure 1o

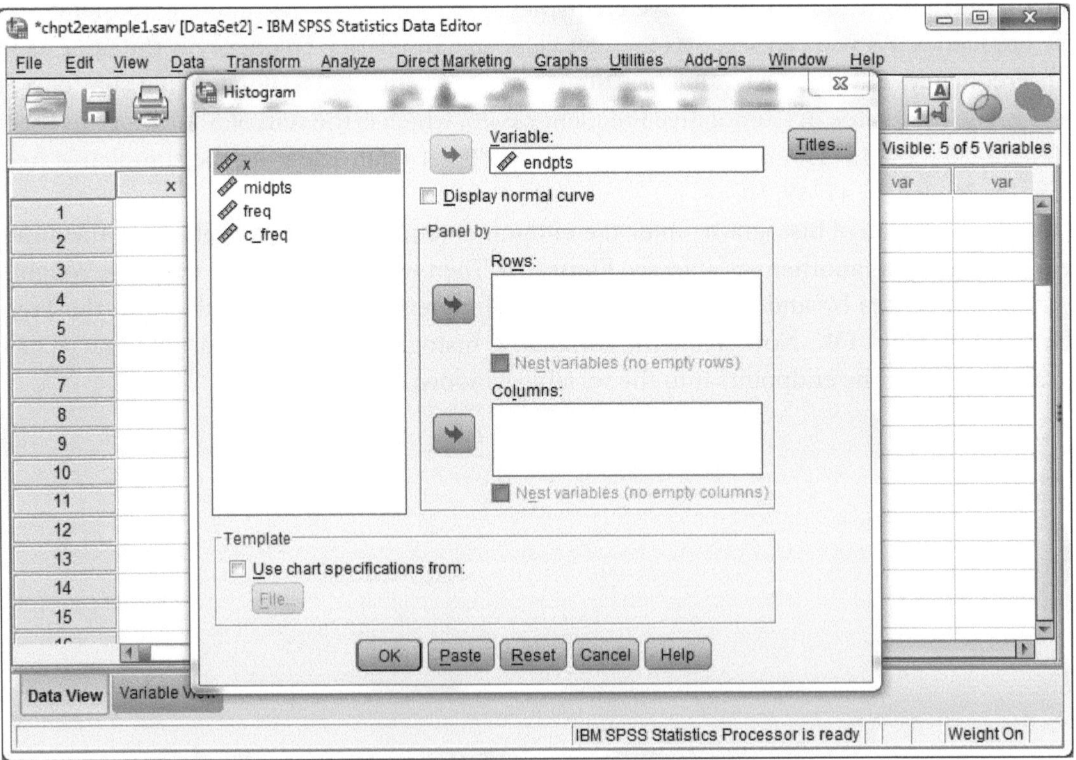

Figure 1p

The graph will be as such; see **Figure 1q.**

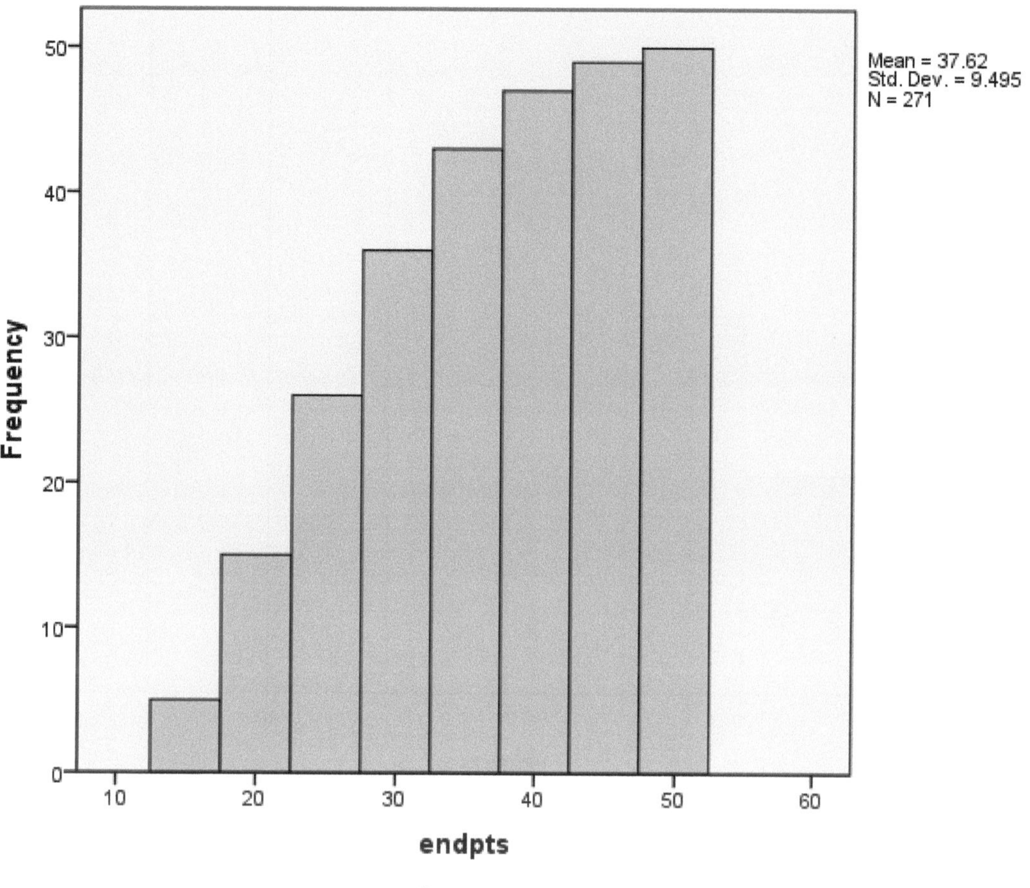

Figure 1q

Again we would like to change our intervals to match the intervals we have for our histogram and our stem-and-leaf plot, so we double click on the graph and use Options → Un-Bin Element → Options → Bin Element and check Custom → Interval Width; set to 5. Also set Custom value for anchor to 10. Click Apply and Close and close the chart editor and the cumulative histogram will look like the one shown in **Figure 1r**.

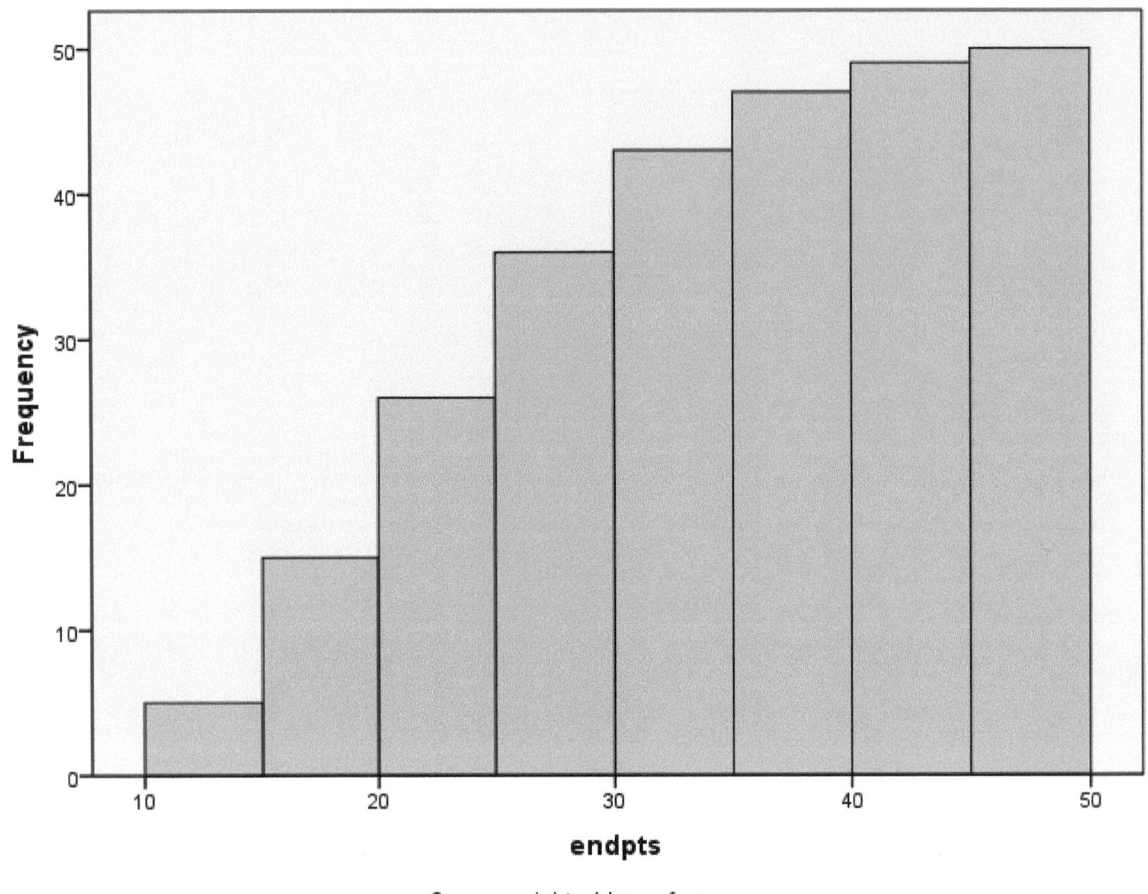

Figure 1r

Relative Frequency and Cumulative Relative Frequency

The frequency column from the stem-and-leaf chart, see **Figure 1d**, is repeated in **Table 1** The second column is called the relative frequency and is the frequency of each row divided by the total number of data values, *n*, which is 50 for this example. The next column is the cumulative frequency, which is the sum of the frequencies of the previous rows and the last row has a cumulative frequency of 50 for the total number of data values. The last column is the cumulative relative frequencies, which is the sum of the relative frequencies of the previous rows. The last cumulative relative frequency is 100%, again the total of all the data values. The cumulative frequency is also available via Analyze → Descriptive Statistics → Frequencies; move *x* into the variable window and click OK. The cumulative frequencies are highlighted in **Figure 1s**.

Frequency	Relative Frequency	Cumulative Frequency	Cumulative Relative Frequency
5	10%	5	10%
10	20%	15	30%
11	22%	26	52%
10	20%	36	72%
7	14%	43	86%
4	8%	47	94%
2	4%	49	98%
1	2%	50	100%

Table 1

x

		Frequency	Percent	Valid Percent	Cumulative Percent
Valid	10	1	2.0	2.0	2.0
	11	1	2.0	2.0	4.0
	12	1	2.0	2.0	6.0
	14	2	4.0	4.0	10.0
	15	2	4.0	4.0	14.0
	16	4	8.0	8.0	22.0
	17	2	4.0	4.0	26.0
	18	1	2.0	2.0	28.0
	19	1	2.0	2.0	30.0
	20	4	8.0	8.0	38.0
	21	4	8.0	8.0	46.0
	23	1	2.0	2.0	48.0
	24	2	4.0	4.0	52.0
	25	2	4.0	4.0	56.0
	26	2	4.0	4.0	60.0
	27	2	4.0	4.0	64.0
	28	1	2.0	2.0	66.0
	29	3	6.0	6.0	72.0
	30	1	2.0	2.0	74.0
	31	2	4.0	4.0	78.0
	32	2	4.0	4.0	82.0
	33	1	2.0	2.0	84.0
	34	1	2.0	2.0	86.0
	35	2	4.0	4.0	90.0
	36	1	2.0	2.0	92.0
	38	1	2.0	2.0	94.0
	41	1	2.0	2.0	96.0
	44	1	2.0	2.0	98.0
	47	1	2.0	2.0	100.0
	Total	50	100.0	100.0	

Figure 1s

Ogive ($\bar{o} - j\bar{i}v$) Cumulative Frequency Polygon:

Another line graph of importance is the cumulative frequency polygon, called the ogive. It is obtained by connecting the endpoints of the bars of the intervals in the cumulative histogram. The first endpoint, 10, will have the value 0. The next endpoint, 15, will have the value 5, the endpoint 20 will have the value 15, and so on.

We will use our example to draw an ogive. Add one more column, one new variable, and give it the values as shown in **Figure 1t**. (To make the ogive, the first frequency is 0; when we make the cumulative histogram, the first frequency is 5.) Follow the same procedures as before to make a line graph. The ogive is shown in **Figure 1u**.

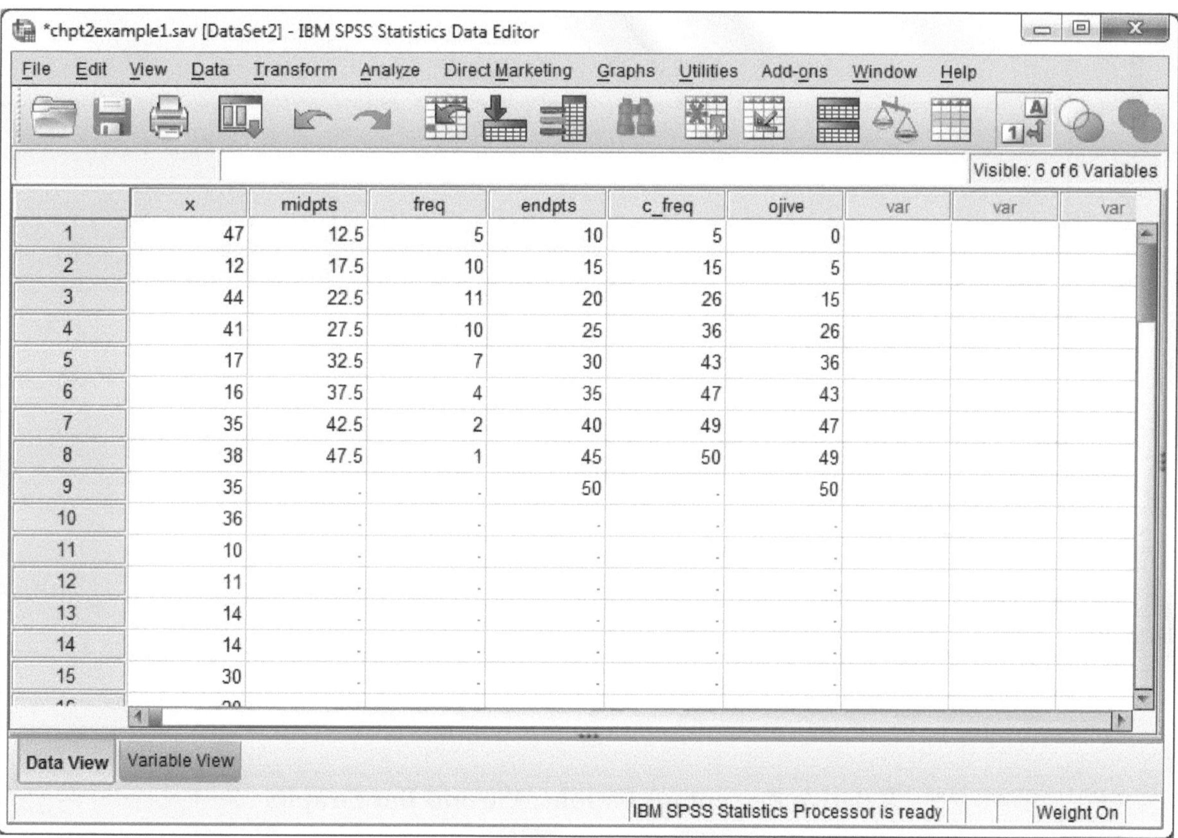

	x	midpts	freq	endpts	c_freq	ojive	var	var	var
1	47	12.5	5	10	5	0			
2	12	17.5	10	15	15	5			
3	44	22.5	11	20	26	15			
4	41	27.5	10	25	36	26			
5	17	32.5	7	30	43	36			
6	16	37.5	4	35	47	43			
7	35	42.5	2	40	49	47			
8	38	47.5	1	45	50	49			
9	35	.	.	50	.	50			
10	36	.	.	.	.	.			
11	10	.	.	.	.	.			
12	11	.	.	.	.	.			
13	14	.	.	.	.	.			
14	14	.	.	.	.	.			
15	30	.	.	.	.	.			

Figure 1t

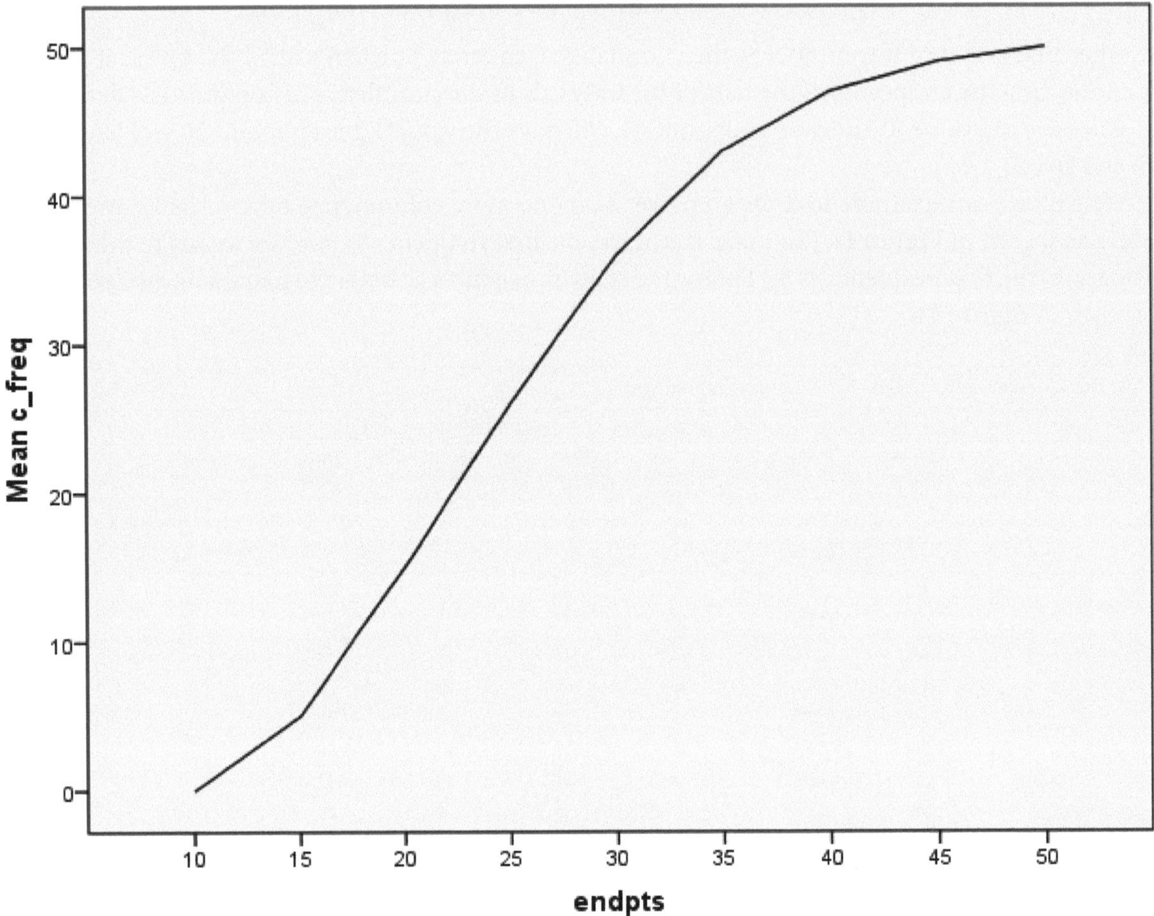

Figure 1u

Relative Percentage Graphs: Pie and Bar Graphs

Two other graphs we want to talk about in this section are bar graphs and pie charts. Bar graphs and pie charts are both used to represent relative frequency of several variables, such as the relative frequency of ethnicity in a class, gender, full-/part-time student, and so on. Bar graphs and pie charts can be used for any variable type, but mostly they are used for the nominal variable. The nominal variable doesn't have numerical data, so a histogram is not appropriate for nominal data. Only a bar graph or pie chart will do. We will use party membership of US Senators in 2012 as our example.

Example 2.

Open the data file PartyMembershipsUSsen2012. Click on Graphs → Legacy Dialogs → Bar, leave Simple as default, check Summaries of separate variables, and click Define. Highlight the three variables,

Democrats, Republicans, and Independents and move them to the box labeled Bars Represent and Change Statistic to Sum of values. Click OK and you should see the graph in **Figure 2a**.

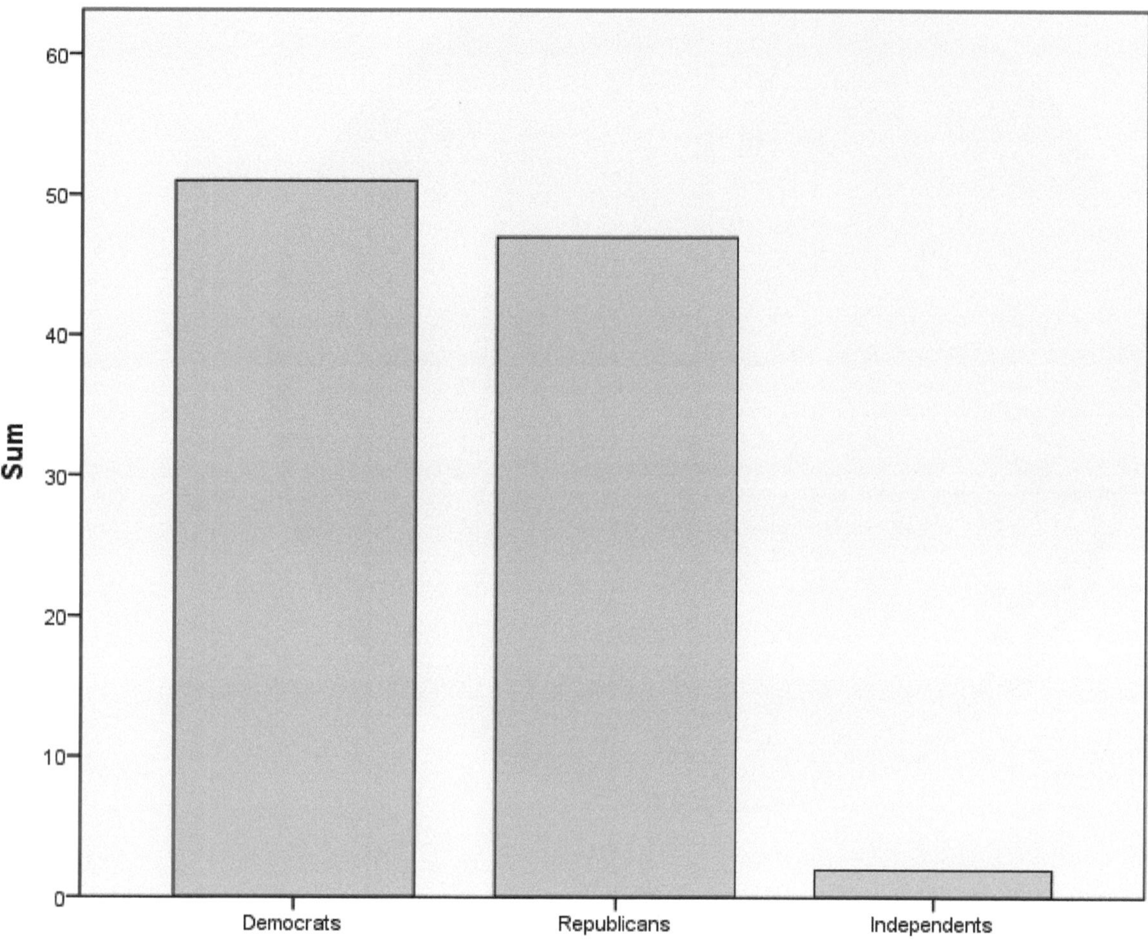

Figure 2a

For a pie chart do the same process except we don't need to change statistic. Use Graphs → Pie, check the circle Summaries of separate variables, and click Define. Hit OK and **see Figure 2b**. To find the number of degrees in a slice of the pie chart, we divide the number of data items that represent the slice, such as the number of democrats—which is 51—by the total number of data items—the total number of senators is 100—and multiply by $360°$. We have $\frac{51}{100} \cdot 360° = 183.6°$. However SPSS will do this calculation for us.

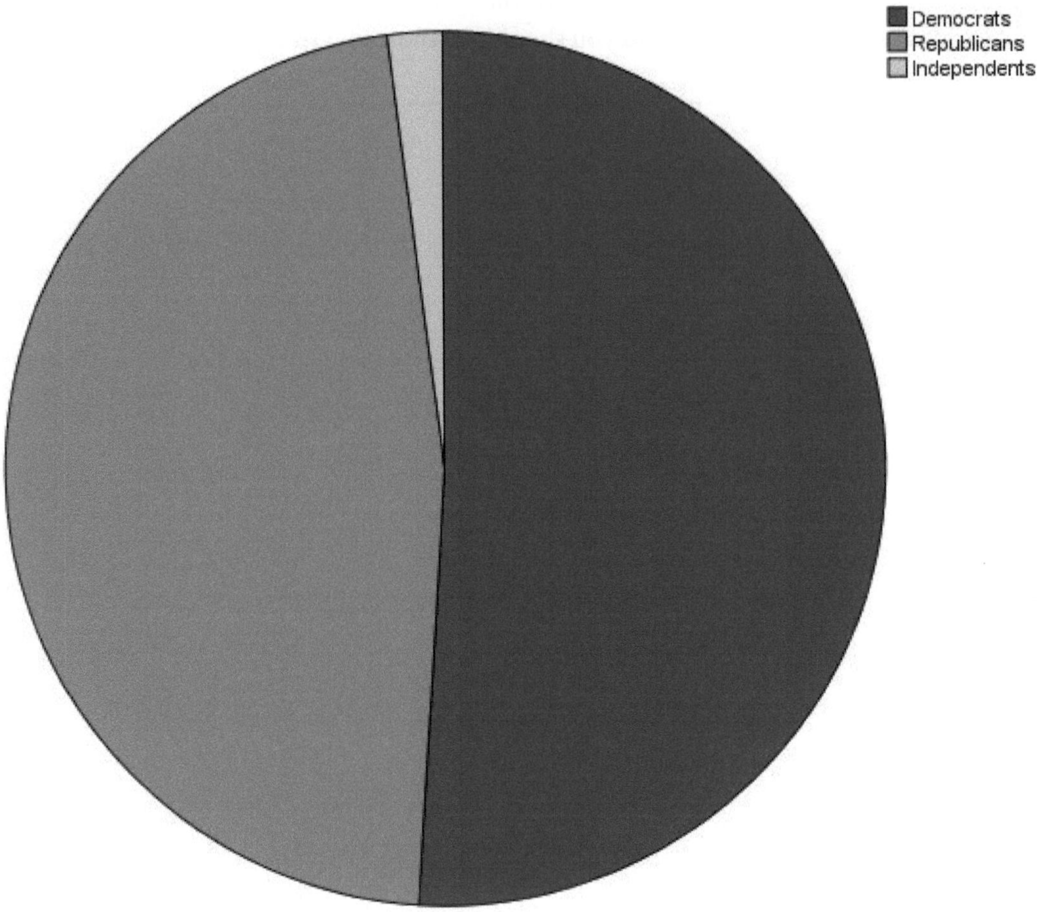

Democrats
Republicans
Independents

Figure 2b

Let's go back to our histograms. Two important statistics we will discuss in the next chapters are the mean (average value) and the standard deviation (average distance from the mean). In Chapter 6 we will discuss the Normal Distribution curve, but we want to use it in this section as well. Essentially a normal distribution curve has only two parameters, mean and standard deviation. We will use the mean and standard deviation of a frequency distribution to draw a series of normal distribution curves that relate to a series of histograms.

Example 3.

Open the data set Sleep.(3) We want to draw several histograms and their associated normal distribution curves.

Skewness

Skewness is a measure of the difference between the mean and the median. If the mean is larger than the median, then the skewness is positive. If the mean is smaller than the median, then the skewness is negative. If the mean equals the median, then the skewness is zero. We will use Analyze → Descriptive Statistics → Explore, move all the variables (body weight, brain weight, non-dreaming sleep, dreaming sleep, total sleep, maximum lifespan, and gestation) to the dependent list. Under plots deselect stem-and-leaf and select histograms. Click continue and click OK, or you can use Analyze → Descriptive Statistics → Frequencies; move all above variables into the Variables window, deselect Display frequencies, click the tab Charts, and select Histogram with normal curve; **see Figure 3a**. Hit continue and OK. Observe the different histograms and the associated normal curves; **see Figure 3c.**

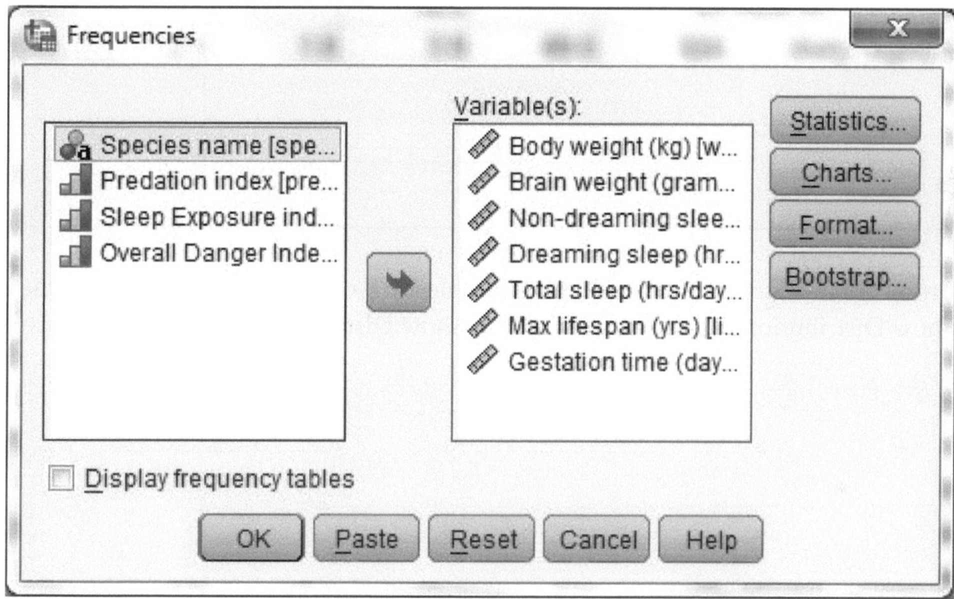

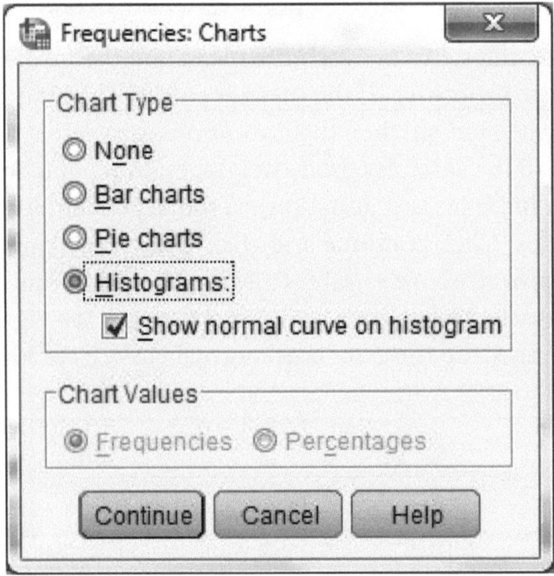

Figure 3a

If the normal curve is not on the graph, then double click on the graph to bring up the chart editor and select Show Distribution Curve; see **Figure 3b**. (Show Distribution Curve is also under Elements.)

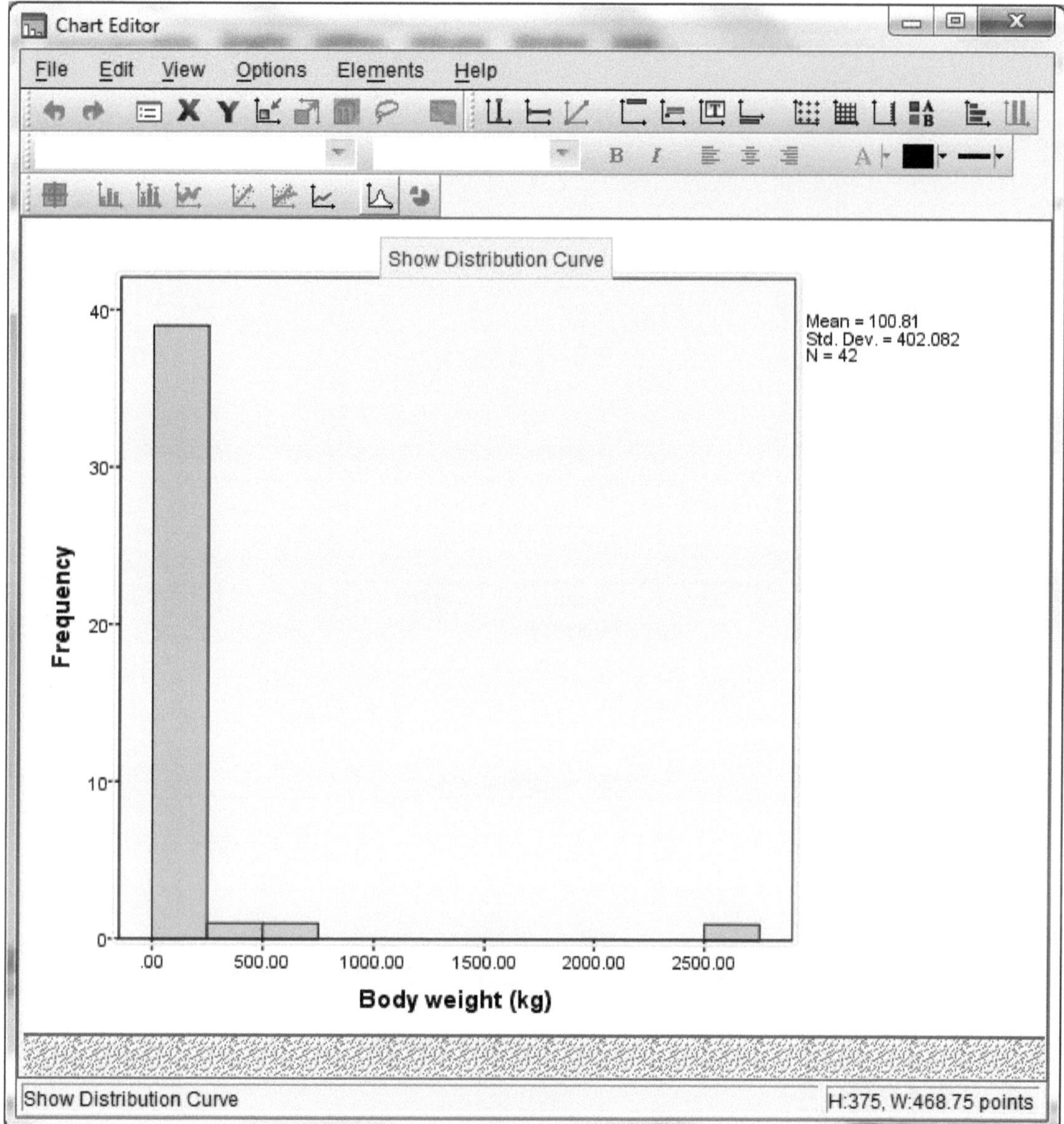
Chart Editor
File Edit View Options Elements Help
Show Distribution Curve
Mean = 100.81
Std. Dev. = 402.082
N = 42
Frequency
40
30
20
10
0
.00
500.00
1000.00
1500.00
2000.00
2500.00
Body weight (kg)
Show Distribution Curve
H:375, W:468.75 points

Figure 3b

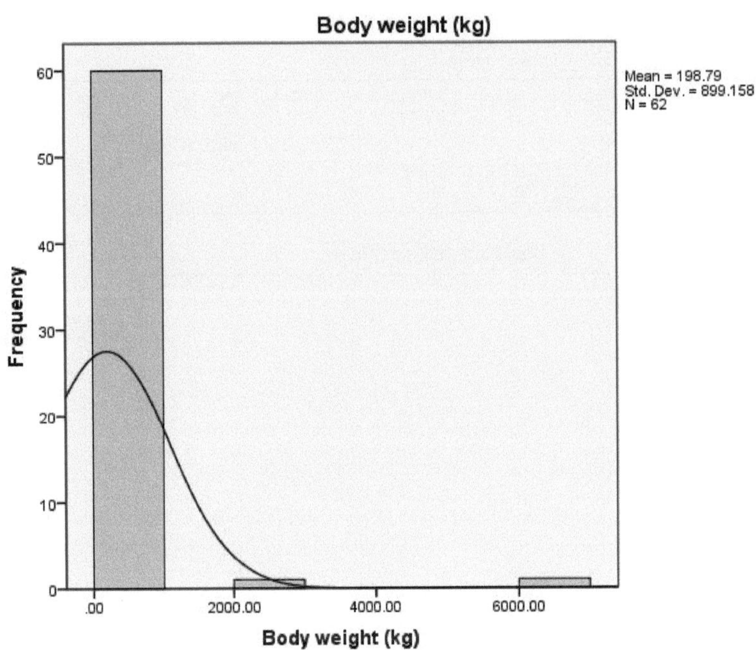

Skewness 5.803, Kurtosis 35.505

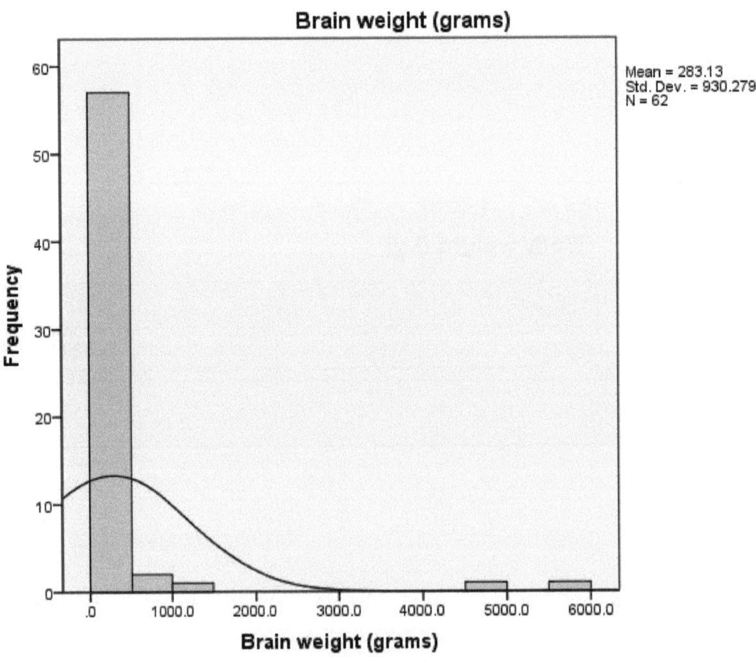

Skewness 5.560, Kurtosis 33.092

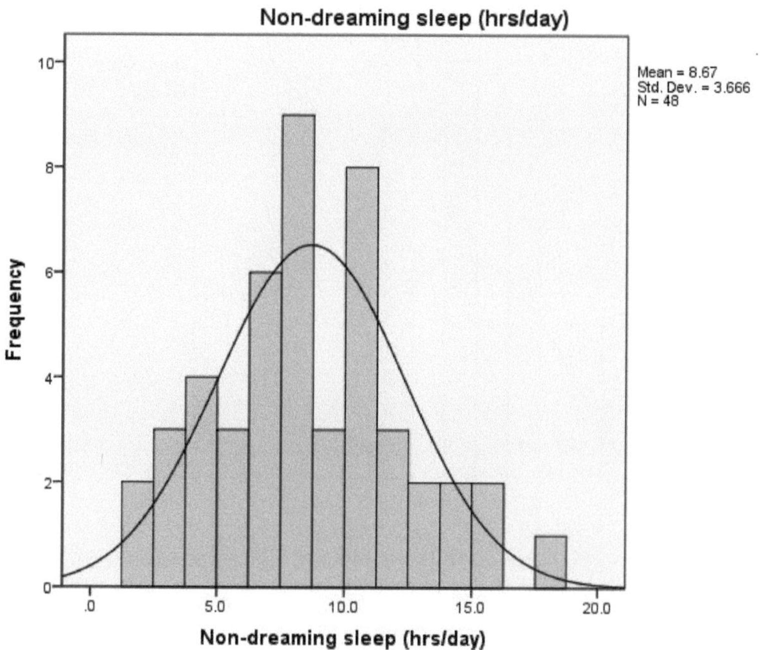

Skewness .253, Kurtosis -.414

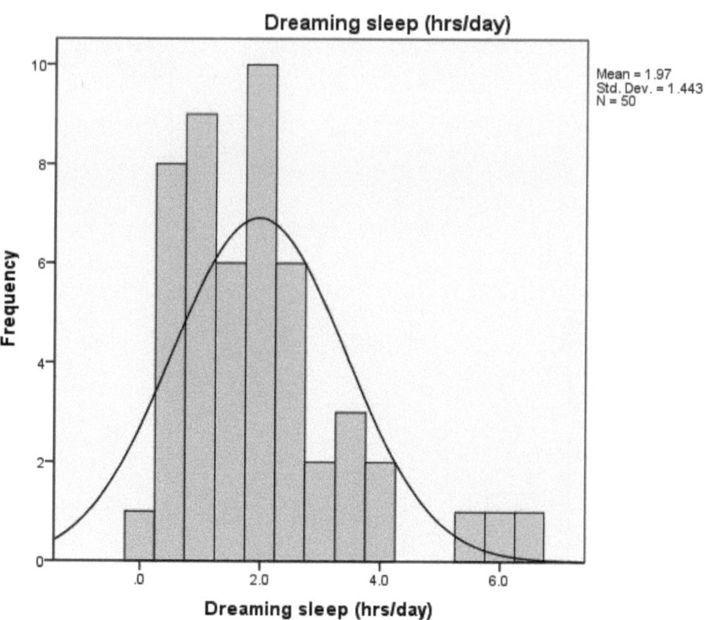

Skewness 1.499, Kurtosis 2.670

Total sleep (hrs/day)

Mean = 10.53
Std. Dev. = 4.607
N = 58

Frequency

Total sleep (hrs/day)

Skewness .362, Kurtosis -.596

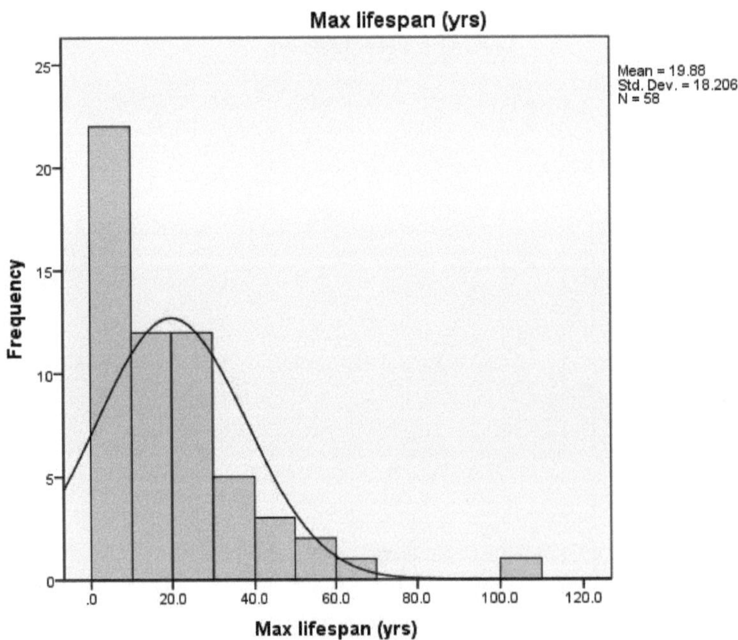

Skewness 2.085, Kurtosis 5.449

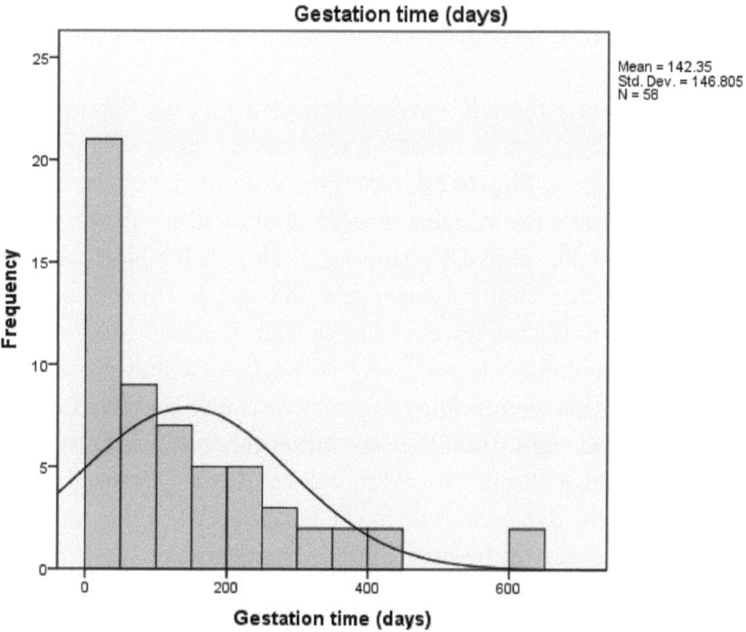

Skewness 1.759, Kurtosis 4.206

Figure 3c

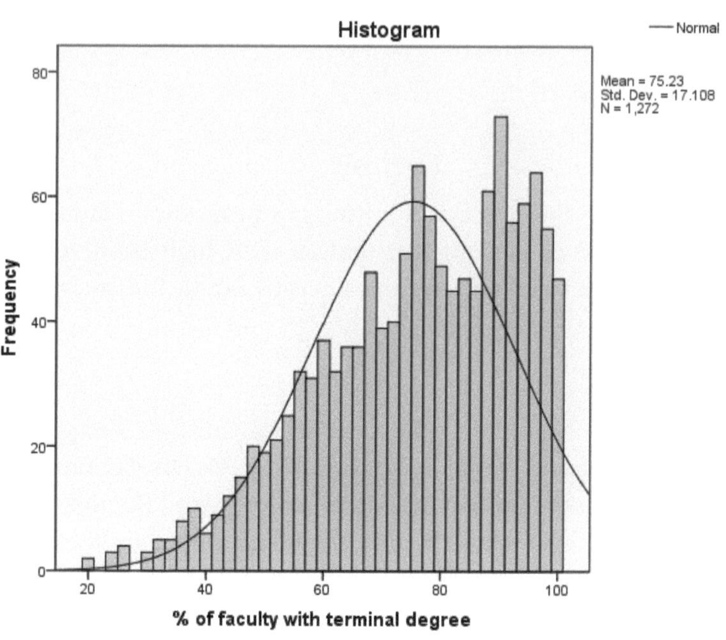

Skewness -.623, Kurtosis -.246

Figure 3d

A standard normal curve will be discussed in Chapter 6 but looks like the normal curve associated with the Total Sleep histogram or the non-dreaming sleep curve; however, the histograms are not normal distributions.

If the peak is shifted to the left and there is a long tail to the right, we say the distribution is skewed right, as in Max lifespan; see **Figure 3c**. A distribution that has the peak on the right and the long tail on the left is said to be skewed left; see **Figure 3d**. Open the data file labeled Colleges, use Graphs → Legacy Dialogs → Histograms, move the variable labeled % of faculty with terminal degree (facterm) to the variable window, and proceed as above. Be sure to check Display Normal Curve. Click OK. The observed histogram and associated normal curve are now skewed to the left; see **Figure 3d**. Note the peak of the curve is on the right and the tail flares out to the left. It is said to be skewed left.

We also say a skewed-right distribution is positively skewed and a skewed-left distribution is negatively skewed. The motivation for this terminology is one measure of skewness is Mean minus Median ($\overline{x}-m$). If a distribution is skewed right, then the few values far to the right will pull the mean value over to the right, a few larger values will raise the mean, hence the mean minus median is positive. The median is the point where 50% of the data values lie to the left and 50% of the data values lie to the right, so the median will be closer to the peak of the distribution. Hence if the peak is to the left and the tail is to the right, then the median will be left of the mean and $\overline{x}-m$ will be positive. On the other hand if the peak is to the right and the tail is to the left, then the mean will be to the left of the median and $\overline{x}-m$ will be negative. So a positive-skewed distribution is right-skewed (positive is to the right on the number line), and a negative-skewed distribution is left-skewed (negative is to the left on the number line).

When SPSS gives the output of descriptive values it will have a skewness value toward the bottom. There are different formulas to calculate a skewness value, but we won't concern ourselves with the actual formula. We'll just note if the skewness value is positive or negative and that a skewness value other than zero indicates a degree of skewness.

Kurtosis

Our next concept in this chapter is the concept of kurtosis or peakness—flatness of a distribution. The kurtosis value tells us how peaked or flat our distribution is. A high positive value means it is very peaked, note body weight and brain weight above, and a negative value indicates it is somewhat flat, note non-dreaming sleep and total sleep above.

Example 4.

Open the data set Sleep again. Use Analyze → Descriptive Statistics → Frequencies and move all the variables we used in Example 4 into the variable window. Uncheck Display Frequency tables. Click on the Statistics tab and put a check in the Distribution box Skewness and Kurtosis and click continue and OK; **see Figure 4a**. We'll concern ourselves for the time being with only the Skewness and Kurtosis values; the standard error involves the sample size and it is not of concern to us at this point.

Statistics

		Body weight (kg)	Brain weight (grams)	Non-dreaming sleep (hrs/day)	Dreaming sleep (hrs/day)	Total sleep (hrs/day)	Max lifespan (yrs)	Gestation time (days)
N	Valid	62	62	48	50	58	58	58
	Missing	0	0	14	12	4	4	4
Skewness		6.564	5.072	.298	1.453	.201	2.014	1.684
Kurtosis		45.741	26.271	-.236	2.320	-.508	5.885	2.852

Figure 4a

Statistics

% of faculty with terminal degree

N	Valid	1272
	Missing	30
Skewness		-.623
Std. Error of Skewness		.069
Kurtosis		-.246
Std. Error of Kurtosis		.137

Figure 4b

In Non-Dreaming Sleep and Total Sleep, the Kurtosis values are negative. The graph of non-dreaming sleep shows several bars that have approximately the same height, hence the distribution is flat over those intervals. The graph of total sleep has alternating intervals whose bars are high and low so that the average value is approximately the same across those intervals, thus fattening the distribution across those intervals; **see Figure 3c**. Follow this same process with Colleges, % of Faculty with Terminal Degrees (**see Figure 4b**), and note the Kurtosis value is again negative, indicating a flatness for the distribution (as can be noted on the intervals with higher bars, they will balance out with the neighboring intervals with low bars).

Although a skewed right distribution has mean – median > 0, that is not always the case; consider the following example.

Example 5.

Consider the following frequency distribution: chpt2example5.sav

X	1	2	3	4	5	6	7	8
Freq	7	7	6	4	2	1	1	1

Show that the histogram is skewed right, but mean minus median is negative.

Solution: Open the data set chpt2example5.sav. Make sure Data is weighted by freq (use Data → Weight Cases, check weight cases by, move freq into the Frequency Variable window, and hit OK). Use Graphs → Legacy Dialogs → Histogram, move x into the variable window, check display normal curve, and hit OK. As we can see in **Figure 5a**, the graph is skewed right. Now use Analyze → Descriptive Statistics → Frequencies, move the x variable into the Variables window, click the tab Statistics, and check mean, median, and skewness, as in **Figure 5b**. The output should read the mean is 2.966 and the median is 3; hence mean minus median is negative, but the skewness is positive; **see Figure 5c**.

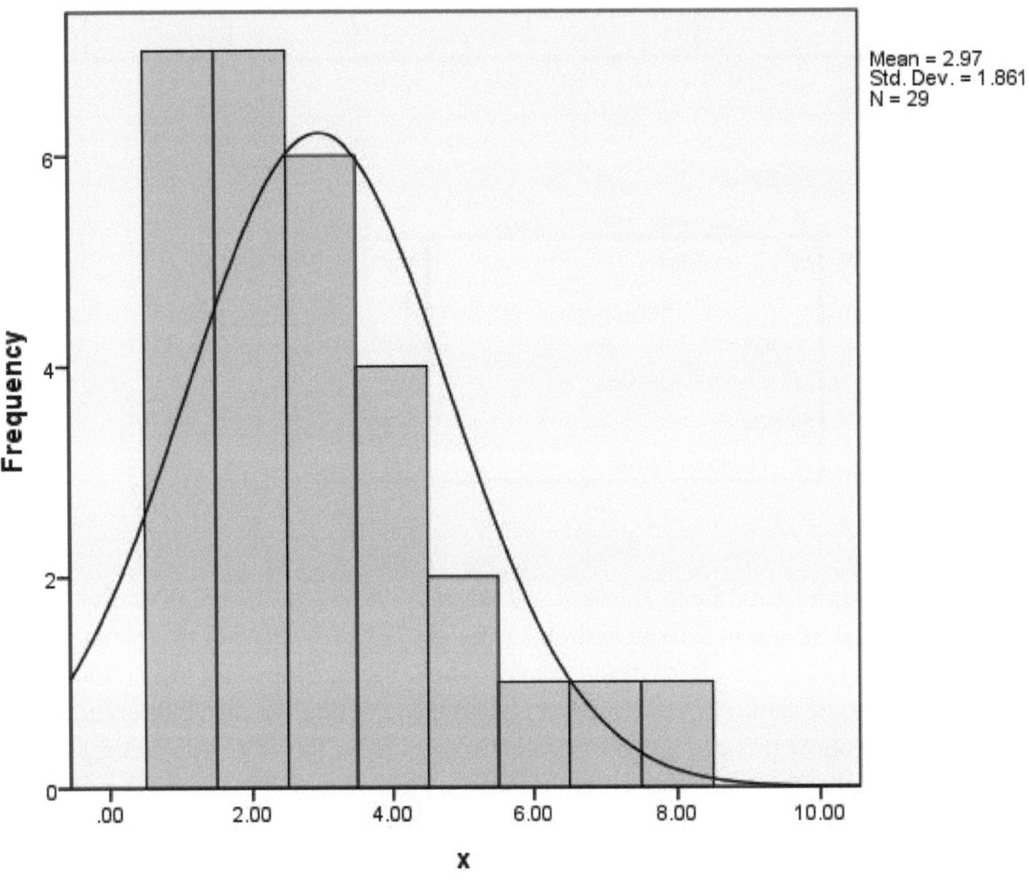

Figure 5a

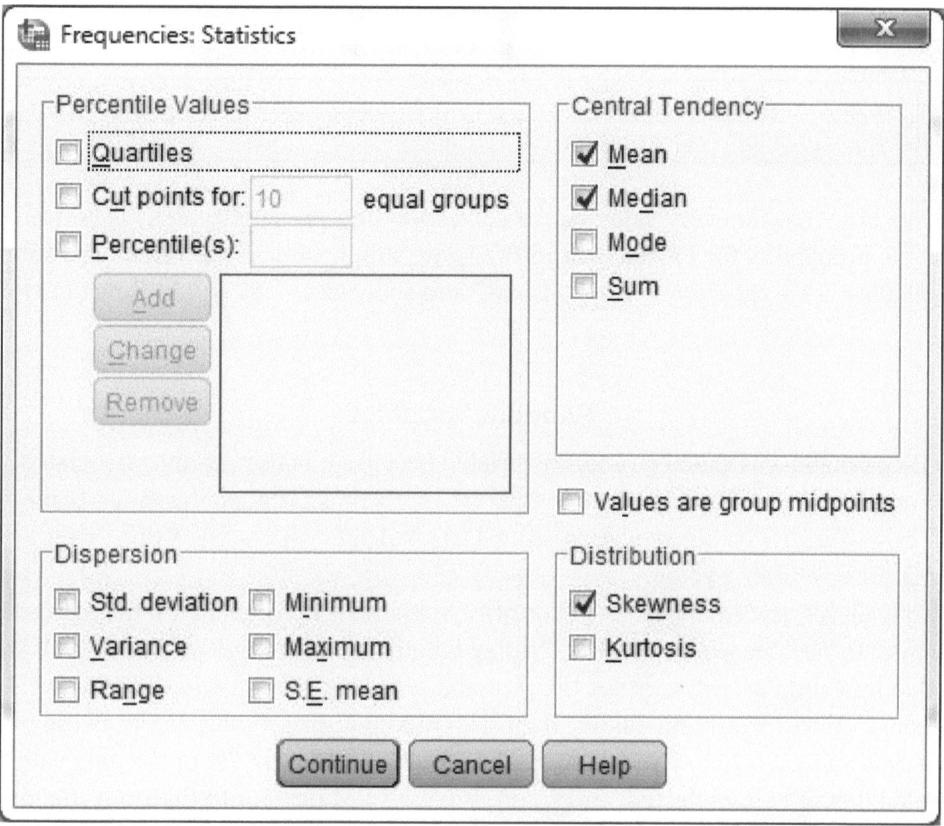

Figure 5b

Statistics

x

N	Valid	29
	Missing	0
Mean		2.9655
Median		3.0000
Skewness		1.089
Std. Error of Skewness		.434

Figure 5c

Probability Distributions

A probability distribution is such that the probability for any value is between 0 and 1 (0% and 100%) and the total of all the probabilities of the values is 1 (100%).

Example 6.

Find the missing probability in the following probability distribution:

X	3	7	11	13	14	17	21
P(X)	.05	.12	.22	?	.14	.23	.11

Solution: We note that for every value X, the corresponding probability P(X) is between 0 and 1, so to determine the probability for 13, that is find P(13), we simply add all the values and subtract from 1. That is we add .05 + .12 + .22 + .14 + .23 + .11 = .87 and subtract 1 – .87 = .13. Thus P(13) = .13.

Random Variables

In the previous example X is called a random variable because it is the variable associated with a probability distribution function. In particular if $x \in X$, then $P(x)$ is assigned the values (called the random variate) $P(3)=.05, P(7)=.12, P(11)=.22, P(13)=.13, P(14)=.14, P(17)=.23, P(21)=.11$, and $P(x)=0$ for any other *x* value.

Go back to Colleges, use Analyze → Descriptive Statistics → Frequencies, move % of Faculty with Terminal degrees to variable window, check Display Frequency tables, and click on OK. If we look at the column headed by Valid Percent, we'll see the probability for each X value in that data set. If we were to consider the data values as random values, then the corresponding valid percent is the probability for selecting that data value. So 20 has a .2% probability of being selected (.2% of the data values are 20), 23 has a .1% probability of being selected, and so on. We use valid percent to discount the missing values, thus making the total of the probabilities of all the values 100% or 1. Therefore for any distribution in SPSS, valid percent is always a probability distribution—the probability for each value is between 0 and 100% and the total of all the probabilities is 100%.

Let's review what we learned in this chapter:

1. Stem-and-Leaf Display—the first step in organizing data
2. Histograms—a graphical representation of the data set
3. Cumulative Histogram—very useful when we talk about Cumulative Distribution Functions
4. Line graphs
    a.) Frequency polygon—a line graph showing the frequency distribution of the data set
    b.) Ogive or Cumulative frequency polygon—a line graph that starts at 0 and ends at 100% or the total frequency on the Y-axis.
5. Bar and Pie Graphs—these graphs are used to demonstrate relative frequency among nominal variables.
6. Normal Distribution
    a.) Skewness—a measure of how skewed the data are
    b.) Kurtosis—a measure of how flat the data are
7. Probability Distribution—the probability for each value is between 0 and 1 (0 and 100%) and the total of the probabilities of all the values is 1 (100%)

Class Work 2

Use the age variable from the survey in Chapter 1 to make a stem-and-leaf plot and from that graph make a histogram, frequency polygon, cumulative frequency histogram, and ogive. (If you don't have a class survey, then you may use the data set labeled class survey 2007.) Are the data skewed? What is the skewness value? Are the data peaked or flat? What is the kurtosis value?

Homework 2

1. Use the following data to make a stem-and-leaf plot for the number of fireflies in twenty-four different ten-meter squares on a June night. (Chpt2problem1.sav)

68	66	22	19	21	47	39	36	21	41	15	36
35	51	22	38	48	26	23	20	56	70	65	47

2. Use the following data to make a stem-and-leaf plot for the heart rates of twenty-two adult men after twenty-minute workouts on an aerobic machine. (Chpt2problem2.sav)

115	169	182	132	145	117	142	139	163	155	145
137	179	169	110	125	124	134	136	119	146	122

3. Use the following data to make a stem-and-leaf plot for the number of unusable prints in a run of twenty 1000 prints. (Chpt2problem3.sav)

90	109	107	102	99	91	101	108	112	94
92	101	102	91	89	94	100	92	95	92

4. Use the following data to make a stem-and-leaf plot for the percentage of instructors who use the text "Statistics with Power Using SPSS" in twenty-two different institutions. (Chpt2problem4.sav)

1.2	3.5	10.4	4.0	6.1	7.8	2.3	9.5	4.7	5.4	3.2
2.1	6.3	4.1	1.7	3.3	8.1	3.5	4.1	5.6	1.8	2.3

5. Create a histogram using the data from problem #1. Experiment with different interval sizes and different numbers of intervals.

6. Create a histogram using the data from problem #2. Experiment with different interval sizes and different numbers of intervals.

7. Create a histogram using the data from problem #3. Experiment with different interval sizes and different numbers of intervals.

8. Create a histogram using the data from problem #4. Experiment with different interval sizes and different numbers of intervals.

9. Create a Cumulative histogram using the data from problem #1. Experiment with different interval sizes and different numbers of intervals. Draw a frequency polygon and a cumulative frequency polygon (ogive). Use Graphs →- Line to draw the line graphs.

10. Create a Cumulative histogram using the data from problem #2. Experiment with different interval sizes and different numbers of intervals. Draw a frequency polygon and a cumulative frequency polygon (ogive). Use Graphs →- Line to draw the line graphs.

11. Create a Cumulative histogram using the data from problem #3. Experiment with different interval sizes and different numbers of intervals. Draw a frequency polygon and a cumulative frequency polygon (ogive). Use Graphs →- Line to draw the line graphs.

12. Create a Cumulative histogram using the data from problem #4. Experiment with different interval sizes and different numbers of intervals. Draw a frequency polygon and a cumulative frequency polygon (ogive). Use Graphs →- Line to draw the line graphs.

13. Find the Kurtosis and Skewness values for the data from problem #1.

14. Find the Kurtosis and Skewness values for the data from problem #2.

15. Find the Kurtosis and Skewness values for the data from problem #3.

16. Find the Kurtosis and Skewness values for the data from problem #4.

17. Find the missing value in the following probability distribution:

x	8	10	13	17	19	21	22	28	30	33
P(x)	.10	.18	.12	.08	.13	?	.14	.09	.02	.06

18. Find the missing values in the following probability distribution if we know P(13) is twice P(26).

x	9	11	13	14	18	22	24	26	27	29
P(x)	.08	.05	?	.11	.12	.04	.01	?	.10	.10

19. Find the missing values in the following probability distribution if we know P(8) is twice P(4) and P(12) is the sum of P(8) and P(4).

x	2	4	6	8	10	12	14	16	18	20
P(x)	.08	?	.10	?	.05	?	.03	.11	.15	.06

20. Fill in the following probability distribution for the sum of the outcomes of the toss of two dice. (One die has 6 sides, hence 6 outcomes 1, 2, 3, 4, 5, 6. Two dice have 6 x 6 = 36 outcomes, 6 choices for the first die and 6 choices for the second die. The possible sums are given below; find the missing probabilities. That is, find the number of outcomes that have the given sum.)

x	2	3	4	5	6	7	8	9	10	11	12
P(x)	1/36	?	3/36	?	5/36	?	?	4/36	?	2/36	?

21. Show that .9999... = 1. This proves there is no space, no matter how small, between .9999... and 1.

Just for fun: Create a data set with 44 data values ranging from 10 to 20, such that the skewness value and the kurtosis value are both equal to 0. (Or make the values as close to 0 as possible.)

Reference

(1) NCTM Mathematics Teacher, January 2006 Volume 99, Issue 5, page 372

(2) Chaotic Elections! A Mathematician Looks at Voting

(3) (Data from which conclusions were drawn in the article "Sleep in Mammals: Ecological and Constitutional Correlates" by Allison, T. and Cicchetti, D. (1976), *Science*, November 12, vol. 194, pp. 732–734. Includes brain and body weight, life span, gestation time, time sleeping, and predation and danger indices for sixty-two mammals. Submitted by Roger Johnson (rwjohnso@silver.sdsmt.edu) [27/Jul/94] (8k)).

Chapter 3

Measures of Central Tendency

There are three measures of central tendency; they are the **mean**, **median**, and the **mode**.

Objectives

After completing this chapter, a student should be able to:

✓ Calculate and understand the definition of the mean
✓ Calculate and understand the definition of the median
✓ Calculate and understand the definition of the mode

Mean:

The mean is the average value of a set of data. It can also be thought of as the center of gravity of data; that is, the sum of the distances from the mean for each data value on one side of the mean will equal the sum of the distances from the mean for each data value on the other side. The mean will balance the data; in fact if the differences of all the data values and the mean are added together, then the sum will be zero. The formula for the mean is $\frac{\sum x}{n}$ where capital sigma, $\sum$ is the summation symbol and simply means "add values together," x represents the individual values, and n is the total number of values.

Example 1.

Find the mean of the following data. Also compute a new variable which is the difference of the mean and the x values and show the sum of that variable is zero. 6, 18, 10, 12, 9, 5, 4, 3, 15, 13, 12, 16, 18, 15,

14, 13, 17, 7, 6, 7 (chpt3example1.sav). (This problem can be done using a hand-held calculator before using SPSS.)

Solution: Enter the data into a SPSS data editor or open the data file chpt3example1.sav. To find the mean use Analyze → Descriptive Statistics → Frequencies, move the *x* variable into the variable window, click on the tab Statistics, and put a check mark in the mean box; **see Figure 1a**.

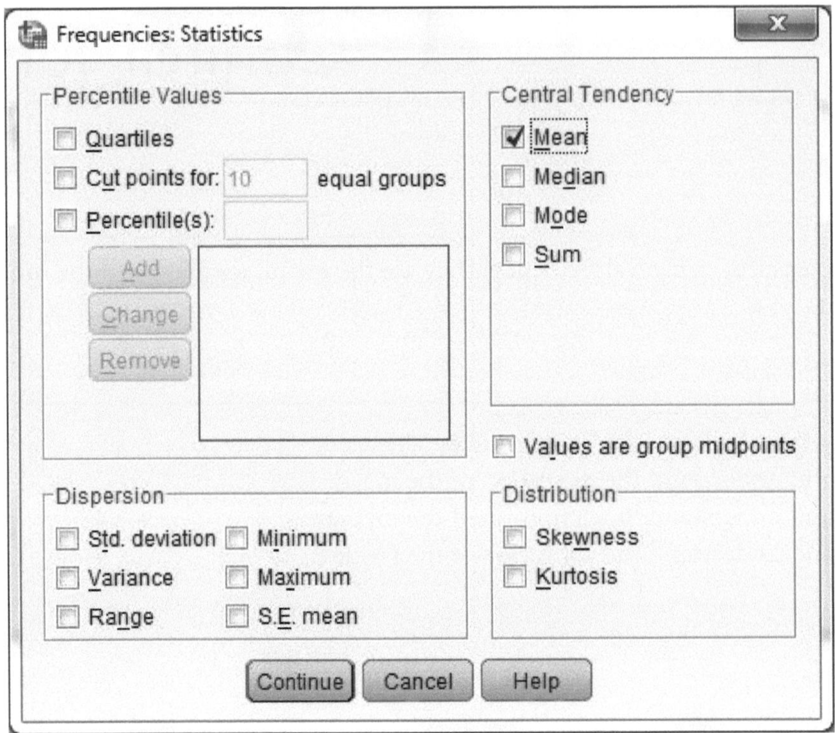

Figure 1a

Click on Continue then OK. The mean is 11; see **Figure 1b**.

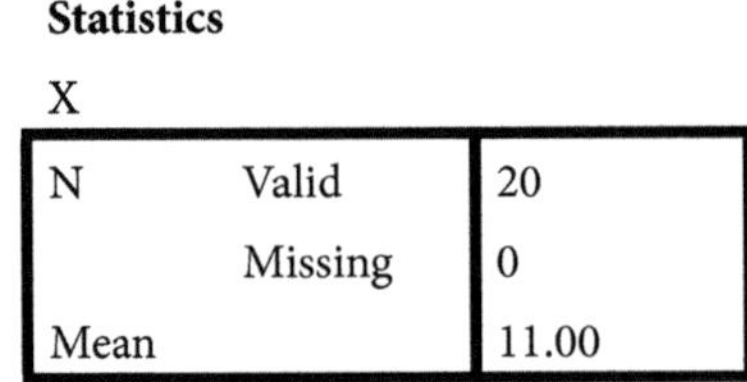

Statistics

X

N	Valid	20
	Missing	0
Mean		11.00

Figure 1b

(To avoid the display frequencies tables, deselect display frequencies tables after Analyze → Descriptive Statistics → Frequencies.)

As a matter of academic understanding of the mean, we'll also find the mean by use of the formula $\bar{x} = \frac{\sum_{i=1}^{n} x_i}{n}$. To find the sum of the data values, use Analyze → Descriptive Statistics → Frequencies, click on the tab Statistics, and put a check mark in the sum box. Hit OK. The sum is 220; see **Figure 1c**. The total number of data values is 20 (note the statistics output box has 20 next to N, which is the number of data values) so the mean is 220/20 = 11.

Statistics

X

N	Valid	20
	Missing	0
Sum		220

Figure 1c

Let's check to make sure the sum of the differences of all the x values and the mean is zero. Go back to the data editor and click on Transform → Compute, call the new variable diff (difference), and create the formula diff = x – 11; **see Figure 1d**.

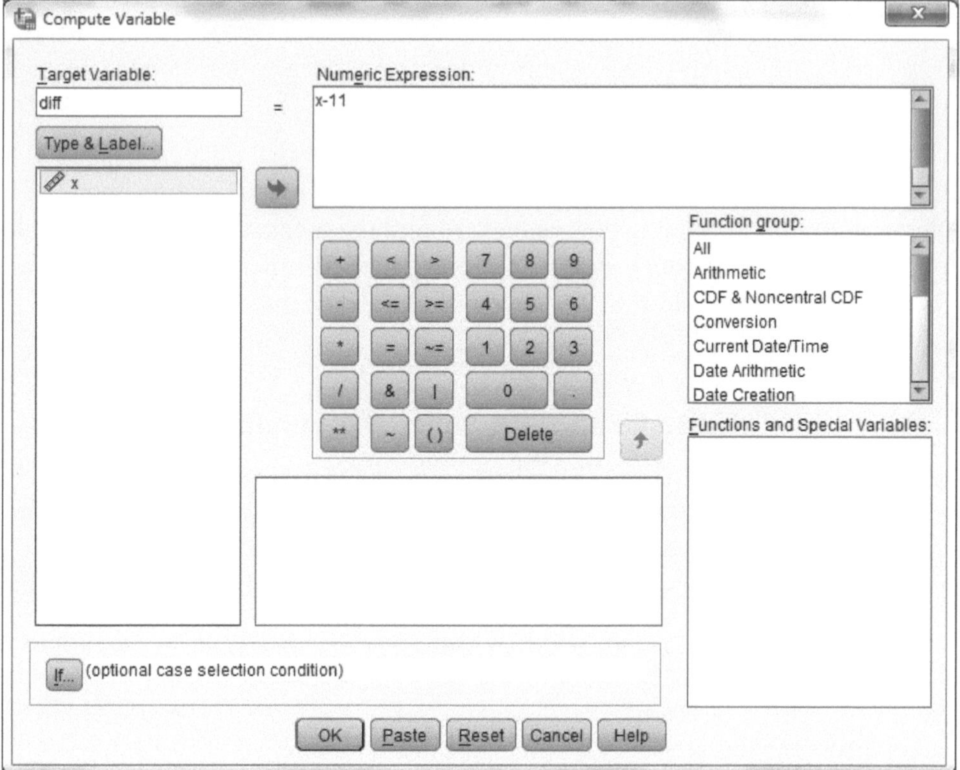

Figure 1d

Click OK and a new variable will be created with the difference of x and the mean. Now find the sum of the diff column (use the procedure above). Is it zero? This illustrates that the mean is the center of gravity. (We will assume everyone is familiar with the concept of center of gravity because we're sure everyone played with their utensils at the dinner table at one time in their past.) (Teaching tip: this data set can also be used to illustrate the concepts of median and mode. A 21st data value can be added for further illustration. These concepts are explained below.)

The symbol for the mean is $\overline{x}$ (x bar), so the equation for the mean can be written as $\overline{x} = \frac{\sum x}{n}$, where the summation symbol means to add all the x values.

Example 2a.

Use Chapter 1 Example 2 data set to find the mean income.

Solution: Open the data file chpt1example2.sav, use Analyze → Descriptive Statistics → Frequencies, move the income variable to the variables window, click on statistics tab, and check the statistic mean. Click OK. Is the answer 51,000?

As we saw in the previous chapters, whenever we use the explore function we get all the descriptive statistics, including the mean. If you need more practice finding the mean, then access data sets from the Internet or elsewhere to find means. (Use Analyze → Descriptive Statistics → Frequencies, put the variable of interest in the variable window, click on statistics, and check mean, continue, and OK.) Sometimes finding the mean is the only job required of a statistician.

Median

The median is the number (which may or may not be a data value) that divides the data in half. That is, half the data values are less than the median and half the data values are greater than the median. We have two different formulas for finding the median. To find the median we first rank the data—that is, put the data in order—normally from lowest to highest, ascending order, but the data can also be put in descending order, from highest to lowest. The key point is the data must be put in order.

Median for an odd number of data values.

If the number of data values is odd, then the median is simply the data value in the middle. For example, if the number of data values is 9 then the median is the 5th data value, the middlemost data value after the data have been ranked. Now you can see why it's important to rank the data values, otherwise we wouldn't know what the fifth data value is if the data aren't in order. In general if the number of data values, n, is odd, then the median is in the position of the $\frac{(n+1)}{2}$ data value of the ranked data values.

Example 2b.

Find the median income in this set.

Gender	Position	Income
Female	President	$100,000
Male	Vice President	$50,000
Male	Manager	$40,000
Female	Worker 1	$35,000
Male	Worker 2	$30,000

Solution: The data set Income is already ranked and has 5 data values, thus the median is the 3rd data value. The median is $40,000. Let's let SPSS find the median for us. Open the data set chpt1example2.sav or the data set you saved from that example, use Analyze → Descriptive Statistics → Frequencies, move the Income variable into the variable window, click on Statistics, check the box Median, click on continue, and hit OK. The output box will have Median 40000.00.

Median for an even number of data values.

If the number of data values is even, then the median is the average of the middle two ranked data values. Thus, if N is even, then the median is the sum of the $\frac{n}{2}$ data value plus the $\frac{(n+1)}{2}$ data value divided by 2. It is the average of the two middlemost data values. In this case the median may or may not be a data value.

Example 3.

Suppose the data set in Example 2 are revised as such:

Gender	Position	Income
Female	President	$100,000
Male	Vice President	$50,000
Male	Manager	$40,000
Female	Worker 1	$35,000
Male	Worker 2	$30,000
Male	Worker 3	$30,000

What is the median of the variable Income?

Solution: Income now has 6 data values, so the median is the average of the third and fourth data values. That is, the median is the average of $40,000 and $35,000, which is $37,500. As we see the median is not a data value, but three of the data values lie below $37,500 and three of the data values lie above $37,500. Thus the median splits the data in half. Let's let SPSS find the median. Add the sixth data value to the data set you used for Example 2 or open chpt3example3.sav. Use Analyze → Descriptive Statistics → Frequencies, move the Income variable into the variable window, click on the Statistics tab, check the Median box, and click on Continue and OK. The median should read 37500.

Example 4a.

Find the median of these data values: 4, 8, 2, 3, 1, 6, 9, 11, 3.
4b. Find the median of these data values: 1, 3, 5, 8, 2, 4, 5, 6, 6, 7.

Solution: Try both of these examples by hand before using SPSS to make sure the concept is clear. To use SPSS open the data file chpt3example4.sav. Use the steps as outlined in the above examples; you can move both variables simultaneously by highlighting both variables before moving them to the variable window. Just hold the shift key down as you click on the variables. The output should read the median for variable *a* is 4 and the median for variable *b* is 5. If we rank the data, we can see 4 is the 5th data value in *a* and 5 is the average of the 5th and 6th data values in *b*.

a ranked: 1, 2, 3, 3, 4, 6, 8, 9, 11
b ranked: 1, 2, 3, 4, 5, 5, 6, 6, 7, 8

The median is that value where 50% of the data values are less than or equal to the median and 50% of the data values are greater than or equal to the median.

Medians are sometimes a more useful tool for describing the average value than the mean. This is because if the data have a few values called outliers (extreme values compared to the rest of the data values), then the mean will be pulled by the extreme values and may not reflect the "true" average of the data. This is particularly the case when discussing income or housing.

Example 5.

Refer back to the data in Example 2.

Gender	Position	Income
Female	President	$100,000
Male	Vice President	$50,000
Male	Manager	$40,000
Female	Worker 1	$35,000
Male	Worker 2	$30,000

Find the mean and the median of Income for this data set.

Solution: Open the data set chpt1example2 and use Analyze → Descriptive Statistics → Frequencies, move income to the variables window, click on the tab Statistics, and check both boxes Mean and Median, Continue, and OK. The mean is 51000 and the median is 40000. Only two people make $51,000 or close or above $51,000. The rest make amounts closer to $40,000. Hence 40000 is actually a better average value for this data set.

In general the median is the most-used measure of central tendency when the data reflect income or the housing market, because most of the income values or house prices are grouped together, but always a few really extreme values shift the mean off the point where the majority of the data values lie.

The third measure of central tendency is the mode.

Mode

The mode is that data value that occurs most often. If two different data values have the same number of occurrences, more than once, then we say the data are bimodal. If no data value occurs more than once or if three or more data values have the same number of occurrences more than once, then we say the data do not have a mode. The data have only one mode, two modes, or no modes. SPSS will always give some value for the mode regardless if a mode exists or not. If all the data values occur only once, then SPSS will give the smallest data value and put a note on the bottom saying multiple modes exist. If one data value occurs more often than any other data value, then SPSS will give that mode. If two or more different data values occur more than once and the same number of occurrences for each data value, then SPSS will give the smaller mode and put a note on the bottom saying multiple modes exist. In these cases the statistician needs to look at the data to see if the mode is significant.

Example 6.

Find the modes for the following data sets, chpt3example6.sav.

6a. 11, 13, 12, 10, 10, 15, 17, 11, 12, 15, 15, 13, 18, 11, 15
6b. 11, 13, 12, 10, 10, 15, 17, 11, 12, 15, 15, 13, 18, 11, 15, 11
6c. 11, 13, 12, 10, 15, 17, 18

Solution: We can see from the data sets that *a* has a mode of 15 (15 occurs four times and no other data value occurs four times), *b* has two modes 11 and 15 (now both 11 and 15 occur four times) and *c* has no mode (because all data values only occur once). With SPSS use Analyze → Descriptive Statistics → Frequencies, move all three variables into the variable window, click on the Statistics tab, check the Mode box, click Continue, and hit OK. You should get the following output (**Figure 2**). SPSS gave a mode for all the variables *a*, *b*, and *c*, but with *b* and *c* there's a footnote attached that says, "a. Multiple modes exist. The smallest value is shown."

Statistics

		a	b	c
N	Valid	15	16	7
	Missing	2	1	10
Mode		15.00	11.00[a]	10.00[a]

a. Multiple modes exist. The smallest value is shown

Figure 2

The mode is not used often for giving the central value of a data set, but if the data set has a mode, it is a number that will stand out in the data set because it the data value that occurs most often. It can be used as a quick measure of the average value without doing any ranking or any summing of the data values. When the data are categorical, nominal data, then the mode is the only measure of any significance. Categorical data can't have a mean or median because they can't be summed or ranked.

Example 7.

Find the mode for the data set PartyMembershipsUSsen2012(1).

Solution: Open the data set PartyMembershipsUSsen2012.sav. Find the sum for the three different parties. Use Analyze → Descriptive Statistics → Frequencies, move all three variables, Democrats, Republicans, and Independents into the variable window. Click on Statistics, check the Sum box, and hit Continue and OK. The output is given in **Figure 3**. We see the Democrats have a sum of 51, the Republicans sum is 47, and the Independents sum is 2. The mode is Democrats. (Modes are data values, not frequency values.) In 2012, the senate had more Democrats than any other party, so the "average" senator was a Democrat. (The number 50 in the top row means all 50 states are accounted for in each party.)

Statistics

		Democrats	Republicans	Independents
N	Valid	50	50	50
	Missing	0	0	0
Sum		51	47	2

Figure 3

Mean, Median and Mode of Weighted Data

Weighted data means each data value has a weight, or frequency attached to it. For example the data set 1, 1, 1, 1, 2, 2, 3, 3, 3, 4, 4, 5, 5, 5, 6, 6, 6, 6, 6, 6 can be tabulated as such:

x value	1	2	3	4	5	6
Frequency	4	2	3	2	3	6

Thus the given *x* values are weighted by the frequency values, so to find the sum we multiply the *x* values by the frequency. Thus the sum of the *x* values is $1\times4+2\times2+3\times3+4\times2+5\times3+6\times6=76$. The frequency values are given because 1 occurs four times, hence the frequency is 4; 2 occurs twice, so the frequency is 2; 3, three times; 4, twice; 5, three times; and 6 occurs six times. To find the mean, median, and mode we need to weight cases as will be shown in the next example.

Example 8.

Find the mean, median, and mode of the weighted data set:

X value	1	2	3	4	5	6
Frequency	4	2	3	2	3	6

Solution: Enter the values into two columns as shown in **Figure 4a** or open the data set chpt3example7.sav. In the tool bar, click on the tab labeled Data and scroll down to weight cases. Click on that. A new window will appear, **see Figure 4b**; check the circle next to "Weight cases by." The Frequency variable window will now be active, so move the freq variable into that window. Click on freq, then the move → button. Hit OK. SPSS has now attached all variables in the data editor to the freq variable. To find the mean, median, and mode use Analyze → Descriptive Statistics → Frequencies, move the x variable into the variable window, click on Statistics, check the boxes Mean, Median, and Mode, and hit Continue and OK. The mean median and mode are given in **Figure 4c**. To avoid clutter in the output always deselect Display frequency tables unless they're needed.

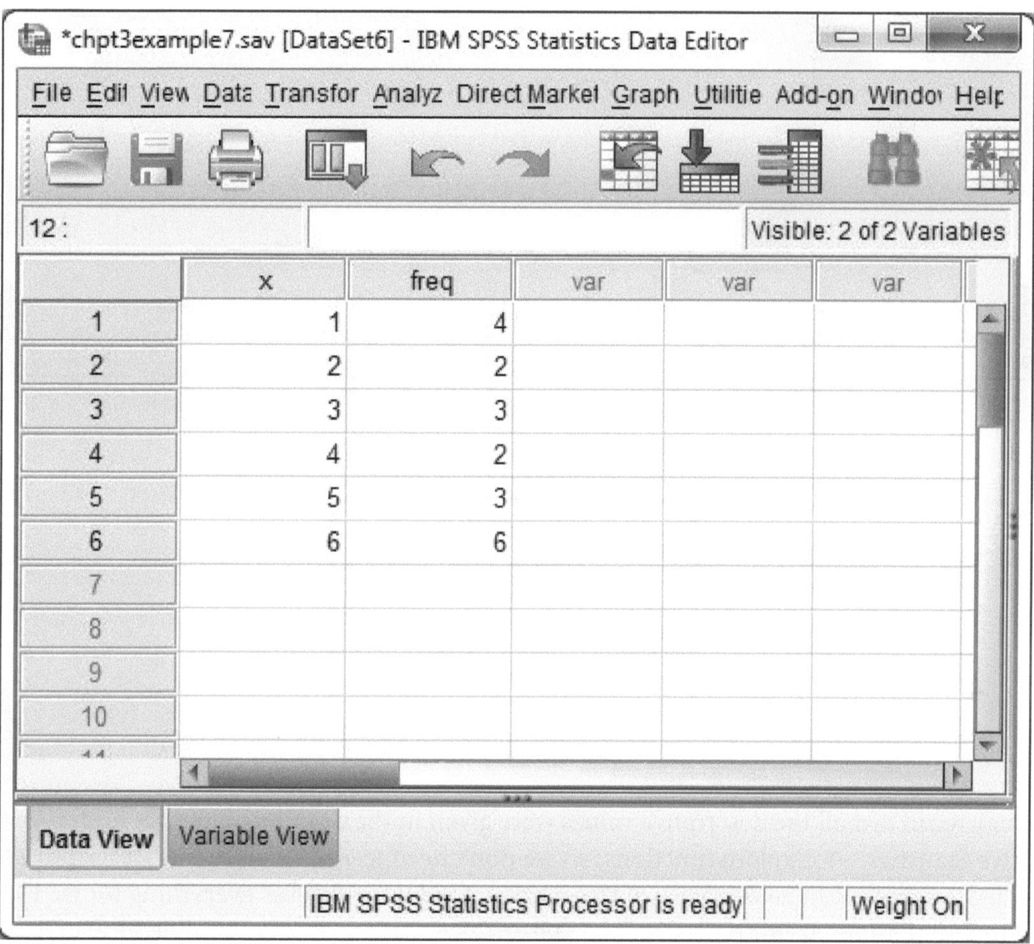

Figure 4a

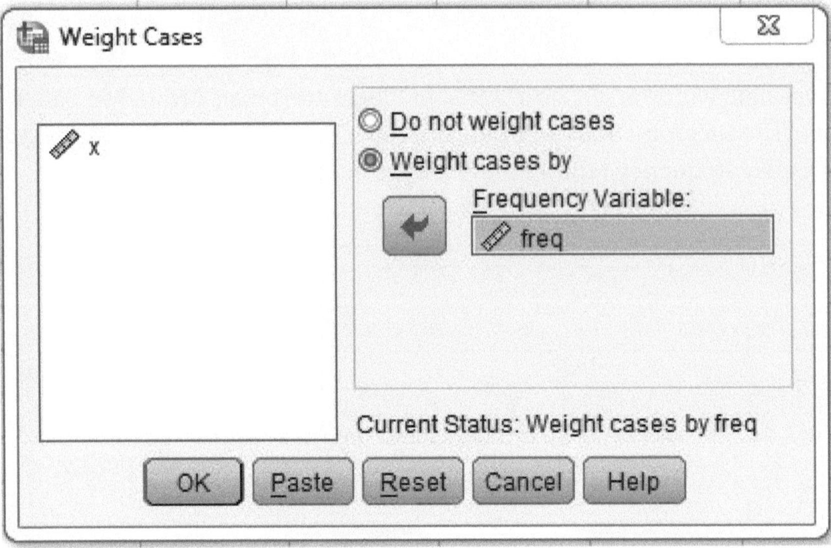

Figure 4b

Statistics

x

N	Valid	20
	Missing	0
Mean		3.80
Median		4.00
Mode		6

Figure 4c

You may recall that all the descriptive values were given in the descriptive output under Analyze → Descriptive Statistics → Explore functions, so we don't need to use Analyze → Descriptive Statistics → Frequencies if we want an overview of the data set; Explore will cover everything for us. However, if we want to find the sum or just one particular statistic, we can use the above procedure.

Quartile

In Explore we also got a chart called a box plot. The old texts used to call it the box-and-whiskers plot, **see Figure 5a**, because the middle is a box and the two ends are straight lines called the whiskers. (This is the box plot for Example 9.) The box represents the middle 50% of the data, and the two lines at the end represent the minimum and maximum values of the data. The graph is divided into four sections called quartiles, which are divided as such: Quartile 1, $Q_1 = 25$, the 25th percentile, 25 % of the data values are less than or equal to Q_1; Quartile 2, $Q_2 = 50$, the 50th percentile, 50% of the data values are less than or equal to Q_2 (Q_2 is the median); and Quartile 3, $Q_3 = 75$, the 75th percentile, 75% of the data values are less than or equal to Q_3.

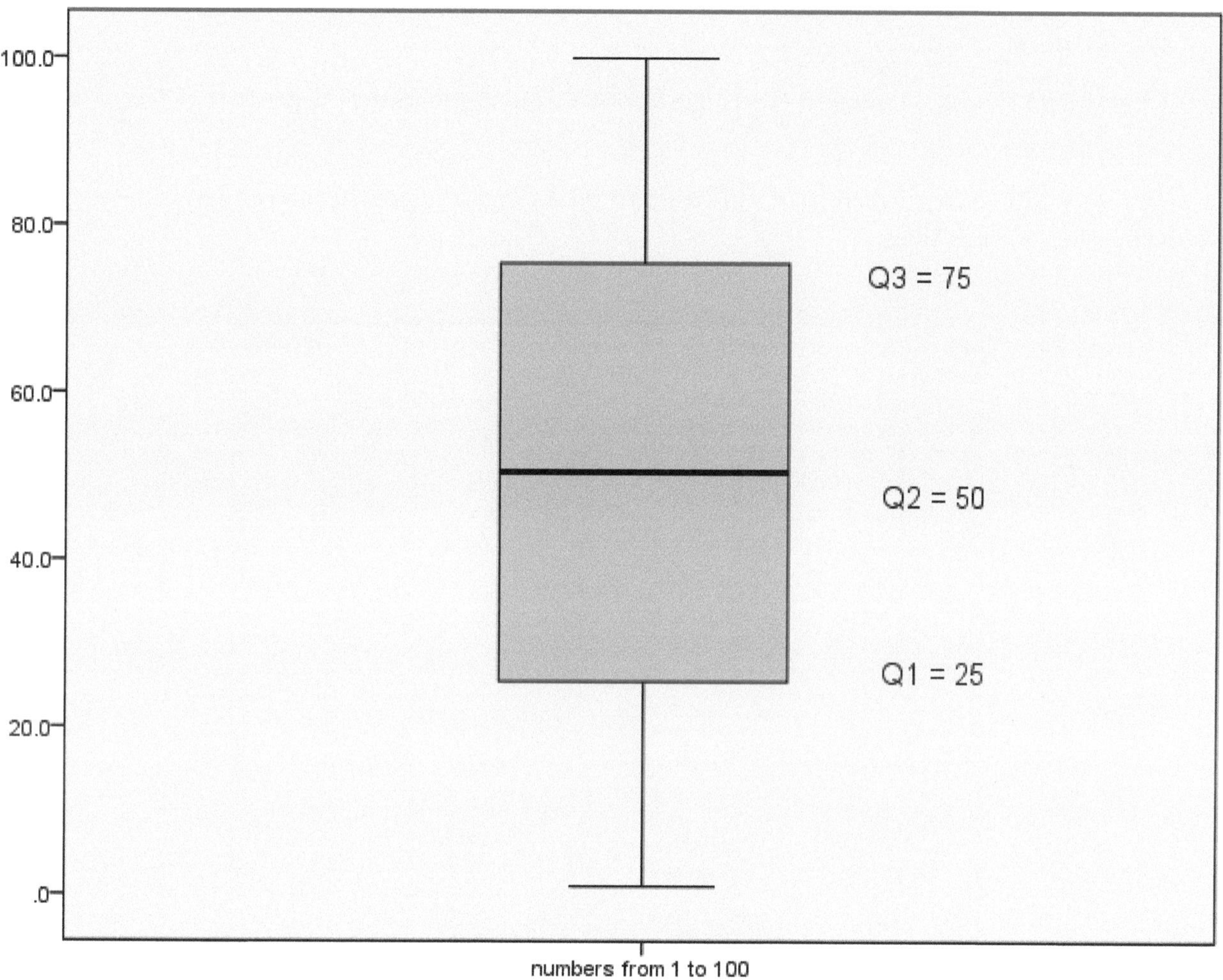

Figure 5a

Example 9.

Input the numbers from 1 to 100 in an SPSS data editor and find the Quartile and box plot. Chpt3example9.

Solution: To find the quartile use Analyze → Descriptive Statistics → Frequencies, move the variable of interest into the variable window, click on the Statistics tab, and check the box marked Quartiles; **see Figure 5b**. The output will be as in **Figure 5c**. To find the quartiles without the computer, just find the medians of the first half and the second half of the data. That is, treat the first half as a set of data and use the methods described above to find the median, which will be Q_1. Likewise, treat the second half as a set of data and find the median, which will be Q_3. Note: SPSS uses a different formula to find the quartiles, but the method described above works for most cases. Also the boxplots drawn by SPSS don't always align with the quartiles.

SPSS method:

$Q_1 = (1-a)x_k + ax_{k+1}$ where $a = \frac{1}{4}\left[(n+1) \bmod 4\right]$ and k equals the integer part of $\frac{(n+1)}{4}$;

$Q_3 = (1-a)x_k + ax_{k+1}$ where $a = \frac{1}{4}\left[(3n+3) \bmod 4\right]$ and k equals the integer part of $\frac{(3n+3)}{4}$, n is the number of data values.

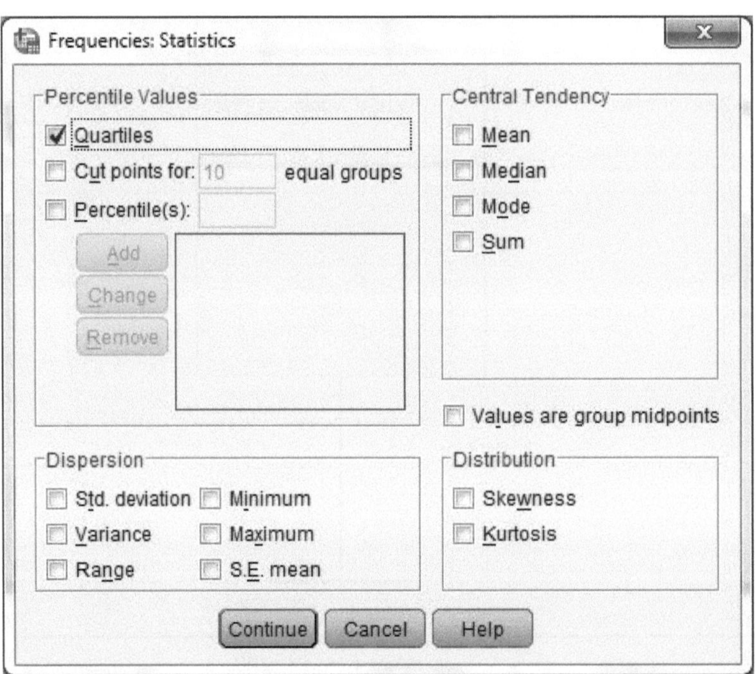

Figure 5b

Statistics

Numbers from 1 to 100

N	Valid	100
	Missing	0
Percentiles	25	25.250
	50	50.500
	75	75.750

Figure 5c

The interquartile range is the difference between the third quartile and the first quartile, which is **Interquartile Range** = $Q_3 - Q_1$.

Box Plot

A box plot (also called a box-and-whiskers plot) illustrates the first three quartiles and the middle 50% of the data; **see Figure 5a**. Since the first quartile is the 25th percentile, then 25% of the data lie at or below that value. The third quartile is the 75th percentile, so 75% of the data lie at or below that value. Thus 50% of the data lie between Q_1 and Q_3. This is represented by the box in the box plot. The whiskers are the 25% of the data that lie at or below Q_1 and at or above Q_3. As we expect the mean to lie somewhere in the middle, the box plot gives a quick range of values for the mean. Input the quartiles into the boxplot by using Options → Text Box in the Chart Editor. The textbox can be moved around inside the chart editor using the cursor.

Example 10.

Use the data set chpt3example4.sav to draw two box plots, and find the quartile and interquartile range. Show that the answers are almost the same by finding the medians of the first half and the second half.

Solution: Let's first draw the box plots. Use Graphs → Legacy Dialog →Boxplot. Check the box marked Summaries of separate variables and leave Simple as default. Click on Define. Move the two variables *a* and *b* into the window Boxes Represent and click OK. The graphs should be as in **Figure 6a**. To find the quartiles use Analyze → Descriptive Statistics → Frequencies, put *a* and *b* in the Variable window, and click on Statistics. Check the box Quartiles, then Continue and OK. The output should be as in **Figure 6b**.

For variable *a*, the minimum value is 1, the first quartile Q_1 is 2.5 (the average of 2 and 3, the second and third data values), the median of Q_2 is 4, and the third quartile Q_3 is 8.5 (the average of 8 and 9, the seventh and eighth data values). The interquartile range is 6, the difference of 8.5 and 2.5. The maximum value is 11.

For variable *b*, the minimum value is 1, the first quartile, Q_1 is 2.75 (the first quartile by the method described above would be 3), the median of $Q_2 = 5$, and the third quartile Q_3 is 6.25 (6 by the method described above, that is, find the median of the second half). The interquartile range is 3.5 (6.25 – 2.75) but we can use 3, which is the difference between 6 and 3. The maximum value is 8.

We can find the interquartile range without evaluating $Q_3 - Q_1$. Use Analyze → Descriptive Statistics → Explore, click on Statistics, and check the box Descriptives. (It should already be checked as it is the default case.) The output is given in **Figure 6c**. The interquartile range is highlighted.

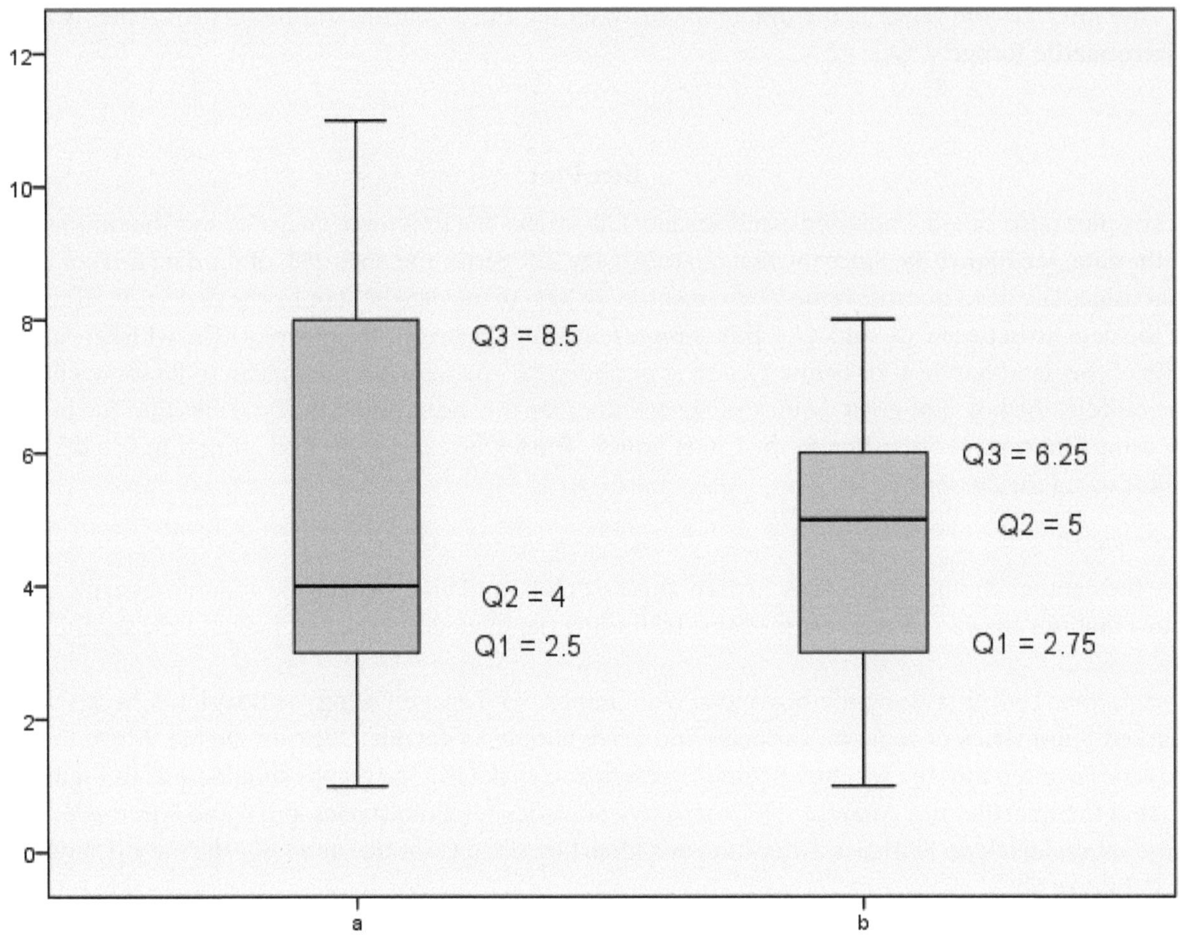

Figure 6a

Statistics

		a	b
N	Valid	9	10
	Missing	1	0
Percentiles	25	2.5000	2.7500
	50	4.0000	5.0000
	75	8.5000	6.2500

Figure 6b

Descriptives

			Statistic	Std. Error
a	Mean		5.2222	1.15202
	95% Confidence Interval for Mean	Lower Bound	2.5656	
		Upper Bound	7.8788	
	5% Trimmed Mean		5.1358	
	Median		4.0000	
	Variance		11.944	
	Std. Deviation		3.45607	
	Minimum		1.00	
	Maximum		11.00	
	Range		10.00	
	Interquartile Range		6.00	
	Skewness		.530	.717
	Kurtosis		-1.089	1.400
b	Mean		4.4444	.72860
	95% Confidence Interval for Mean	Lower Bound	2.7643	
		Upper Bound	6.1246	
	5% Trimmed Mean		4.4383	
	Median		5.0000	
	Variance		4.778	
	Std. Deviation		2.18581	
	Minimum		1.00	
	Maximum		8.00	
	Range		7.00	
	Interquartile Range		3.50	
	Skewness		-.089	.717
	Kurtosis		-.434	1.400

Figure 6c

Because this text is using SPSS for all the concepts, we'll use the quartile found by SPSS and not concern ourselves with finding the quartile by hand. The interested reader can look up the definition for modular arithmetic (basically a mod b is the remainder when a is divided by b) and check if the above formulas for SPSS are indeed accurate. (They are, we checked.)

Percentiles

The formulas for percentiles are also quite involved so we'll skip them and simply use the SPSS output to find percentiles. Percentiles have the same definition as quartiles; for example the 30th percentile is that value (which may or may not be a data value) such that 30 percent of the data values in the data set are at or below that value. As a short reference to find percentiles without SPSS, simply multiply the total number of data values by the percentile (as a decimal value, not a percent) and find that data value in the ranked data set.

Example 11.

Open the data set StateSATscores1990.sav. Find the 10, 20, 30, ..., 90, 100 percentiles for both the math and verbal scores. (2)

Solution: Use Analyze → Descriptive Statistics → Frequencies, move both variables into the variable window, and click on the tab Statistics. In the Frequencies: Statistics window, check the box Cut points for 10 equal groups (see **Figure 7a**). If you need to add other percentiles (such as 68 and 75), then put a check mark in the Percentiles box and add the Percentile you want by putting the number in the box next to the Percentile(s) line and click Add; see **Figure** 7a After you added all the percentiles of interest, click Continue and OK. Your output should look like **Figure 7b**.

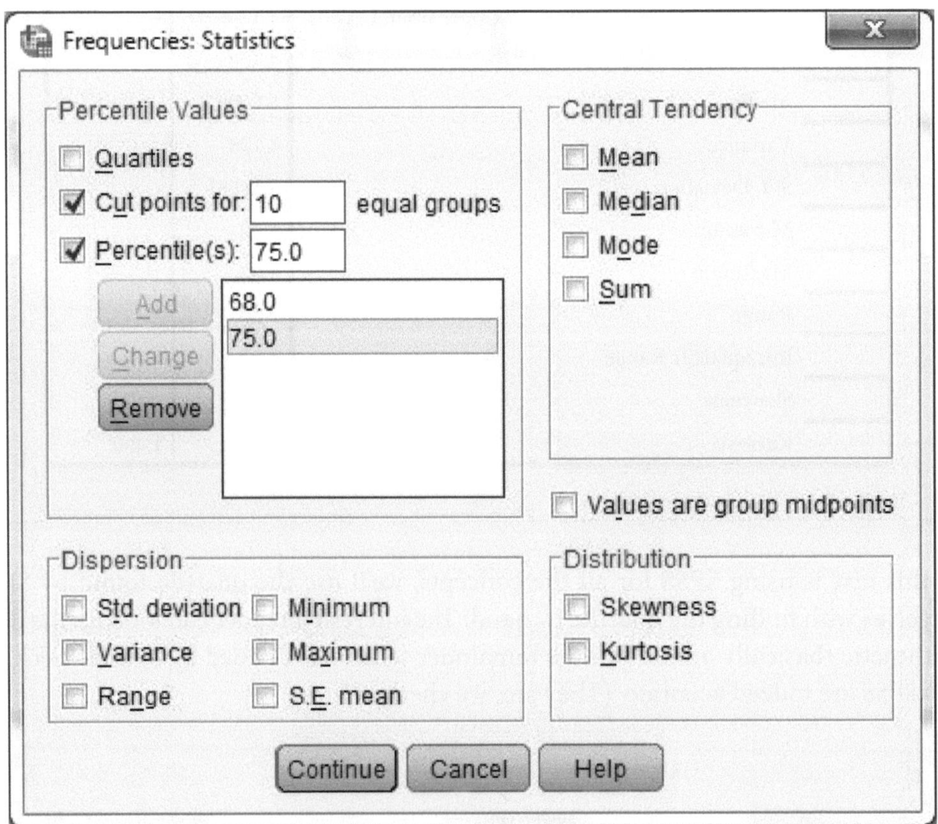

Figure 7a

Statistics

		Mean SAT I Verbal by State	Mean SAT I Math by State
N	Valid	51	51
	Missing	0	0
Percentiles	10	408.2000	459.4000
	20	418.4000	466.0000
	30	425.8000	472.2000
	40	433.8000	483.4000
	50	443.0000	490.0000
	60	459.2000	513.2000
	68	470.0000	519.0000
	70	471.2000	519.8000
	75	476.0000	523.0000
	80	477.0000	526.2000
	90	490.4000	545.4000

Figure 7b

Let's review what we've learned:

1. Mean—measure of central tendency used most often, particularly dealing with scale data that have a normal distribution. (The data are not skewed.)
2. Median—measure of central tendency used with scale data that are skewed.
3. Mode—measure of central tendency used with nominal (categorical) data.
4. Quartile—divides the data into four sections such that a quarter of the data values are in each section, used to make the box plot.
5. Percentiles—divide the data into Percent. Percentiles can range anywhere from 1 to 100 percent.

Class Work 3

Find the mean, median, and mode of the following grouped data set:

X	1	2	3	4	5	6
Frequency	10	12	18	20	16	9

These are weighted data so you need to enter the data into two columns, one for the variable and one for the frequency. Then you need to weigh cases. Use Data → Weigh Cases, click on the Weigh cases by button, and move frequency into the case window, Continue. Also draw the box plot, find the quartiles and inter-quartile range, and label those on the box plot.

Homework 3.

The following data sets are found in chpt3problems1-4.sav

1. Find the mean, median, and mode of the following data set:

   37.8 26.5 29.2 31.5 32.5 50.4 62.3 39.1 38.2 26.5

2. Find the mean, median, and mode of the following data set:

   26 22 31 24 22 32 25 31 48 42 23 22
   32 35 37 29 22 24 19 36 31 50

3. Find the mean, median, and mode of the following data set:

   6 10 5 8 11 6 14 6 5

4. Find the mean, median, and mode of the following data set:

   191 230 295 311 262 245 269 222 240 240 250
   274 236 190 251 242 215 210 216 321

5. Draw a histogram for problem 1 and decide which measure is best to describe the central tendency of the data.

6. Draw a histogram for problem 2 and decide which measure is best to describe the central tendency of the data.

7. Draw a histogram for problem 3 and decide which measure is best to describe the central tendency of the data.

8. Draw a histogram for problem 4 and decide which measure is best to describe the central tendency of the data.

9. Find the mean, median, and mode of the following grouped data: chpt3problem9.sav; be sure to weight cases. This is the outcome set for the toss of two dice.

x	2	3	4	5	6	7	8	9	10	11	12
Freq	1	2	3	4	5	6	5	4	3	2	1

10. Find the mean, median, and mode of the following grouped data: chpt3problem10.sav; be sure to weight cases.

x	7	8	9	10	11	12	13	14	15	16
Freq	34	22	56	75	36	68	91	26	45	28

11. Draw a box plot for problem 1 and find the three quartiles and the interquartile range. Label those values on the box plot.

12. Draw a box plot for problem 2 and find the three quartiles and the interquartile range. Label those values on the box plot.

13. Draw a box plot for problem 3 and find the three quartiles and the interquartile range. Label those values on the box plot.

14. Draw a box plot for problem 4 and find the three quartiles and the interquartile range. Label those values on the box plot.

15. Draw a box plot for problem 9 and find the three quartiles and the interquartile range. Label those values on the box plot.

16. Draw a box plot for problem 10 and find the three quartiles and the interquartile range. Label those values on the box plot.

17. Find the 20th and 80th percentiles for problem 9.

18. Find the 30th and 70th percentiles for problem 10.

19. Use the data set chpt3problem19.sav to find the mode for the variable age.

20. The data set chpt3problem19.sav has the two variables ethnicity and gender as Nominal variables and they were input as string variables. SPSS cannot find the mode of string variables, which is why we always enter the data as numeric and change the measure and labels after we enter the data. However, we can still find the mode of string variables by using Analyze → Descriptive Statistics → Frequencies; move the variables of interest into the variable window and click on charts. Either Pie chart or Bar chart will give the relative frequencies (click on Show Data Labels in the chart editor), and from those numbers pick the mode for ethnicity and gender.

21. Open the data set classsurvey2007.sav and find the mode of all the variables, Gender, Age, Ethnicity, Cost, and Status. These variables were all entered numerically and the measure and values were changed to reflect qualities.

Just for fun. Guess which digit appears most often in a 365-day calendar and then use SPSS to find the answer. Did you correctly guess the answer?

References

(1) Created from table of US Senators 2012
(2) The Math Forum @ Drexel.

Chapter 4

Measures of Dispersion (Variability)

Objectives

After completing this chapter, a student should be able to:

- ✓ Calculate and understand the definition of the Standard Deviation
- ✓ Calculate and understand the definition of the Variance
- ✓ Calculate and understand the definition of the Range
- ✓ Calculate and understand the Coefficient of Variation

There are three measures of dispersion: standard deviation, variance, and range.

Standard Deviation

The standard deviation is the most often-used measure of dispersion and it gives the average distance the data values are from the mean. The formula for the standard deviation is $s=\sqrt{\frac{\sum(x-\overline{x})^2}{n-1}}$. We need to square all the differences of the mean and the x values because as we saw in the last chapter, $\sum(x-\overline{x})=0$, and it makes no sense to say the standard deviation is 0. Because the square of a real number is always nonnegative, the sum of the squares will also be nonnegative and the square root will be a real number. We take the square root so that the standard deviation will have the same units as the data values. (Just as it makes no sense to talk about area—which is a square unit—in the same context as length, which is a linear unit. An example is if the area of a square equals the perimeter of the square, then what is the length of a side of the square? This example doesn't make sense because we can't equate length with area, but to say "the numerical value of the area of a square is the same as the numerical value of the perimeter, then what is the length of a side of the square" is a valid question. The answer is 4.) We divide by $n - 1$ to get an average measure of dispersion. Why don't we divide by n? The answer is that otherwise this standard deviation is biased toward the sample mean, $\overline{x}$, which is being used in place of the population mean μ. If we have μ then indeed we would divide by n, which is called a population standard deviation. But normally in our samples we estimate μ with the sample mean $\overline{x}$. Hence our sample standard deviation is biased toward the sample mean. This is not good if we divide by n because then it is called a biased standard deviation. By dividing by $n - 1$ we make the standard deviation a little bit larger and we call it an unbiased estimate of the population standard deviation. However, in most practical applications, the sample size is very large (in the thousand range) so the difference between a biased standard deviation and an unbiased standard deviation is minuscule (less than .0001). SPSS by default will always use an unbiased estimate for the standard deviation. SPSS will also treat all data sets as small sample sets and by default always perform t tests (more about that later). But the dividing line used by statisticians to distinguish between a small sample and a large sample is 30. There are two different techniques to evaluate statistics, one for a large sample (sample size greater than 30) and one for small samples (sample size less than or equal to 30). SPSS ignores all that and treats everything as if it's a small sample.

Example 1.

Find the standard deviation of the following sample set: 6, 18, 10, 12, 9, 4, 3, 2, 15, 13, 12, 16, 18, 15, 14, 13, 17, 7, 5, 6 (chpt4example1.sav).

Solution: Enter the data into an SPSS data editor, or open chpt4example1.sav. Use Analyze → Descriptive Statistics → Frequencies, move the variable x into the variable window, click Statistics, and check the box Standard Deviation; see **Figure 1a**. The standard deviation is 5.180, as shown in the output Statistics table in **Figure 1b**.

Frequencies: Statistics

Percentile Values
Quartiles
Cut points for: 10 equal groups
Percentile(s):
Add
Change
Remove

Central Tendency
Mean
Median
Mode
Sum

Values are group midpoints

Dispersion
Std. deviation
Minimum
Variance
Maximum
Range
S.E. mean

Distribution
Skewness
Kurtosis

Continue Cancel Help

Figure 1a

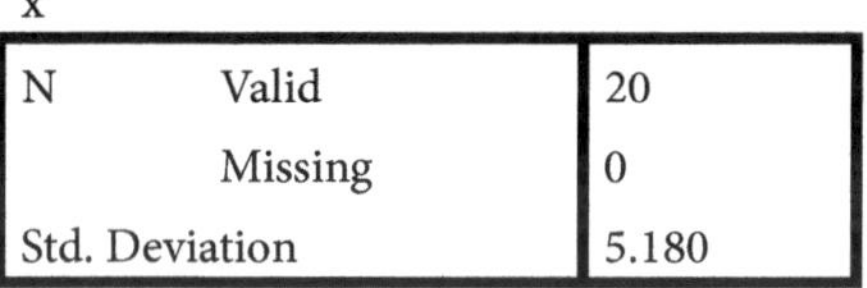

Statistics

x

N	Valid	20
	Missing	0
Std. Deviation		5.180

Figure 1b

Variance

Variance is the next measure of dispersion. Variance is the square of the standard deviation and is used extensively with analysis of variance; see chapters 9 and 15. The formula for variance is $s^2 = \frac{\sum(x-\bar{x})^2}{n-1}$. It's the same as the standard deviation formula except it's missing the square root. We have this relationship: Variance $= s^2$ or Standard Deviation $= \sqrt{Variance} = \sqrt{s^2} = s$. (Standard deviation is always nonnegative so we don't need to worry about a negative value for *s*. Recall in algebra $\sqrt{x^2} = |x|$).

Example 2.

Find the variance for the following data set: 6, 18, 10, 12, 9, 4, 3, 2, 15, 13, 12, 16, 18, 15, 14, 13, 17, 7, 5, 6 (chpt4example1.sav).

Solution: Open the data set for Example 1. If it's not already open, click on File on the SPSS Data Editor, scroll down to most Recently Used Data, and find the data set there. Use Analyze → Descriptive Statistics → Frequencies, move the x variable into the variable window, click on the tab Statistics, check the box next to Variance, then Continue and OK. The output should show the value for the variance as 26.829; **see Figure 2.**

Statistics

x

N	Valid	20
	Missing	0
Variance		26.829

Figure 2

As with Standard Deviation, Variance can either be biased (divide the sum of the differences squared by n) or unbiased (divide the sum of the differences squared by $n - 1$). SPSS, by default, will always give an unbiased variance and Standard Deviation. However, the need may arise when a biased estimate is needed, such as we have a small population and the sample set is the entire population. Thus the Standard Deviation and Variance from the sample are the population Standard Deviation and Variance. To have SPSS give us the biased Standard Deviation and Variance, simply add the mean to the data set.

Example 3.

Find the population (biased) Standard Deviation and Variance for the data set 6, 18, 10, 12, 9, 4, 3, 2, 15, 13, 12, 16, 18, 15, 14, 13, 17, 7, 5, 6 (chpt4example1.sav).

Solution: The mean for this data set is 10.75, so to now find the Standard Deviation and Variance of the population, we add 10.75 to the old data set and form this new data set 6, 18, 10, 12, 9, 4, 3, 2, 15, 13, 12, 16, 18, 15, 14, 13, 17, 7, 5, 6, 10.75. Enter the data into a data editor, use Analyze → Descriptive Statistics → Frequencies, move the variable x into the variable window, click on the tab Statistics, check the boxes Standard Deviation, Variance, and hit Continue and OK. (Or use Analyze → Descriptive Statistics → Descriptives, move the variable x into the variable window, click on the tab Options, check the boxes Standard Deviation and Variance, and hit Continue and OK.) The answer should be Standard Deviation 5.049 and Variance 25.488; **see Figure 3.** Note that the population Standard Deviation and Variance are less than the sample Standard Deviation and Variance. An unbiased estimate is larger than a biased estimate. (If numerators are equal, the bigger the denominator, the smaller the fraction.)

Statistics

x

N	Valid	21
	Missing	0
Std. Deviation		5.049
Variance		25.488

Figure 3

Range

Our last measure of dispersion is the range. The range is simply the difference between the highest data value and smallest data value. The formula for the range is Range = Maximum value – Minimum value.

Example 4.

Find the range of the following data set: 6, 18, 10, 12, 9, 4, 3, 2, 15, 13, 12, 16, 18, 15, 14, 13, 17, 7, 5, 6 (chpt4example1.sav).

Solution: Open the data set given above, use Analyze → Descriptive Statistics → Frequencies, move the variable x into the variable window, click on the tab Statistics, check the box Range, Maximum and Minimum, and click on Continue and OK. (Or use Analyze → Descriptive Statistics → Descriptives, move the variable x into the variable window, click on the tab Options, check the box Range, Maximum, and Minimum, and click on Continue, OK.) The output should give the range as 16; **see Figure 4**. Note the maximum value in the data set is 18 and the minimum value is 2: 18 – 2 = 16.

Statistics

x

N	Valid	20
	Missing	0
Range		16
Minimum		2
Maximum		18

Figure 4

We can always find the minimum and maximum values by clicking the Minimum and Maximum boxes in the Statistics window, and thus find the range by subtracting. However, the goal of this text is to be able to find all statistics using SPSS. Nonetheless we still want you to be aware of what the statistics mean and how they were obtained.

The range found above is called the exclusive range because it excludes the minimum value (the minimum value was subtracted out). If we wish to include the minimum value in the range of data values, then we need to increase the range by 1. This is called the inclusive range because we include the minimum value.

Example 5.

Find the inclusive range of the data set 6, 18, 10, 12, 9, 4, 3, 2, 15, 13, 12, 16, 18, 15, 14, 13, 17, 7, 5, 6 (chpt4example1.sav).

Solution: Because we found the exclusive range of this data set to be 16, then the inclusive range is 17: 16 + 1 = 17.

Estimate for the Standard Deviation Using the Range

As we will see in Chapter 6, a normal distribution is a distribution where approximately 95% of the data values fall within two standard deviations of the mean, so a rough estimate for the standard deviation is to divide the range by four, that is, standard deviation $\frac{\max - \min}{4}$.

Example 6.

Use the data set from Example 1, chpt4example1.sav, to approximate the standard deviation. How well does the approximate compare to the actual standard deviation?

Solution: The approximate standard deviation is $\frac{18-2}{4} = 4$. The actual standard deviation is 5.180. The approximate standard deviation is close enough to use for an estimate of the actual standard deviation.

Standard Deviation of a Weighted Sample

The formula for the standard deviation of a weighted sample is $s = \sqrt{\frac{\sum (x_i - \overline{x})^2 f_i}{\sum f_i - 1}}$ where $\overline{x} = \frac{\sum x_i f_i}{\sum f_i}$. A weighted sample is given by a frequency distribution. As we did in Chapter 3, we can find weighted statistics by using the procedure Data $\rightarrow$ Weight Cases. The following will illustrate this procedure.

Example 7.

A highway engineer knows that his crew can lay 5 miles of highway on a clear day, 2 miles on a rainy day, and only 1 mile on a snowy day. Suppose the probabilities are as follows:

Outcome	Clear	Rain	Snow
Probability	.6	.3	.1
Random Variable (miles of highway)	5	2	1

Find the mean, standard deviation, and variance of this distribution.

Solution: Before we use SPSS we need to change the probabilities to frequencies. This is easily accomplished by multiplying each frequency by 10, so that the frequencies are 6, 3, and 1, respectively. The data set is chpt4example6. Because this is a population distribution, then we need to find the mean, add that number to the data set, and set its frequency to 1. The population mean is 3.7, the population variance is 2.61, and the population standard deviation is 1.616.

Coefficient of Variation

When we compare two data sets with different means and different standard deviations, we can tell which data set has more variation by computing the coefficient of variation. (This is useful for an investor who is trying to determine between two companies in which to invest; see problem 21.) The formula is $CV = \frac{s}{\overline{x}} \times 100$ for a sample and $CV = \frac{\sigma}{\mu} \times 100$ for a population. SPSS does not yet have a CV command (that is, SPSS cannot compute the coefficient of variation for a given data set), but we can find the mean and standard deviation of a data set and use a calculator to compute the coefficient of variation. (This is the only statistic we have to calculate by hand.)

Example 8.

Compute the coefficient of variation for the data set 6, 18, 10, 12, 9, 4, 3, 2, 15, 13, 12, 16, 18, 15, 14, 13, 17, 7, 5, 6 (chpt4example1.sav).

Solution: The mean was found to be 10.75 and the sample standard deviation was found to be 5.180, hence the coefficient of variation for the sample is $CV = \frac{5.180}{10.75} \times 100 = 48.186$. The population coefficient of variation is $CV = \frac{5.049}{10.75} \times 100 = 46.9674$. Note the population statistic is smaller than the sample statistic.

When we compare the coefficient of variation between two different samples, we can say the sample with the larger coefficient of variation has a higher degree of variance. Also note that because both the standard deviation and the mean have the same units—dollars, pounds, gas mileage, etc.—then the coefficient of variation will cancel those units and enable us to compare two different coefficients from two different categories. Thus we can see if the variation in wheat production is any different from the variation in the standard of living. Statisticians use all kinds of categories to see if there is a correlation anywhere by comparing variances. This is a useful technique in determining criminal patterns or terrorist activities.

Let's review what we've learned in this chapter.

1. Standard Deviation—the most-used measure of dispersion, it has the same units as the data values.
2. Variance—the second most-used method of dispersion, variance is the square of the standard deviation. Another formula for the variance is $s^2 = \frac{\sum x^2 - n\overline{x}^2}{n-1}$, which is useful if doing calculations by hand.
3. Range—gives the range of all the data values; together with the standard deviation it can be determined if the data are grouped close together or if the data are spread out or if the data have a few extreme values called outliers.
   - i.) If the range is small and the standard deviation is small, then the data are grouped close together.
   - ii.) If the range is large and the standard deviation is small, then the data have a few extreme values (outliers).

iii.) If the range is large and the standard deviation is large, then the data are spread out across the entire range.

4. Standard Deviation of Weighted Data—each data value is assigned a frequency and the data value and frequency are multiplied together to calculate the standard deviation.
5. Coefficient of Variation—used to compare the variation between two data sets.

Class Work 4

Find the standard deviation, variance, range, and coefficient of variation for the following sample data set: 38.7 25.8 38.5 32.5 31.5 50.6 64.5 39

Homework 4

The following data sets are found in chpt3problems1-4.sav.

1. Find the standard deviation, variance, range, and coefficient of variation for the following sample data set:

   37.8 26.5 29.2 31.5 32.5 50.4 62.3 39.1 38.2 26.5

2. Find the standard deviation, variance, range, and coefficient of variation for the following sample data set:

26	22	31	24	22	32	25	31	48	42	23
22	32	35	37	29	22	24	19	36	31	50

3. Find the standard deviation, variance, range, and coefficient of variation for the following sample data set:

   6 10 5 8 11 6 14 6 5

4. Find the standard deviation, variance, range, and coefficient of variation for the following sample data set:

191	230	295	311	262	245	269	222	240	240
250	274	236	190	251	242	215	210	216	321

5. Find the population standard deviation, variance, inclusive range, and coefficient of variation for the population data set:

   106 108 110 111 91 96 97 113 104

6. Find the population standard deviation, variance, inclusive range, and coefficient of variation for the population data set:

   2 3 4 5 6 7 8 9 10 11 12

Another rule for dispersion of data is Chebyshev's theorem: For any data set (population or sample) and for any constant k greater than 1, the proportion of the data set that must lie within k standard deviations on the left side of the mean and k standard deviations on the right side of the mean is at least $1-\frac{1}{k^2}$. For example if $k = 2$, then $1-\frac{1}{2^2}=1-\frac{1}{4}=\frac{3}{4}=.75$ shows that 75% of the data must lie within two standard deviations of the mean. For $k = 3$, then 89% of the data must lie within three standard deviations of the mean and for $k = 4$, then 93.75% of the data must lie within four standard deviations of the mean.

7. Verify Chebyshev's theorem for the data set in problem 1 with $k = 2$. That is, show that at least 75% of the data set is within two standard deviations of the mean. Show the theorem is also true if we use the population standard deviation.

8. Verify Chebyshev's theorem for the data set in problem 2 with $k = 2$. That is, show that at least 75% of the data set is within two standard deviations of the mean. Show the theorem is also true if we use the population standard deviation.

9. If we know the mean is 10.2 and the standard deviation is 1.7, then by Chebyshev's theorem, what would be the two boundary values such that at least 75% of the data lie between the two boundary values?

10. If we know the mean is 32.3 and the standard deviation is 3.1, then by Chebyshev's theorem, what would be the two boundary values such that 89% of the data lie between the two boundary values?

11. If we know the mean is 52.5 and the standard deviation is 4.6, then by Chebyshev's theorem, what would be the two boundary values such that 93.75% of the data lie between the two boundary values?

12. Explain the difference between a sample statistic and a population statistic. What do the terms "biased" and "unbiased" refer to?

13. Compute the coefficient of variation for these two samples: chpt4problem13.sav.

Sample A

13.3 6.5 17.9 16.4 21.3 17.5 26.5 16.9 23.9 32.4 41.0
12.3 17.8 28.4

Sample B

11.9 15.3 21.5 17.5 26.9 11.2 14.9 46.7 24.5 25.9 49.6
34.0 38.8 41.3 32.5

If these numbers represent magnetic susceptibility from sections over two different grids and the coefficient of variation is the measure of variability per unit of expected magnetic susceptibility, then the higher CV might indicate more artifacts are buried in that region. Under this assumption, in what region should the archeologists dig for buried artifacts?

14. Find the standard deviation, variance, range, and coefficient of variation for the following frequency sample data set: chpt3problem9.sav.

X	2	3	4	5	6	7	8	9	10	11	12
Freq	1	2	3	4	5	6	5	4	3	2	1

15. Find the standard deviation, variance, range, and coefficient of variation for the following frequency sample data set: chpt3problem10.sav.

X	7	8	9	10	11	12	13	14	15	16
Freq	34	22	56	75	36	68	91	26	45	28

16. Find the standard deviation, variance, range, and coefficient of variation for the data set: chpt4problem16.sav.

X	12	13	14	15	16	17	18	19	20	21	22
Freq	1	2	3	4	5	6	5	4	3	2	1

This data set is the same as in problem 14 except each data value was increased by 10 points. Does that change the answers as given in problem 14?

17. Find the standard deviation, variance, range, and coefficient of variation for the following frequency sample data set: chpt4problem17.sav.

X	17	18	19	20	21	22	23	24	25	26
Freq	34	22	56	75	36	68	91	26	45	28

This data set is the same as in problem 15 except each data value was increased by 10 points. Does that change the answers as given in problem 15?

18. Find the standard deviation, variance, range, and coefficient of variation for the following frequency sample data set: chpt4problem18.sav.

X	8	12	16	20	24	28	32	36	40	44	48
Freq	1	2	3	4	5	6	5	4	3	2	1

This data set is the same as in problem 14 except each data value was multiplied by 4. How will that change the answers as given in problem 14?

19. Find the standard deviation, variance, range, and coefficient of variation for the following frequency sample data set: chpt4problem19.sav.

X	63	72	81	90	99	108	117	126	135	144
Freq	34	22	56	75	36	68	91	26	45	28

This data set is the same as in problem 15 except each data value was multiplied by 9. How will that change the answers as given in problem 15?

20. 20. Start with the formula $s^2 = \sum \frac{(x-\overline{x})^2}{n-1}$ and derive the shortcut formula for the variance: $s^2 = \frac{\sum x^2 - n\overline{x}^2}{n-1}$.

21. An investor is trying to determine which of these two companies he should invest in. Company A has quarterly returns of 101.6, 230.5, 151.7, and 283.9 dollars. Company B has quarterly returns of 263.4, 335.7, 569.8, and 421.5. Which company has the lower risk? (Hint: the company with the smaller CV will have the smaller risk.)

Just for fun. In a class of 6 students, each guesses the number of pennies in a jar. The six guesses are 85, 92, 73, 64, 59, and 81. The guesses are off by -13, 15, -18, 8, 4, and -4. (Not in the same order.) How many pennies are in the jar?

Chapter 5

Probability

Objectives

After completing this chapter, a student should:

- ✓ Understand the principle of Probability
- ✓ Compute the binomial probability
- ✓ Understand Independent and Dependent events
- ✓ Compute Conditional Probability
- ✓ Understand Inclusion–Exclusion Formula
- ✓ Understand Mutually Exclusive
- ✓ Understand Counting
- ✓ Understand Odds
- ✓ Understand Expected Value
- ✓ Compute geometric probabilities
- ✓ Compute Hypergeometric probabilities
- ✓ Compute Poisson probabilities
- ✓ Understand tree diagrams

Outcomes

As we mentioned in the first chapter, probability is the foundation for statistics. As we also mentioned earlier that the probability of an event is the number of ways the event can occur divided by the total number of possible outcomes.

Example 1.

Find the total number of outcomes for three tosses of a coin. Find the probability of two heads and one tail.

Solution: The toss of a coin has two outcomes: either the coin shows heads on top or the coin shows tails on top. Thus in three tosses there will be $2 \cdot 2 \cdot 2 = 8$ possible outcomes. They are (using H for heads and T for tails)

HHH
HHT
HTH
THH
TTT
TTH
THT
HTT

Thus the first outcome represents heads, heads, heads, or three heads. The second outcome represents heads, heads, tails or two heads and a tail, and so on. How many outcomes have the event of two heads and one tail? The answer is three, as we can see in the above listing of all outcomes for tossing a coin three times; hence the probability for a coin tossed three times to come up with two heads and one tail is 3/8. We say $P(2H,1T) = \frac{3}{8} = .375$.

This experiment can also be thought of as a binomial experiment. Consider the binomial expansion $(H+T)^3 = H^3 + 3H^2T + 3HT^2 + T^3$. Since the probability of *H* is ½ and the probability of *T* is also ½, the term $3H^2T$ will give the probability of 2 heads and 1 tail when we replace *H* and *T* with their probabilities. Thus $P(2H,1T) = 3 \cdot \left(\frac{1}{2}\right)^2 \cdot \left(\frac{1}{2}\right) = \frac{3}{8}$.

Binomial Experiment, Bernoulli Trials

Any experiment that only has two outcomes for each trial is called a binomial experiment. In particular a binomial experiment is composed of Bernoulli trials. A Bernoulli trial is a trial in which

1. There are only two fixed outcomes, called a success and a fail.
2. The probability of a success is always the same; we use the letter p for a success.
3. The probability of a fail is always the same; we use the letter q for a fail.
4. $p + q = 1, q = 1 - p$
5. Each trial is independent; the outcome of one trial will not affect the outcome of a second trial.

A binomial experiment is a finite number, n, of Bernoulli trials. Example 1 is a binomial experiment with 3 trials and $p=\frac{1}{2}$ and $q=\frac{1}{2}$. The terms "success" and "fail" are relative terms: one day we might say getting a head is a success and another day we might say getting a tail is a success, but their probabilities never change and the sum of their probabilities is 1.

SPSS can find the probability of a binomial experiment.

Example 2.

Use SPSS to find the probability of getting two heads if a coin is tossed three times.

Solution: Open an Untitled SPSS Data Editor (open a new data editor). Enter 2 into the first column. In variable view, name the variable x and change the measure to scale. Use Transform → Compute Variable, and enter *prob1* into the Target Variable window. In the Function Group window, see **Figure 1a**, scroll down until you find PDF & Noncentral PDF, and click on that. In the Functions and Special Variables window, double click on Pdf.Binom. The function should move to the Numerical Expression window; see **Figure 1b**. The explanation for the numeric expression is as follows: PDF.BINOM(quant, n, prob) Numeric. This returns the probability that the number of successes in n trials, with probability prob of success in each, will be equal to quant. When n is 1, this is the same as PDF.BERNOULLI.

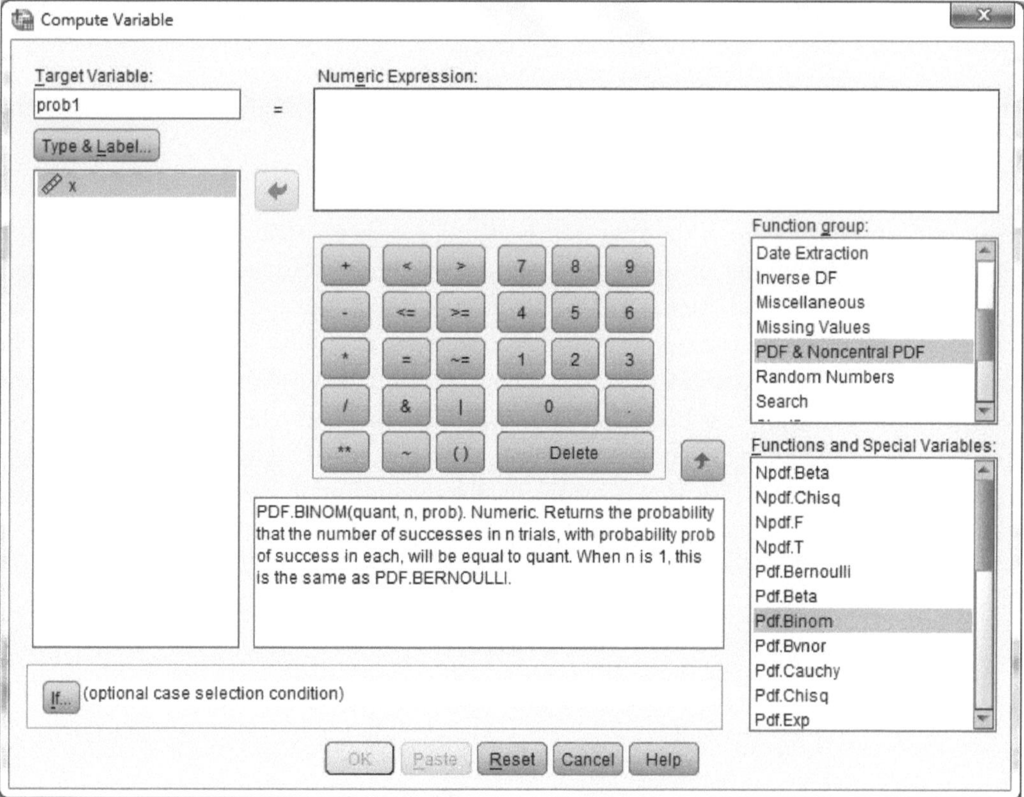

Figure 1a

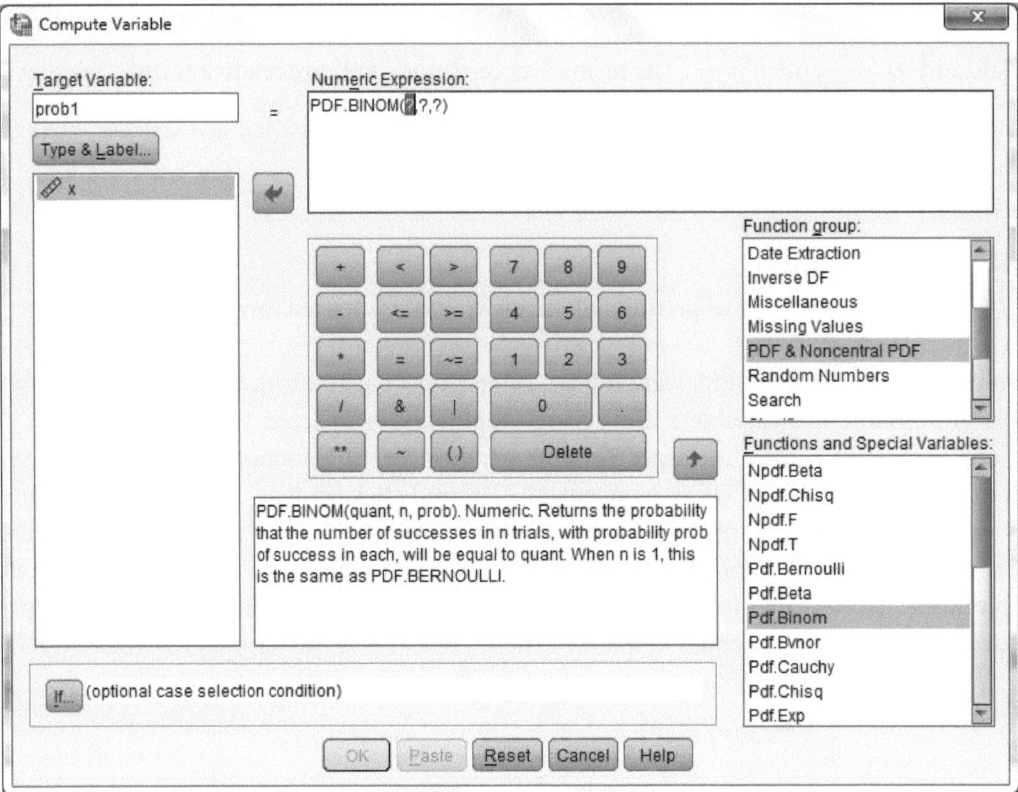

Figure 1b

The three question marks mean the function is asking for three parameters; the function is asking for three parameters, the number of successes (quant), the number of trials (*n*), and the probability of a success (prob). Highlight *x* and use the move button to place *x* in for quant; see **Figure 1c**. Next move the cursor behind the second question mark and backspace to delete it. Then type in 3 for *n* because 3 is the number of trials in our experiment (toss a coin 3 times). Likewise replace the last question mark with .5, the probability of a success because the probability of a head when flipping a coin is 50%. Hit OK and the variable prob will be in a new column of the data editor. Change the number of decimals to 4 and the answer should read .3750. This is a numeric data value and it is the probability for obtaining 2 heads in the toss of a coin 3 times, see **Figure 1d.**

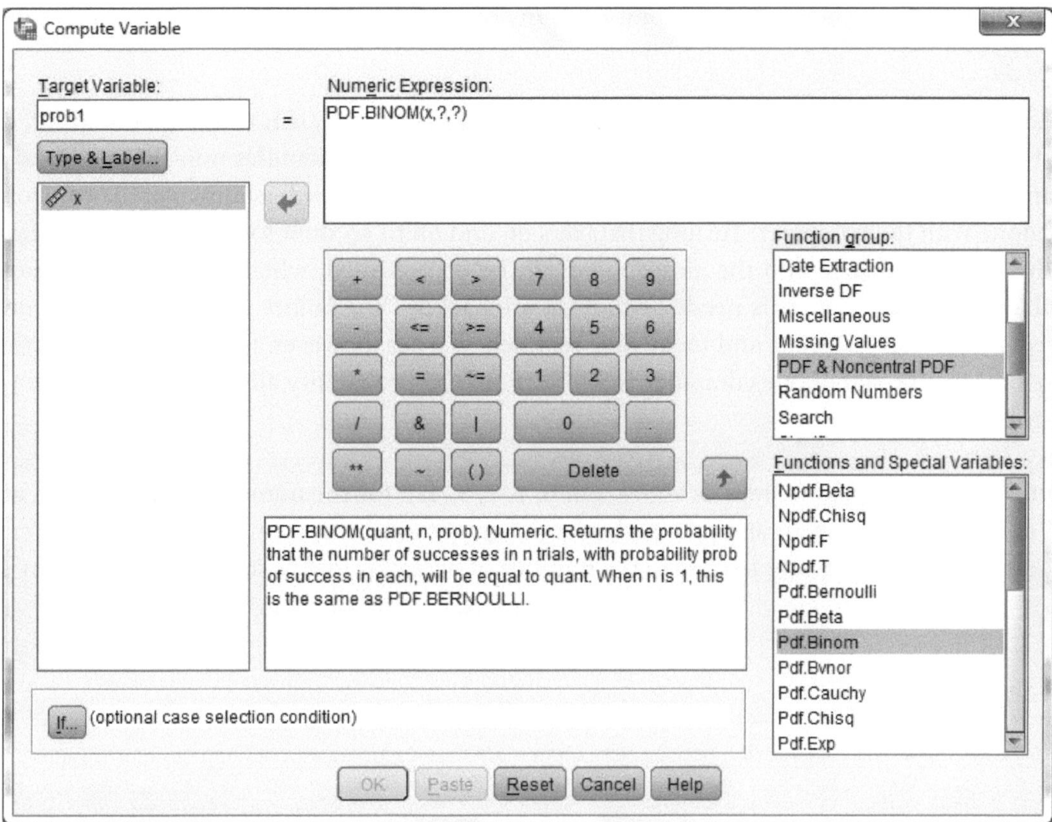

Figure 1c

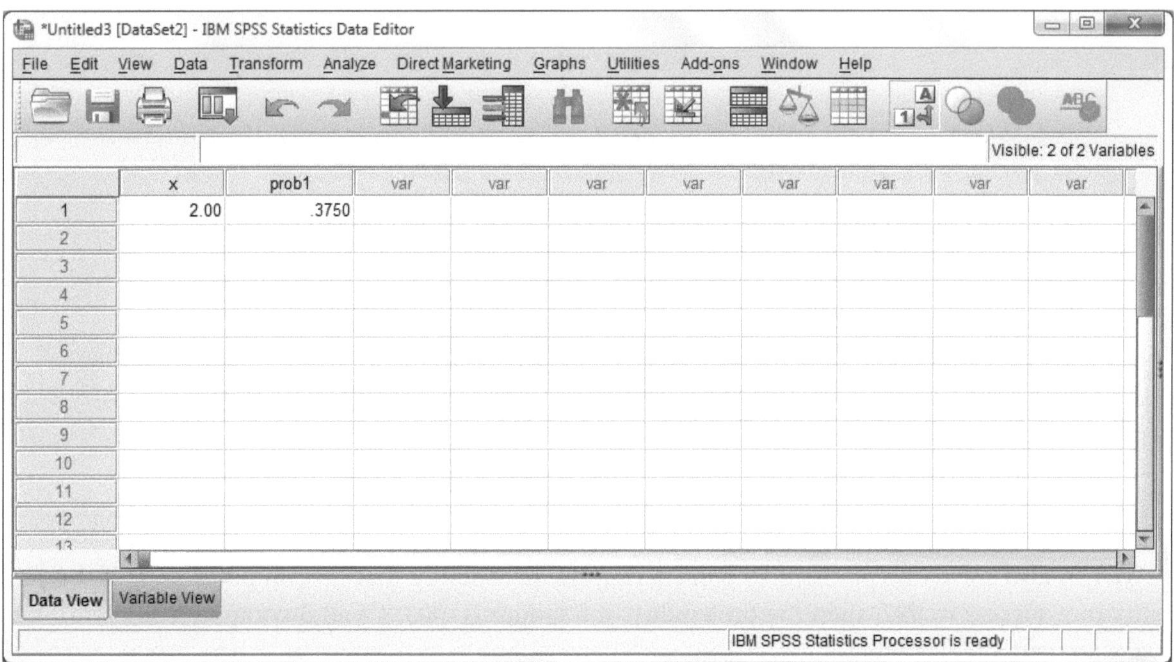

Figure 1d

We can now find the binomial probabilities of any binomial experiment.

Example 3a.

The space shuttle had five computers on board; any one of four is sufficient to ascend into orbit, descend into atmosphere, and control the payload, while the fifth computer handles non-flight-critical avionic functions. (1) The descent into the atmosphere was once performed with a human at the controls, but is usually done with the computer. Turning thrusters on and off to account for changes in the heat shields while the shuttle impacts with the atmosphere can be complicated, which is why the computers are invaluable; however, only one is needed—the rest are backup. The failure rate for the early computers was once in every 6,000 hours, and today they're much sharper; however, for the sake of argument let's suppose the success rate of one computer is 99.7%. Find the probability all five computers are working successfully.

Solution: Follow the same steps as above, put 0, 1, 2, 3, 4, 5 for the number of successes in an x column, 5 as the number of trials and .997 as the probability of a success. The probability of all 5 successes is .9851; see **Figure 2a**. Be sure to change the number of decimals to 4 in the prob1 variable in variable view.

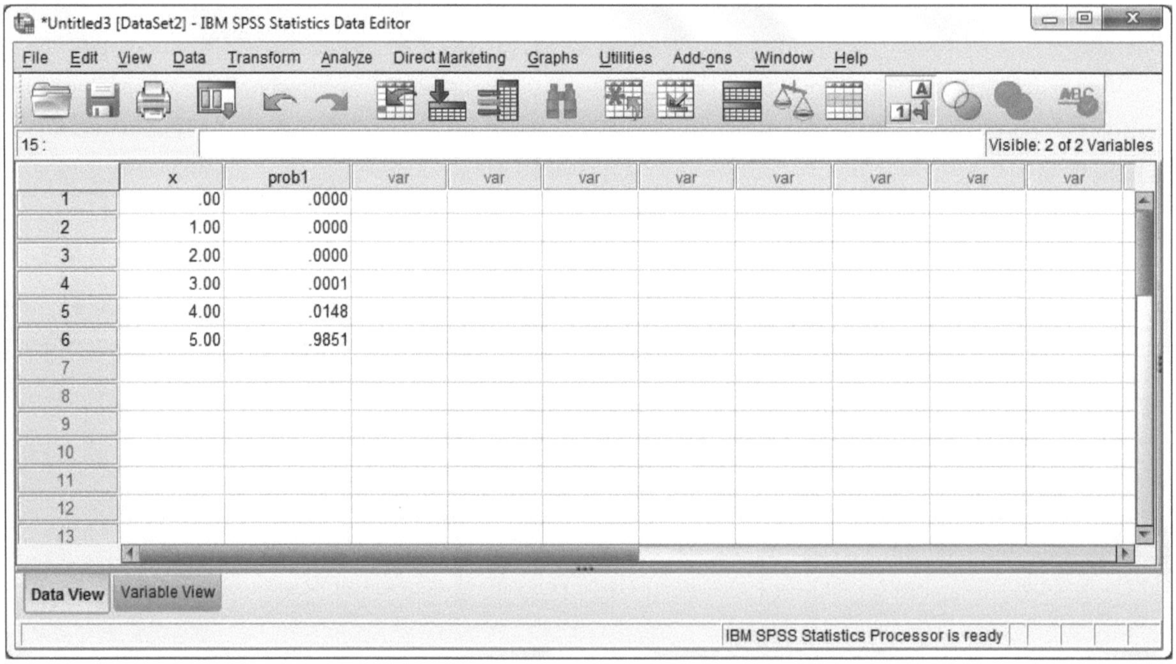

*Untitled3 [DataSet2] - IBM SPSS Statistics Data Editor

File Edit View Data Transform Analyze Direct Marketing Graphs Utilities Add-ons Window Help

15: Visible: 2 of 2 Variables

	x	prob1
1	.00	.0000
2	1.00	.0000
3	2.00	.0000
4	3.00	.0001
5	4.00	.0148
6	5.00	.9851

Data View Variable View

IBM SPSS Statistics Processor is ready

Figure 2a

We can also solve this problem by considering the probability that no computer fails. If the probability of a success is .997, then the probability of a failure is .003. (If all 5 computers are working, it means no computer is failing.)

Example 3b.

Find the probability that no computer out of five fails, where the probability of a failure is .003.

Solution: Use the same data editor and use Transform → Compute Variable, call the target variable *prob2*, and use the Pdf.Binom function with the same *x* values, 5 for *n* and .003 for the probability. The output will read the same as above, except backwards; see **Figure 2b**. The probability that no computers out of 5 fails is .9851. If we want the probability that at least one computer fails, then we add the probabilities for $x = 1, 2, 3, 4, 5$, which is .0149. This is called the complement of the event that no computer will fail; see **Figure 2c**. The probability of the complement $P(\bar{A})$ is $1 - P(A)$. Likewise, the probability of an event is $P(A) = 1 - P(\bar{A})$. Sometimes it's easier to compute the complement of an event rather than the event itself. Then to find the probability of the event, we subtract the probability of the complement from 1. In this case we know the probability that no computer fails out of 5 computer is .9851, so the probability that at least one computer will fail is 1 – .9851 = .0149. This way we subtract only one number rather than add five numbers.

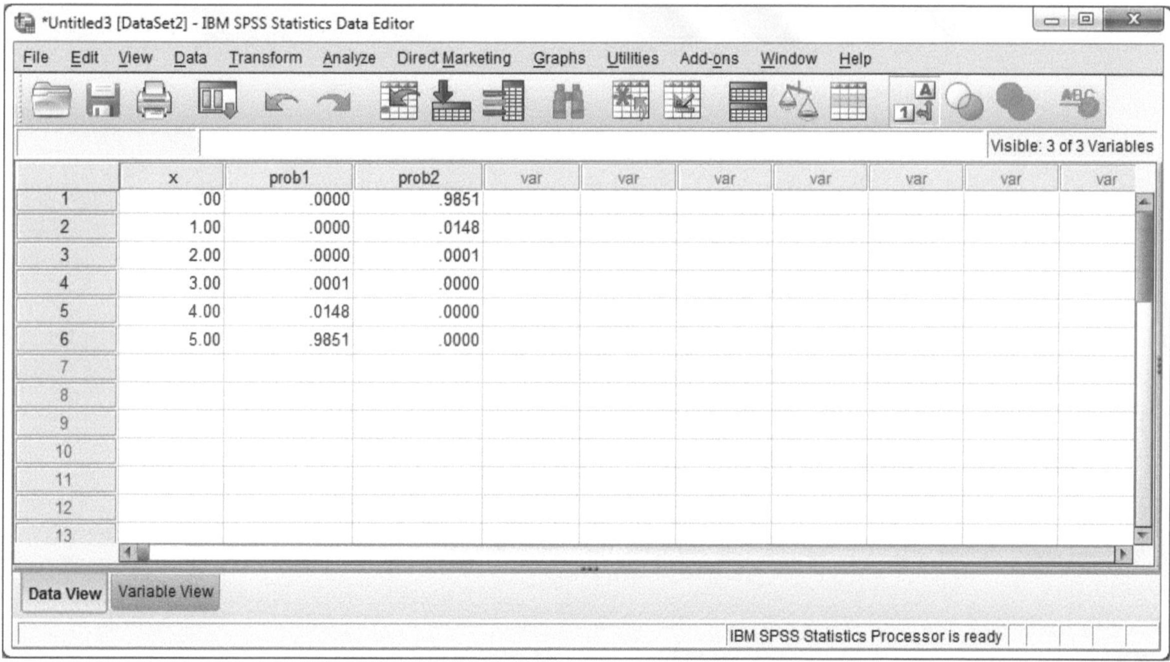

	x	prob1	prob2	var	var	var	var	var	var	var
1	.00	.0000	.9851							
2	1.00	.0000	.0148							
3	2.00	.0000	.0001							
4	3.00	.0001	.0000							
5	4.00	.0148	.0000							
6	5.00	.9851	.0000							
7										
8										
9										
10										
11										
12										
13										

Figure 2b

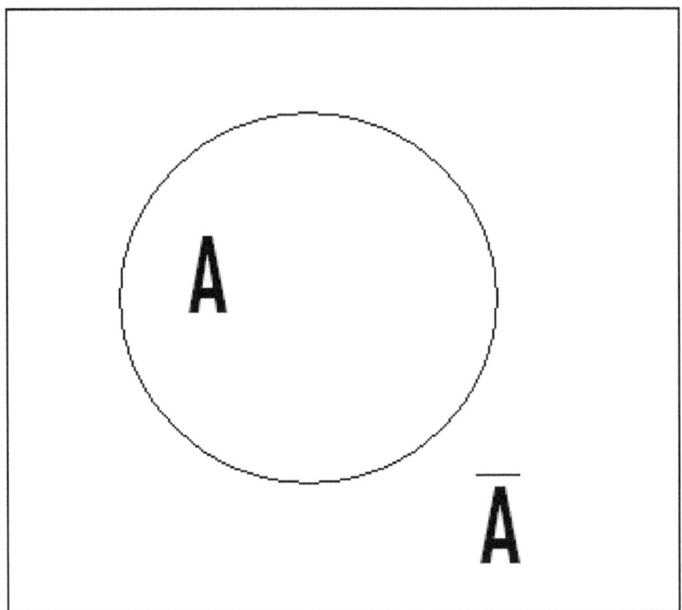

Figure 2c

Examples 2 and 3 are called equivalent events since they represent the same concept. The complement of a success is a failure; hence if the probability of a success is .997, then the probability of a failure is .003, because the sum of the probabilities of a success and a failure is 1. So the event all five computers work is the same as the event none of the five computers fail. Usually with binomial distributions we count from zero successes to *n* successes, but we can just as easily count from zero failures to *n* failures. In this particular problem we don't want five failures, we can tolerate anywhere from 1 to 4 failures. What we really want is the cumulative distribution function from zero to 4 failures. Rather than add each binomial probability by hand, we can use SPSS to find the cumulative sum.

Complementary Events

The complement of an event is the set of all outcomes not in the event; see **Figure 2c**. If A denotes an event, then $\overline{A}$ denotes the complement of the event. Since $A \cup \overline{A}$ is the entire sample space, S, $A \cap \overline{A} = \varnothing$ and since $P(S) = 1$, we have $P(A) + P(\overline{A}) = 1$ or $P(\overline{A}) = 1 - P(A)$; likewise $P(A) = 1 - P(\overline{A})$.

Example 3c.

Find the probability that at least one computer out of 5 is successful given a success rate for a single computer is .997.

Solution: The cumulative distribution function in Transform → Compute Variable will add only the first set of values from 0 to where we want the sum to stop. SPSS cannot add, say, values 1 through 5. Therefore we'll use the complement event and have SPSS add the first five values in *prob2* column.

This is the same as adding the last five values in prob1 column. The complement of a success is a failure with a probability of .003. Hence we want to find the cumulative probability of 4 failures. Use the same data set as above or enter 4 into an SPSS data editor and call the column *x*. Use Transform → Compute Variable. In the Target Variable window, type in *prob3* or any convenient letter. In the Function Group window, locate the CDF & Noncentral CDF, highlight that. Double click the Cdf.Binom function in the Functions and Special Variables window. It should move into the Numeric Expression window. Move the *x* variable into the first position, and place 5 in the second position and .003 into the third position. Hit OK. Remember to change the decimal places to 4. The answer should read 1.0000, which is great because this tells us the probability that zero to 4 computers will fail is 100%, which means out of 5 computers, there's a 100% probability at least one computer is functioning.

Another way to find the solution is to subtract from 1 the probability all 5 computers fail, or subtract from 1 that 0 computers will work. That will be 1 – PDF.BINOM(*x*, 5, .003) where $x = 5$, or 1 – .003^5. Since .003^5 is very small (actually it's .000000243), 1 – .003^5 is close to 1. Probability is almost never 100% or 0%, but can be so close to those numbers that we allow ourselves to use the values 1 or 0. Nonetheless there's always a small chance that extreme values can occur.

Independent Events

In calculating the above probabilities we were using an important concept, the concept of independent events. Two events are independent if the outcome of one event does not affect the outcome of another event. In the previous example, the computers are independent because the success of one computer will not affect the success of another computer. In mathematical terms we say two events are independent if the probability of their intersection is the product of their probabilities. That is, if *A* and *B* are two independent events, then $P(A \cap B) = P(A) \cdot P(B)$. The $\cap$ symbol means intersection; see the Venn diagram in **Figure 3**.

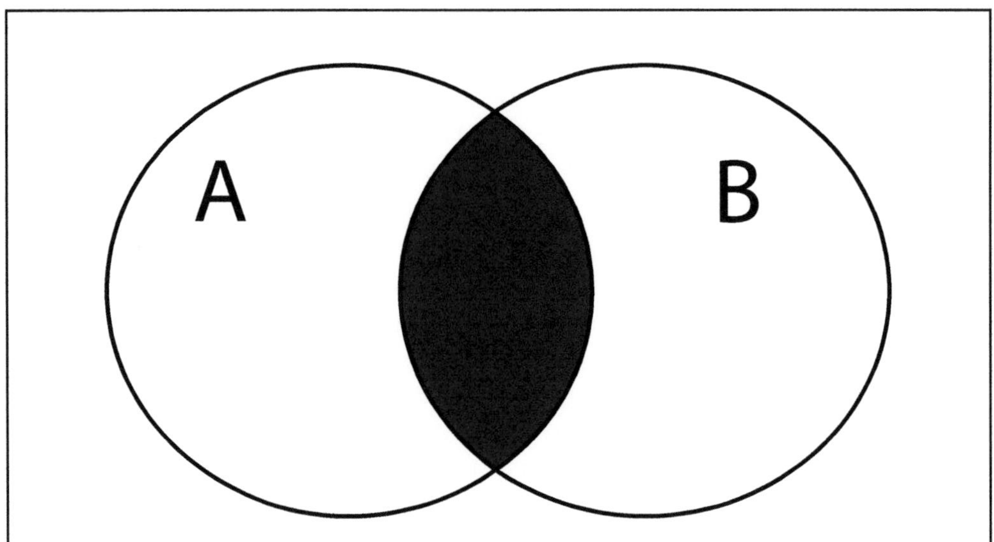

Figure 3

The intersection of two events *A* and *B* are all the outcomes that satisfy both events *A* and *B*. In the Venn diagram, the intersection is the shaded region that includes both *A* and *B*. Two events are independent if the outcome of one event, *A*, does not affect the outcome of a second event, *B*, such as the event of flipping a coin, rolling a die, or having children. (Regardless of whether the first child was a boy or girl, the probability the next child will be a boy or girl is still 50%.)

Example 5a.

Find the probability that out of seven children, the first child is a boy and the next six are all girls.

Example 5b.

Find the probability one of seven children is a boy and the other six are all girls.

Example 5c.

Find the probability that out of eight children, two are boys.

Example 5d.

Find the probability out of eight children, there are at least two boys.

Solution: **5a**. The solution is simply $\left(\frac{1}{2}\right)^7 = .0078125$, because the first child has to be a boy, with probability ½, and the next six have to be girls with probability ½, and there's only one way that can happen. But we'll use SPSS to find the answer. Put 6 in an open column of an SPSS data editor, call the variable *x*, or put 6 in the *x* column of the data editor we used previously. Use Transform → Compute Variable, call the target variable *p*, and move the cursor into the Numeric Expression box. Now the probability of a boy for the first child is .5, so place .5 in the Numeric expression and click on the times (*) button. Find Pdf.Binom, double click on that, put *x* in the first position, 6 in the second position and

.5 in the last position. This means we will multiply the probability of a boy, .5, with the probability of six girls out of six children; see **Figure 4a**. Hit OK and click on the value to see the full number under the tools bar; it should read .0078125. See **Figure 4b**.

5b. In the binomial distribution we like to think of the different trials as slots to be filled with either a *p* for success or a *q* for failure; see **Figure 4c**. (Remember the terms success and failure are relative terms and have no significance as to whether the outcome is good or bad.) We can also use the letters *b* and *g* for boy or girl. So we have 7 slots that we want to fill with 6 *g*'s and one *b*. In how many ways can this be done? The answer is 7, because the one *b* can occupy any of the 7 slots and the 6 *g*'s must occupy the remaining 6 slots. To find the probability of one boy and 6 girls in seven children, we multiply $7\times(.5)^7$ =.0547 (we replace each *b* and *g* with .5). We can use SPSS to find this value using PDF.BINOM(*x*,7,.5). Put 6 in a variable labeled *x* and use Transform → Compute Variable. Call the target variable *p* and click on PDF & Noncentral PDF and double click on Pdf.Binom. Move *x* into the first parameter, and put 7 in the next parameter and .5 as the last parameter. Click OK. The answer will be under *p* in the row with *x* equal to 6; see **Figure 4d**.

5c. From algebra we recall the binomial coefficient of a term in the expansion of $(p+q)^n$ can be found using the function combinations of n choose r, *C*(*n*, *r*) on a calculator, where *n* is the exponent of the binomial and *r* is the exponent of *q*, while *n* – *r* is the exponent on *p* (3). SPSS can generate binomial coefficients using .5 for both *p* and *q* and multiplying PDF.BINOM(*r*,*n*,.5) by 2^n; see note below. Thus to find the probability that out of 8 children, 2 are boys, we multiply *C*(8, 2) with $.5^8$. However, we can skip finding the binomial coefficient and use SPSS directly to find the probability of 2 *b* and 6 *g* in 8 trials. Put 6 (or 2, or both) in a variable labeled *x* and use Transform → Compute Variable. Call the target variable *p*, click on PDF & Noncentral PDFf, and double click on Pdf.Binom. Move *x* into the first parameter, 8 in the next parameter, and .5 as the last parameter. Click OK. The output will be the same for both values, .1094, because the probability is the same for both values *b* and *g*.

5d. We need to find the cumulative probability of 2, 3, 4, 5, 6, 7, and 8 boys since the terminology "at least 2 boys" means we have 2 or more boys. Thus we need to use the Cumulative Distribution Function. The complement of that event is 0 or 1 boy, which is equivalent to 8 or 7 girls. We can find the cumulative probability of 1 boy and subtract the answer from 1 to get the probability of more than 1 boy or we can find the cumulative probability of 6 girls, because that is the complement event of 8 or 7 girls, and get our answer directly. We'll use that event to find our desired probability. Open a data editor and put 6 in an *x* column. Use Transform → Compute variable, label the target variable *p*, place CDF.BINOM(*x*, 8, .5) in the numeric expression box, and click OK; see **Figure 4e**. The return in the data editor window should read .9648. This means the probability of 6 or fewer girls in a family of eight children is 96.48%, which is equivalent to the probability of 2 or more boys in a family of 8 children, hence our solution.

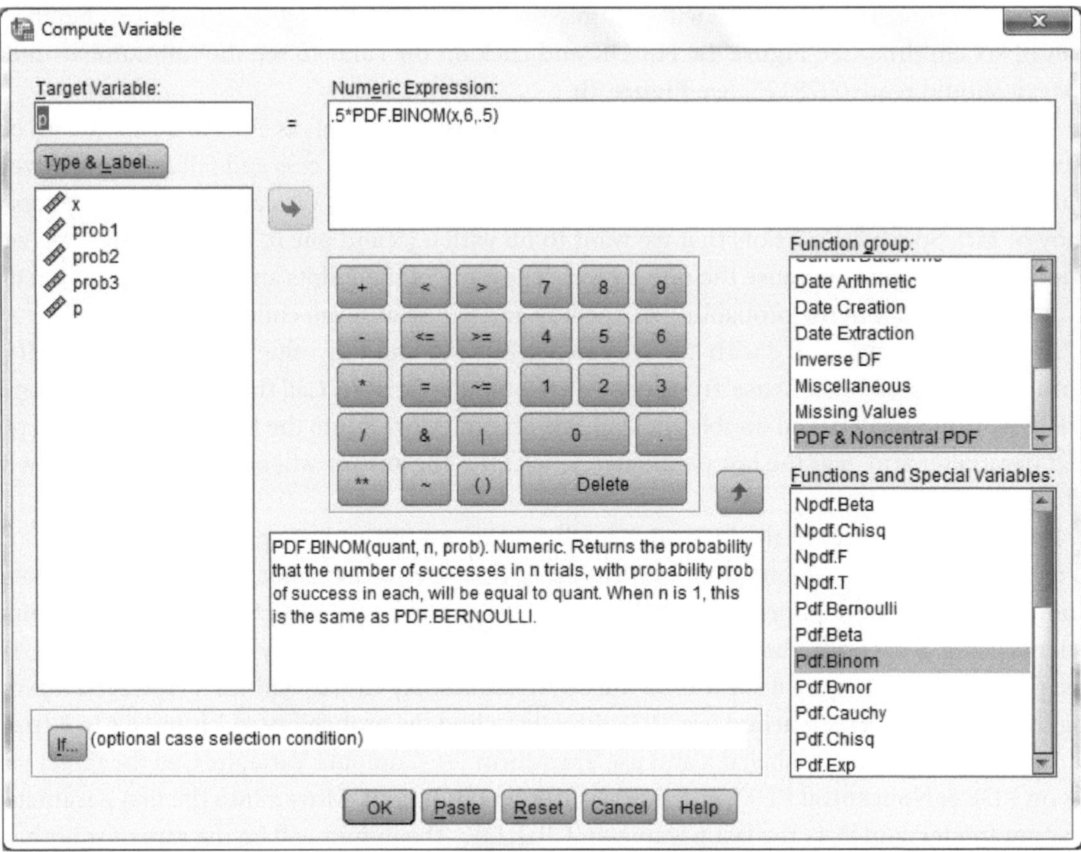
Compute Variable
Target Variable:
p
Type & Label...
x
prob1
prob2
prob3
p
Numeric Expression:
=
.5*PDF.BINOM(x,6,.5)
Function group:
Date Arithmetic
Date Creation
Date Extraction
Inverse DF
Miscellaneous
Missing Values
PDF & Noncentral PDF
Functions and Special Variables:
Npdf.Beta
Npdf.Chisq
Npdf.F
Npdf.T
Pdf.Bernoulli
Pdf.Beta
Pdf.Binom
Pdf.Bvnor
Pdf.Cauchy
Pdf.Chisq
Pdf.Exp
Delete
PDF.BINOM(quant, n, prob). Numeric. Returns the probability that the number of successes in n trials, with probability prob of success in each, will be equal to quant. When n is 1, this is the same as PDF.BERNOULLI.
If... (optional case selection condition)
OK
Paste
Reset
Cancel
Help

Figure 4a

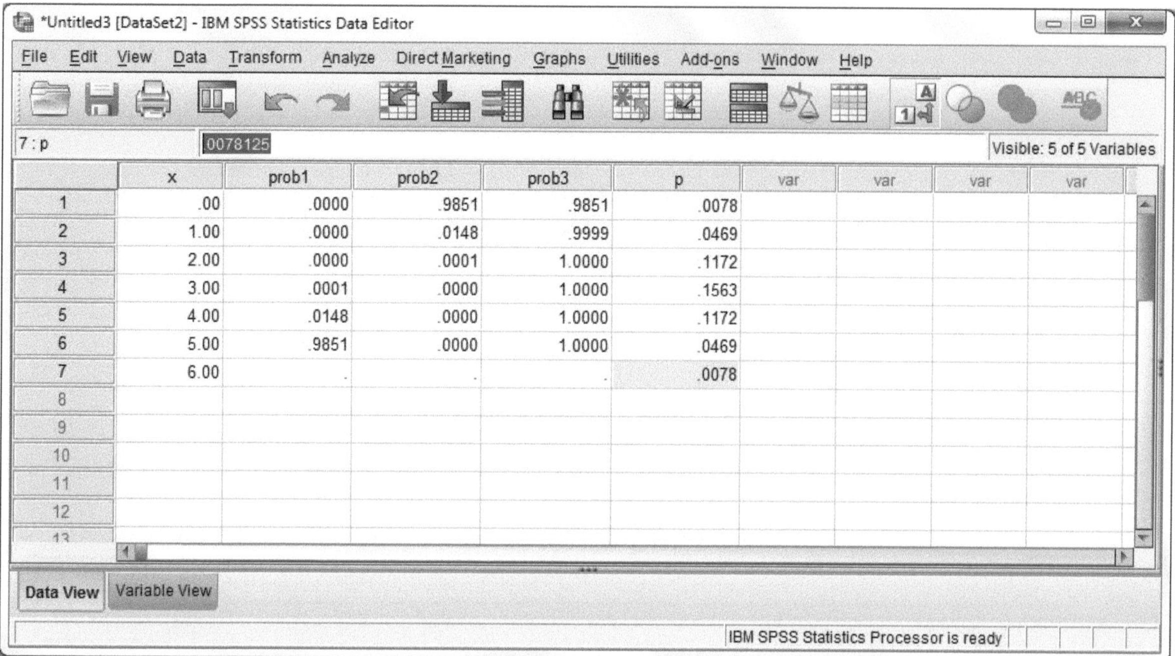

	x	prob1	prob2	prob3	p	var	var	var	var
1	.00	.0000	.9851	.9851	.0078				
2	1.00	.0000	.0148	.9999	.0469				
3	2.00	.0000	.0001	1.0000	.1172				
4	3.00	.0001	.0000	1.0000	.1563				
5	4.00	.0148	.0000	1.0000	.1172				
6	5.00	.9851	.0000	1.0000	.0469				
7	6.00	.	.	.	.0078				
8									
9									
10									
11									
12									

Figure 4b

b g g g g g g

g b g g g g g

g g b g g g g

g g g b g g g

g g g g b g g

g g g g g b g

g g g g g g b

Figure 4c

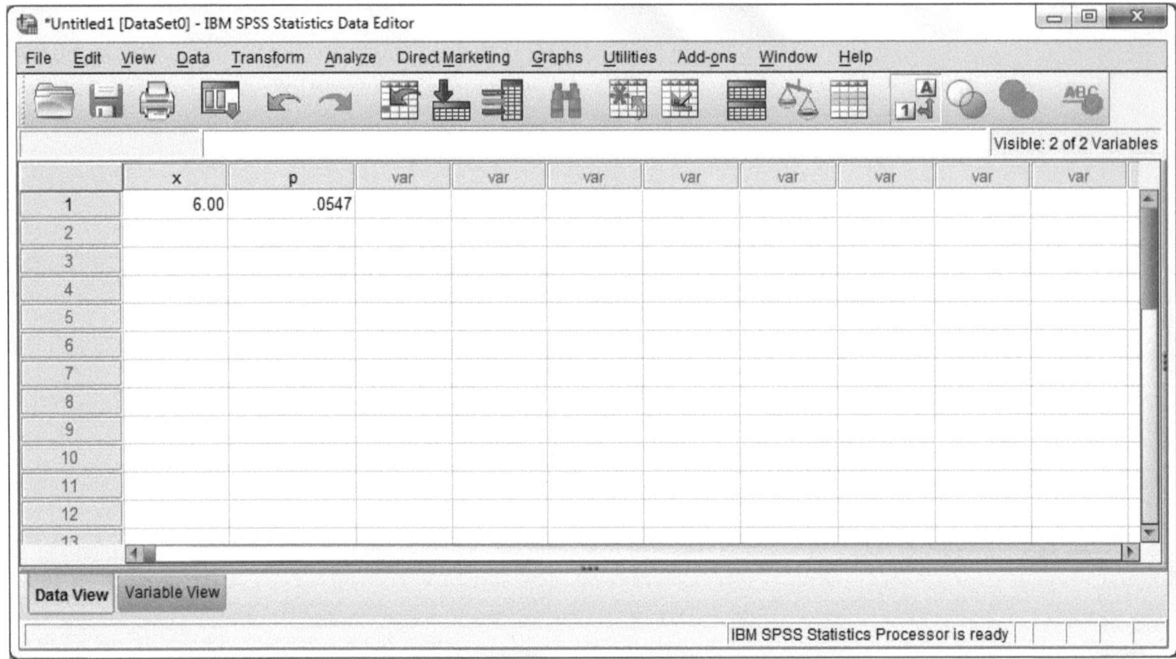

Figure 4d

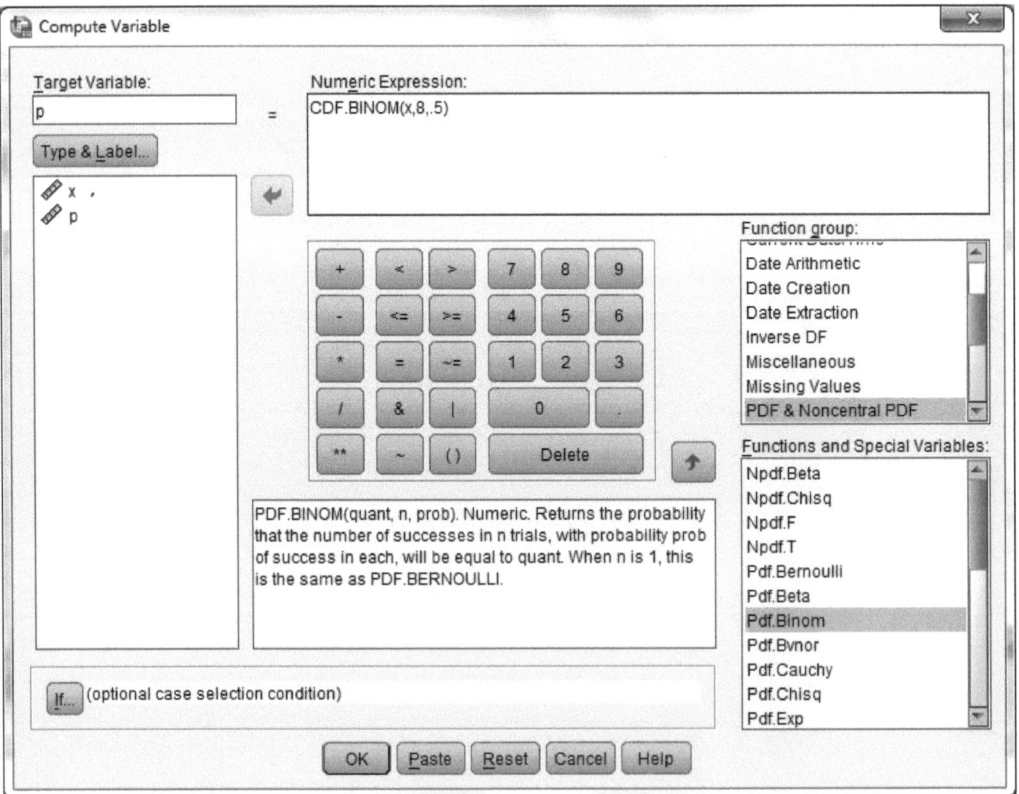

Figure 4e

Note: $C(n, r)$ is called a binomial coefficient and it is the number in the $(r + 1)$th position in the nth row of Pascal's Triangle; see **Figure 4f**. The numbers in Pascal's triangle can be generated by adding the number immediately above with the number immediately above and to the left. They can also be generated with the formula $C(n,r)=\frac{n!}{(n-r)!r!}$, where $n!$ (read n factorial) is the product of the first n positive integers and $0! = 1$. $C(n, r) = C(n, n-r)$ because $C(n,r)=\frac{n!}{(n-r)!r!}=\frac{n!}{r!(n-r)!}=\frac{n!}{(n-(n-r))!(n-r)!}=C(n,n-r)$.

1	1									
1	2	1								
1	3	3	1							
1	4	6	4	1						
1	5	10	10	5	1					
1	6	15	20	15	6	1				
1	7	21	35	35	21	7	1			
1	8	28	56	70	56	28	8	1		
1	9	36	84	126	126	84	36	9	1	
1	10	45	120	210	252	210	120	45	10	1

Figure 4f

Example 5e.

From Pascal's triangle find: $C(5, 3)$, $C(6, 0)$, $C(7, 7)$, $C(8, 5)$, $C(9, 4)$, and $C(10, 6)$.

Solution: $C(5, 3) = 10$. Move down to the 5th row and count to the 4th term. The third term is also 10, which verifies $C(5, 3) = C(5, 5\text{-}3) = C(5, 2)$. $C(6, 0) = 1$, $C(7, 7) = 1$, $C(8, 5) = 56$, $C(9, 4) = 126$ and $C(10, 6) = 210$.

If n is much larger than 10, then we can use SPSS to find the binomial coefficient.

Example 5f.

Find the number of ways to draw a five-card poker hand from a deck of 52 cards.

Solution: This problem asks us to find $C(52, 5)$. We can use a calculator with the formula, but we'll use SPSS instead. Open a new data editor, put 5 in a column, then label the variable x. Use Transform → Compute Variable, call the target variable *bc* for binomial coefficient, move the cursor into the numeric expression, and create the expression EXP(52*LN(2)). To choose EXP() and LN(), click on Arithmetic under Function group and these two functions can be found under Functions and Special Variables. What we need to do is create 2^{52}, but SPSS does not have a power function for arbitrary bases. It does have a power function for the base e, and as we all know from algebra, we can change bases by taking logs of both sides and then applying the inverse log function, which is what EXP(52*LN(2)) does. Now enter a * sign and click on PDF & Noncentral PDF and double click on Pdf.Binom. Put x in the first position, 52 in the second position, and .5 as the probability. This will give us the probability of choosing 5 cards out of 52 but each with a probability of .5 (keep in mind .5 is not the correct probability for choosing a card from a deck of cards). By making all their probabilities equal to .5, then the value will be the binomial coefficient times .5 to the 52nd power. By multiplying by 2 to the 52nd power, we eliminate all the fractions and are left with the binomial coefficient; see **Figure 4g**. Click OK and the binomial coefficient will be 2598960 as in **Figure 4h**.

In general $C(n, r)$ = EXP(n *LN(2))*PDF.BINOM(x,n,.5).

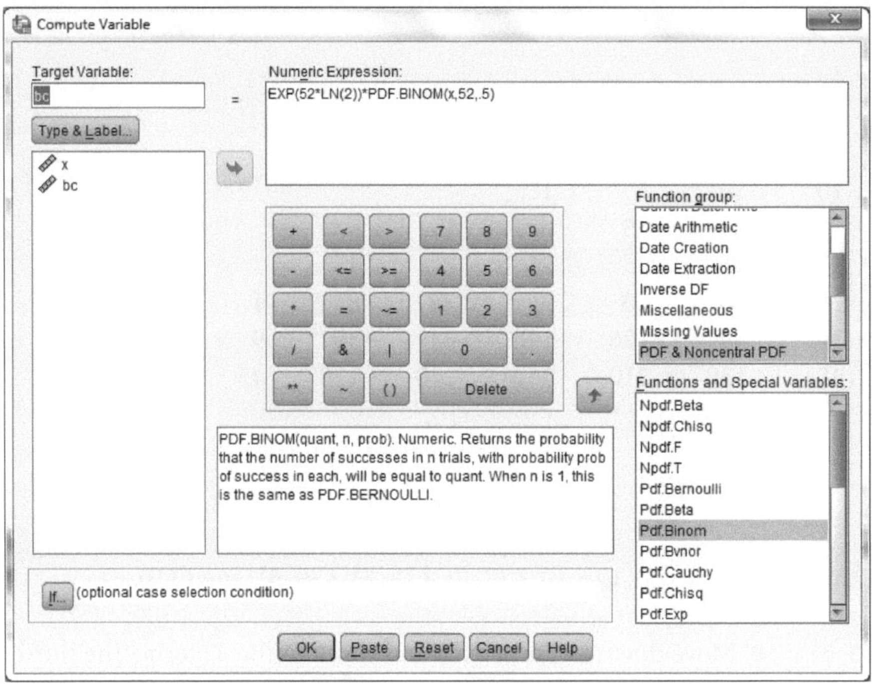

Figure 4g

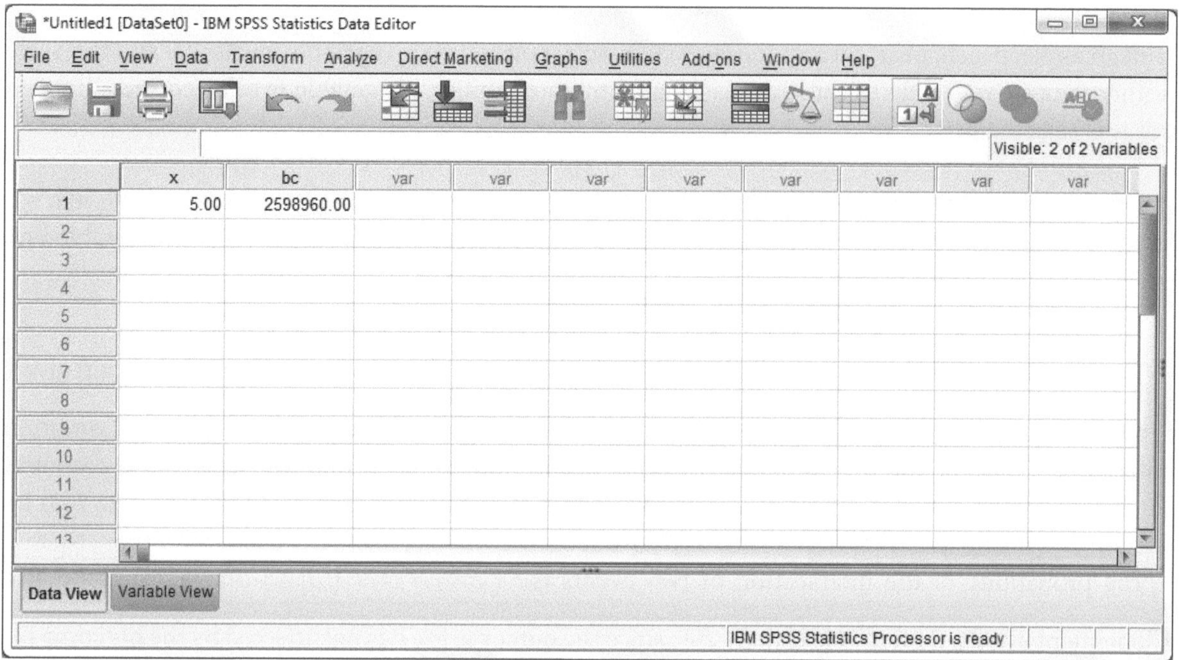

Figure 4h

Conditional Probability

Two events are not independent if the outcome of one event affects the outcome of the second event. An example is to select one card from a single deck of cards and then select a second card from the same deck. Another example—from an apple, peach, pear, carrot, or celery stick—is to select one fruit or vegetable to eat first and then select another fruit or vegetable to eat second. The probabilities of the first event (choose the first card, eat the first fruit) and the second event (choose the second card, eat the second fruit) are different depending on the outcome of the first event.

Let's consider the first example. Suppose the first event is to draw a red card from a standard deck of cards. The probability of that event is $\frac{1}{2}$ because there are 26 red cards and 26 black cards in a standard deck of cards. Now suppose the second event is to draw a black card from the remaining cards. The probability of drawing a black card in the second event will change depending on what card was drawn in the first event. If a red card was drawn in the first event, then the probability of a black card in the second event is $\frac{26}{51}$ because 26 black cards are left in the deck and 51 cards are left in the deck. On the other hand, if a black card was drawn in the first event, then the probability of a black card in the second event is $\frac{25}{51}$ because 51 cards are left in the deck and only 25 cards are black.

Example 6.

From an apple, peach, pear, carrot, or celery stick, select one fruit or vegetable to eat first and then select another fruit or vegetable to eat second. What is the probability a vegetable is selected second given a fruit was selected first?

Solution: If a fruit is selected first, then there are only 2 pieces of fruit left and 2 vegetables left, so of the four pieces of fruit and vegetables, 2 of them are vegetables, thus the probability a vegetable being selected second given a fruit was selected first is $\frac{2}{4}=\frac{1}{2}$.

The formula for conditional probability is $P(B|A)=\frac{P(A\cap B)}{P(A)}$, read as the probability of *B* given *A* is equal to the probability of their intersection divided by the probability of *A*; see **Figure 3**. Conditional probability is that the event *A* has already occurred and now we want to know what is the probability that event *B* will also occur. In essence we treat event *A* as the sample space and then seek to find how much of *B* is inside *A*.

The probability of the intersection of two events is $P(A\cap B)=P(B|A)\cdot P(A)$. If *A* and *B* are independent events, then $P(B|A)=P(B)$, since the outcome for *B* is not affected by the outcome of *A*.

Example 7.

In a hat with 10 numbers, 1 through 10, two numbers are drawn at random. What's the probability that both numbers are odd?

Solution: This is conditional probability. Since 5 numbers are odd and five numbers are even, if the first number drawn is odd, with probability ½, then the probability the next number is odd is 4/9. So the probability that both are odd is $\frac{1}{2}\cdot\frac{4}{9}=\frac{2}{9}$

Conditional probability is also used in contingency tables. An example of a contingency table is the following: A random sample of 100 students was categorized as to male/female and right-handedness/left-handedness. The table is given in **Figure 5a.**

	Right-Handed	**Left-Handed**	**Totals**
Male	42	9	51
Female	44	5	49
Totals	86	14	100

Figure 5a

The probability that a person selected at random is a left-handed female is $\frac{5}{100}=.05$. The probability a person selected at random is female given the person is left-handed is $\frac{5}{14}=.3571$.

We can use SPSS to make contingency tables and find conditional probabilities.

Example 8a.

Using the data set above, find the probability that a person selected at random will be a right-handed male.

Example 8b.

Find the probability that a person selected at random is male given the person is right-handed.

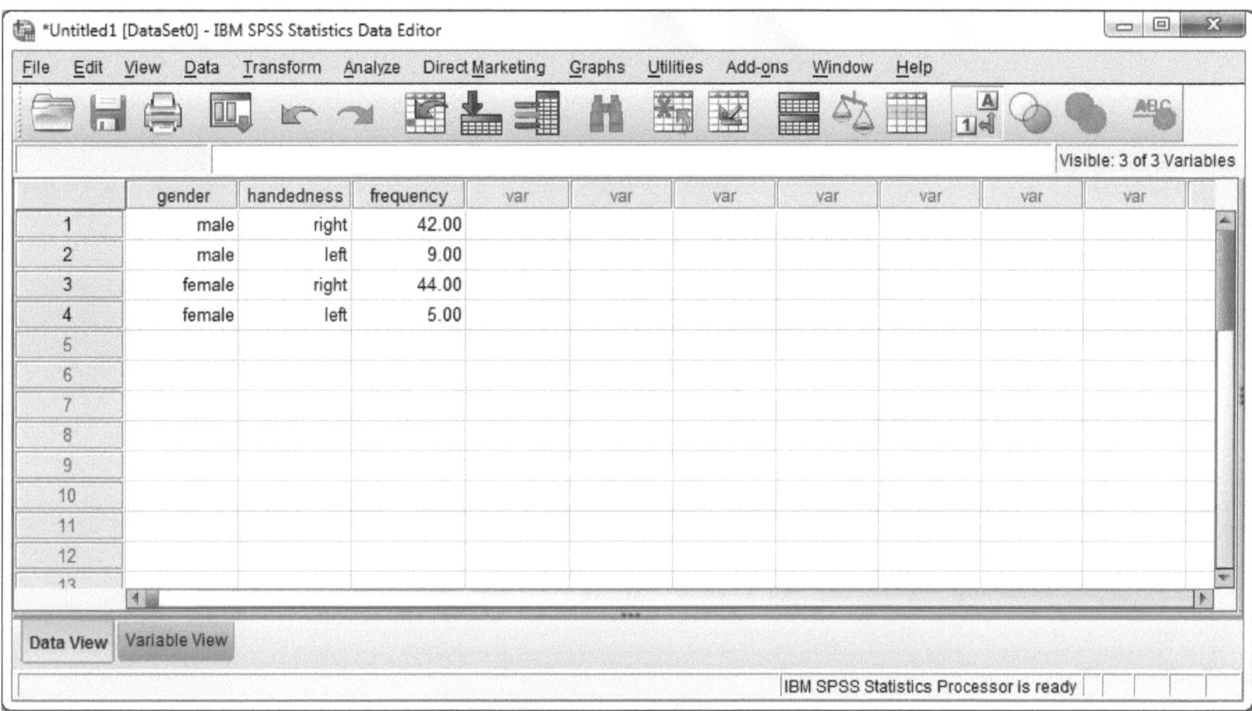

Figure 5b

Solution. Input the data above into a new SPSS data editor or open chpt5example8a.sav. To input the data into a new data editor, use the first variable as gender and the second variable as handedness. The third variable will be the frequency. Use 1 for male, 2 for female, 1 for right-handedness, and 2 for left-handedness and change the value labels to their names; see **Figure 5b**. Change gender and handedness to a nominal variable and frequency to a scale variable. Weight the data using frequency as the frequency variable.

8a. We'll make a contingency table and from that we'll find the desired probability. Use Analyze → Descriptive Statistics → Crosstabs, put gender in the Row(S) window, and put handedness in the Column(s) window. Click on cells and check Total under Percentages; see **Figure 5c**. Click Continue and then OK and the following contingency table will appear in the output; see **Figure 5d**. The probability a person selected at random will be a right-handed male is $\frac{42}{100} = .42 = 42\%$.

8b. To find probability a person selected at random will be male given the person is right-handed, use Analyze → Descriptive Statistics → Crosstabs, put gender in the Row(S) window, and put handedness in the Column(s) window. Click on cells and check Column under Percentages; see **Figure 5e**. Click Continue and then OK and the following contingency table will appear in the output; see **Figure 5f**. The probability a person selected at random is male given the person is right-handed is $\frac{42}{86} = .4884 = 48.8\%$.

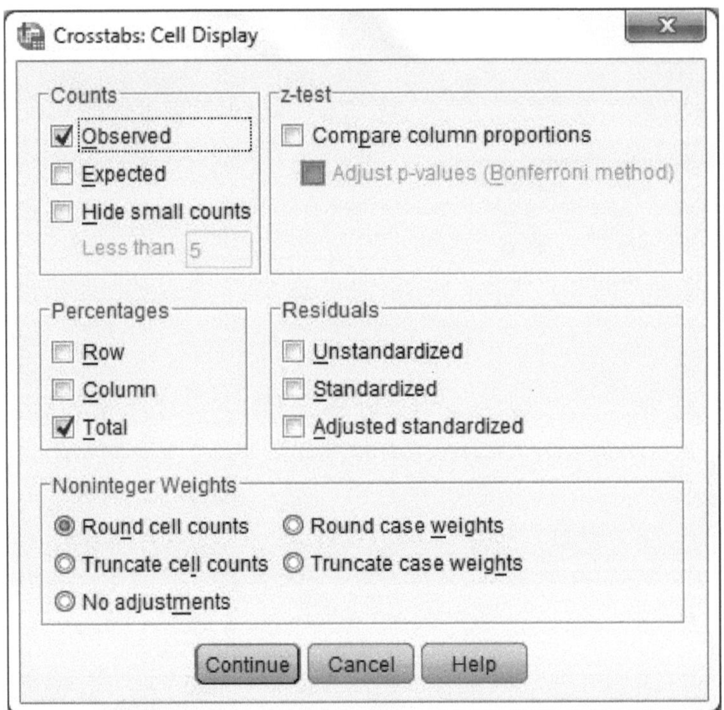

Figure 5c

Gender * Handedness Cross-Tabulation

			Handedness		Total
			Right	Left	
Gender	Male	Count	42	9	51
		% of Total	42.0%	9.0%	51.0%
	Female	Count	44	5	49
		% of Total	44.0%	5.0%	49.0%
Total		Count	86	14	100
		% of Total	86.0%	14.0%	100.0%

Figure 5d

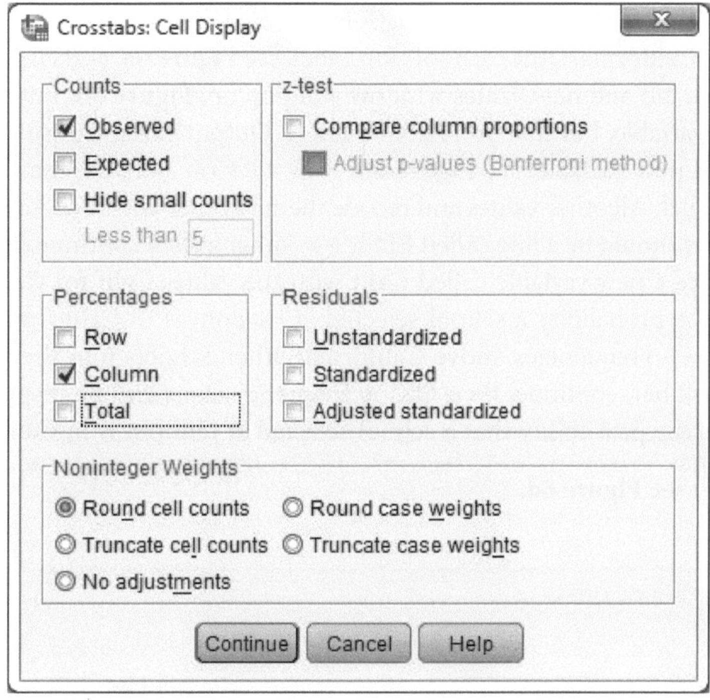

Figure 5e

Gender * Handedness Cross-Tabulation

			Handedness		Total
			Right	Left	
Gender	Male	Count	42	9	51
		% within handedness	48.8%	64.3%	51.0%
	Female	Count	44	5	49
		% within handedness	51.2%	35.7%	49.0%
Total		Count	86	14	100
		% within handedness	100.0%	100.0%	100.0%

Figure 5f

We'll work an example with a large data set that will demonstrate the power of SPSS.

Example 9.

Open the file called Colleges.sav (2)
What is the probability a college chosen at random is a California college?

Solution: Rather than go through the list and count the number of colleges in California, we'll recode the data so that SPSS will tell us which colleges are in California. Use Transform → Recode Into Different Variables, and move the nominal variable State from the variable list into the Input Variable

→ Output Variable window. SPSS will recognize it is a string variable. In the output variable box, type in Calif for Name and California/Other schools for Label, see **Figure 6a**, and click change. Click on Old and New values and an old and new values window will pop up, **Figure 6b**. Put CA as the old variable and Calif as the new variable, but first you need to check Output variables are strings so that you can type in a string for the new variable; see **Figure 6b**. Now click on Add; the recoding from CA to Calif has been completed. Click All other values and recode them to other and click Add; see **Figure 6c**. In the Old → New box, there should be a line called ELSE → "other"; click continue and OK. In the Colleges data editor you'll notice a new variable called Calif with the values Calif for California and Other for other states. To find the probability a school selected at random is in California, we use Analyze → Descriptive Statistics → Frequencies, move California/Other Schools into the variable box and click on plots (charts), check bar, continue, then OK. A frequency chart will be displayed along with a bar plot and you can read the probability that a school selected at random is in California from the chart: $\frac{70}{1302} = .054 = 5.4\%$; see **Figure 6d.**

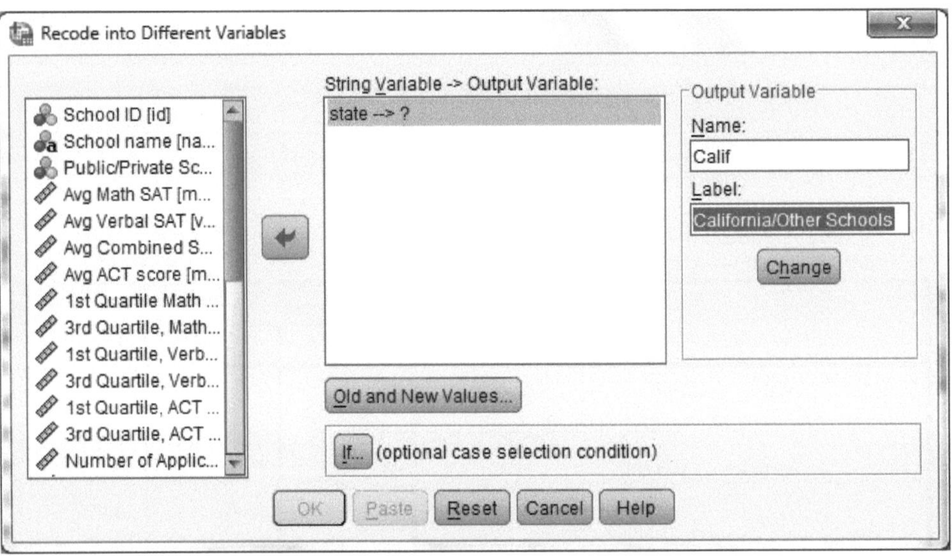

Figure 6a

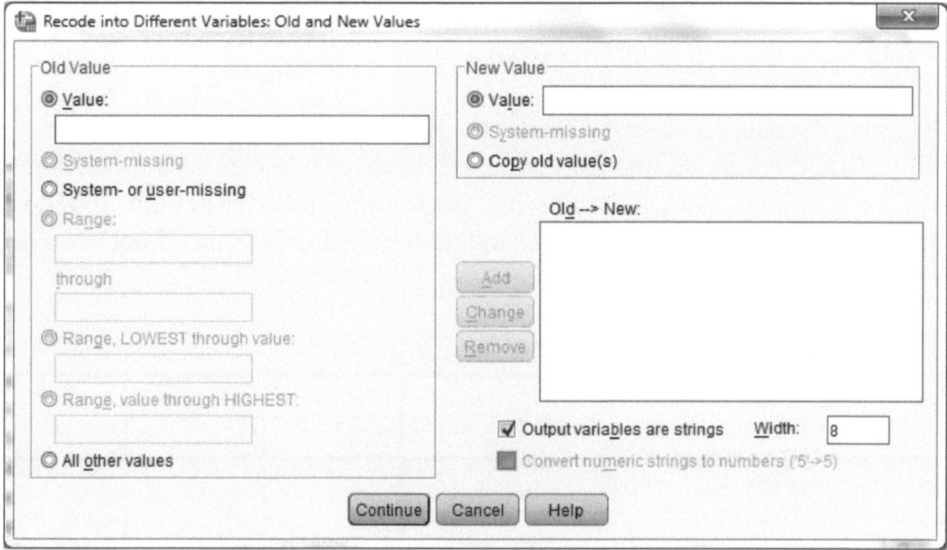

Figure 6b

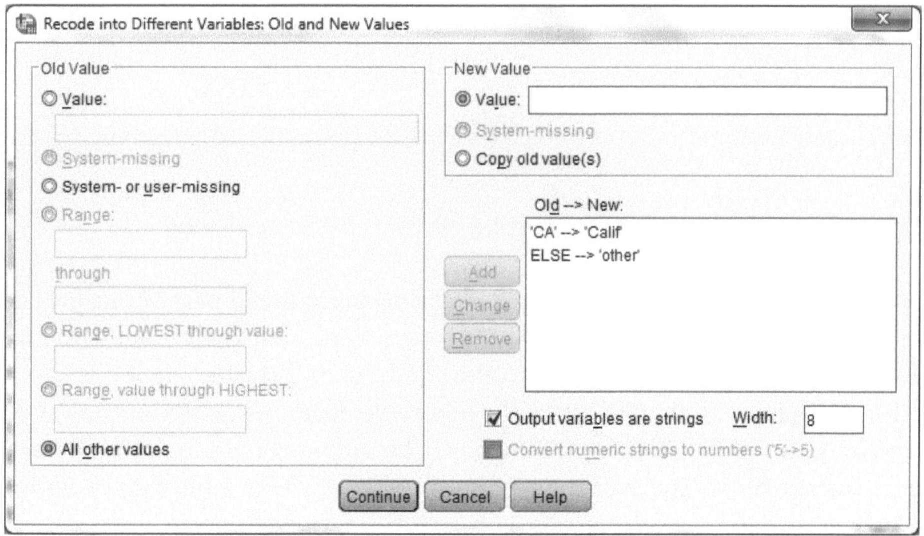

Figure 6c

California/Other Schools

		Frequency	Percent	Valid Percent	Cumulative Percent
Valid	Calif	70	5.4	5.4	5.4
	Other	1232	94.6	94.6	100.0
	Total	1302	100.0	100.0	

Figure 6d

Example 6b.

Find the probability a school is in California and is a private school.

Solution: Keeping the data set above the same, use Analyze ⟶ Descriptive Statistics ⟶ Crosstabs, and move California/Other schools into Row and Public/Private school into column. Click on cells and check the box Percentages ⟶ Total, Continue, OK. The contingency table in **Figure 6e** should appear in the output. The answer that a school selected at random is both a California school and a private school is 3.2% = .032 = 42/1302.

California/Other Schools * Public/Private School Cross-Tabulation

			Public/Private School		Total
			Public	Private	
California/Other Schools	Calif	Count	28	42	70
		% of Total	2.2%	3.2%	5.4%
	Other	Count	442	790	1232
		% of Total	33.9%	60.7%	94.6%
Total		Count	470	832	1302
		% of Total	36.1%	63.9%	100.0%

Figure 6e

Example 6c.

Given the school is in California, what is the probability it is a private school?

Solution: We know the school is in California, that's the conditional part of our probability, so now we want to know what the probability is that it's a private school. Use Analyze ⟶ Descriptive Statistics ⟶ Crosstabs and move California schools into Row and Public/Private school into column. Click on cells, deselect Percentages ⟶ Total, and check the box Percentages ⟶ Row, Continue, OK. We're using rows because we moved California schools into Row. The output should look like **Figure 6f**. Note the row percentages sum to 100%. The probability the school is a private school given it's in California is $\frac{42}{70} = .6 = 60\%$. Likewise we can find the conditional probability a school is in California given it's a public (or private) school, just check Percentages ⟶ Columns in the cells tab of the cross-tabs window; see **Figure 6g**. The probability a school is in California given it is a private school is 5%.

California/Other Schools * Public/Private School Cross-Tabulation

	Public/Private School		Total
	Public	Private	

California/Other Schools	Calif	Count	28	42	70
		% within California/Other Schools	40.0%	60.0%	100.0%
	Other	Count	442	790	1232
		% within California/Other Schools	35.9%	64.1%	100.0%
Total		Count	470	832	1302
		% within California/Other Schools	36.1%	63.9%	100.0%

Figure 6f

California/Other Schools * Public/Private School Cross-Tabulation

			Public/Private School		Total
			Public	Private	
California/Other Schools	Calif	Count	28	42	70
		% within Public/Private School	6.0%	5.0%	5.4%
	Other	Count	442	790	1232
		% within Public/Private School	94.0%	95.0%	94.6%
Total		Count	470	832	1302
		% within Public/Private School	100.0%	100.0%	100.0%

Figure 6g

Are the events "California schools" and "public/private schools" independent events? We know for independent events $P(A) \cdot P(B) = P(A \cap B)$, and for conditional probability $P(A \cap B) = P(B|A) \cdot P(A)$, so if two events are independent, then $P(A) \cdot P(B) = P(B|A) \cdot P(A)$; canceling $P(A)$ on both sides, we have $P(B) = P(B|A)$, which says the probability of *B* is unaffected by the outcome of *A*. That is, given that *A* has occurred, the probability for *B* will not be affected. Now the probability of a California school is 5.4%, but the probability the school is in California given it's a private school is 5%. Since these two numbers are not equal, these events are not independent.

As another example of two events being dependent, consider that the probability a school is a public school is 36.1% and the probability the school is public given it's not in California is 35.9%; **see Figure 6f**. Since these two numbers are not equal, these events are not independent. (Although they do pass a chi-square test of independence because they are very close; see Chapter 12.)

Inclusion–Exclusion Formula

The probability of the union of two events is $P(A \cup B) = P(A) + P(B) - P(A \cap B)$, which is pretty straightforward; see **Figure 3**. If *A* and *B* have a nonempty intersection, then when we count all the outcomes in *A* with all the outcomes in *B*, we'll have counted all the outcomes in both *A* and *B* twice, so we need to subtract those outcomes once. This is called the inclusion–exclusion formula.

Mutually Exclusive Events

If the events A and B have no outcomes in common, then we say A and B are disjoint or mutually exclusive. In that case $P(A \cup B) = P(A) + P(B)$ because $P(A \cap B) = 0$; see **Figure 7a.**

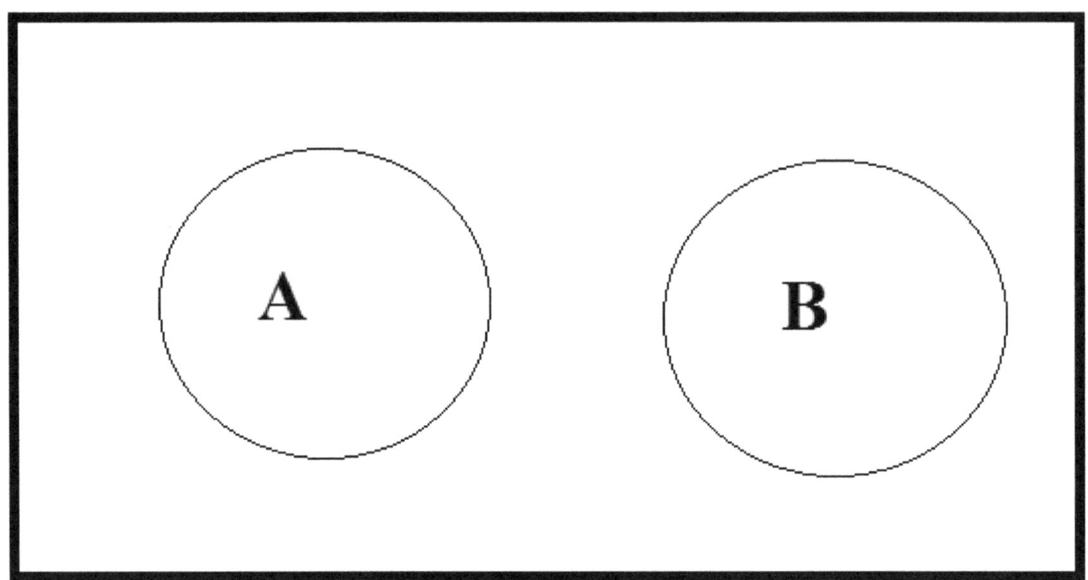

Figure 7a

Bayes' Theorem

If A is a subset of B and B is sliced into disjoint subsets, B_i such that the sum of the B_i's equal B, that is $\sum B_i = B$, then $P(B_j|A) = \frac{P(B_j \cap A)}{P(A)} = \frac{P(A \cap B_j)}{\sum P(A \cap B_i)} = \frac{P(A|B_j)P(B_j)}{\sum P(A|B_i)P(B_i)}$; see **Figure 7b.**

Example 7.

To illustrate Bayes' Theorem, we'll use the colleges' data set, Colleges.sav to find the probability that a school is in Hawaii given that it is a public school.

Solution: Rather than recode everything, we'll use Bayes' Theorem. Open the data set Colleges.sav. We want the probability that a school is in Hawaii given that it is a public school. The set A will be public schools and the B_i's will be the states. If B_h represents the state of Hawaii, then we need to find $P(B_h|A)$. Use Analyze → Descriptive Statistics → Crosstabs and move State into Row and Public/Private into Column. Click on Cells and check Percentages → Column. The output will be quite lengthy, but the part we want is shown in **Figure 7c**. The probability the school is in Hawaii given it's a public school is .4% =.004.

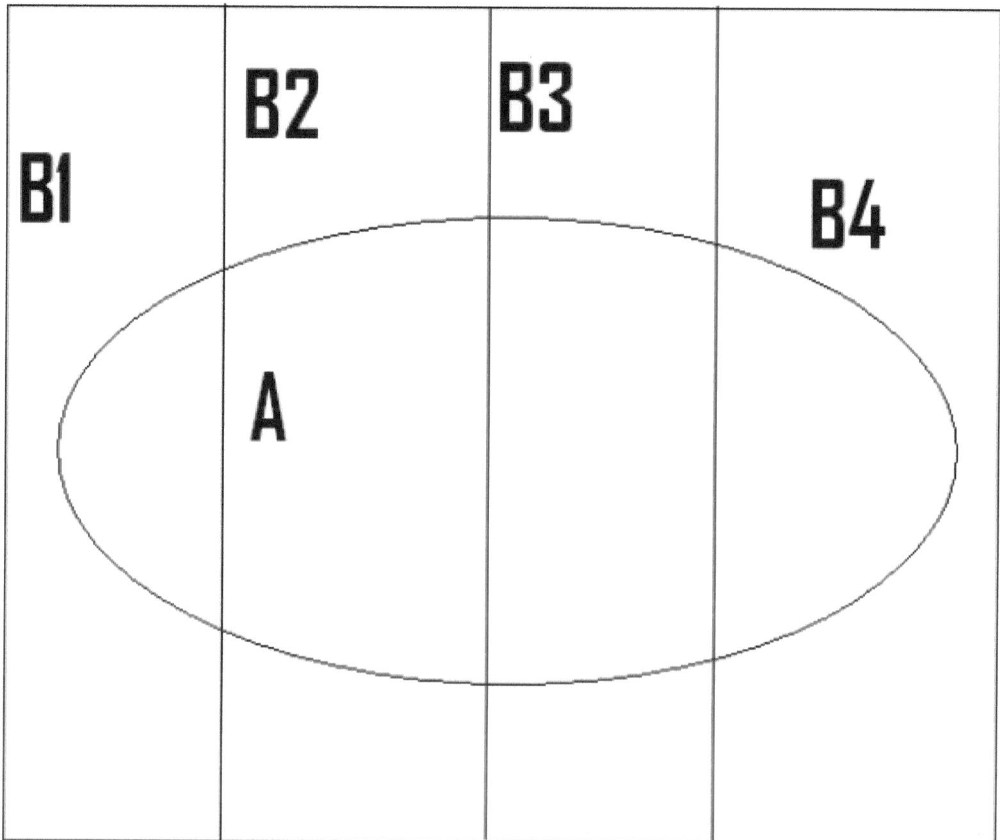

Figure 7b

		Public	Private	Total
HI	Count	2	3	5
	% within Public/Private School	0.4%	0.4%	0.4%
Total	Count	470	832	1302
	% within Public/Private School	100.0%	100.0%	100.0%

Figure 7c

Counting

A big part of probability is the ability to count, which is not always so simple. For instance: 1) how many five card poker hands are possible; 2) how many license plates have a number followed by three letters followed by three numbers; 3) how many arrangements of the word Mississippi are possible; and 4) how many ways can you buy 8 cookies given that at least 3 must be chocolate chip, 2 must be shortbread, and at least one a gingersnap? The rest of the cookies can be any of these three. These types of problems are interesting in and of themselves and are used to find probabilities based on how many ways an event can occur divided by how many outcomes are possible, but the intricacies of counting do get complicated and it takes time to carefully understand the techniques. Thus we will not delve here into counting techniques; however, here are the solutions to the above questions.

1. There are 52 cards in a deck of cards and we want to choose 5 of them. There are 52 ways to choose the first card, 51 ways to choose the second card (one has already been chosen), 50 ways to choose the third card (two have already been chosen), and similarly 49, 48 ways to choose the fourth and fifth cards. The product of these five numbers is the number of ways we can choose 5 cards from a deck of 52 cards in order. But since the order doesn't matter about how we choose the cards, we will divide out the number of ways to arrange 5 cards, which is 5! (read 5 factorial). 5! = $5 \cdot 4 \cdot 3 \cdot 2 \cdot 1 = 120$. So the number of ways to choose 5 cards out of 52 cards is $C(52,5) = \frac{52 \cdot 51 \cdot 50 \cdot 49 \cdot 48}{120} = 2{,}598{,}960$. In general we say $C(n,k) = \frac{n!}{(n-k)!k!}$, where k is the number of items we want to choose and n is the number of items we can choose from. This is the formula for combinations of n things taken k at a time.

2. This is the most basic form of counting; it is called the multiplication principle. Essentially we have 7 slots and we can place 10 numbers in the first slot; the numbers 0–9, 26 letters from A to Z in the second, third, and fourth slots; and again 10 digits from 0 to 9 in the remaining three slots. The total number of license plates with this scheme is $10 \cdot 26 \cdot 26 \cdot 26 \cdot 10 \cdot 10 \cdot 10 = 175{,}760{,}000$.

3. Here the order of the letters matter, except for those letters that are repeated. If we consider all letters as distinct, then we have 11! ways to arrange the eleven letters. However the *s* is repeated 4 times, the *i* is repeated 4 times, and the *p* is repeated twice, so we need to divide by $4! \cdot 4! \cdot 2! = 1152$. Thus the number of different arrangements for the word Mississippi is $\frac{11!}{1152} = 34{,}650$. In general the permutation of n things taken k at a time is $P(n,k) = \frac{n!}{(n-k)!}$.

4. This is called the cookies-and-bins problem. We already have 6 cookies—3 chocolate chip, 2 shortbread, and one gingersnap—so we need two more cookies from any of the three bins. We could take one chocolate and one gingersnap, or two chocolate chip and zero of the other two, or one shortbread and one gingersnap, etc. We can consider the cookies as two slots and the dividers between the bins as two more slots (we don't need a divider before the first bin or after the third bin) so we have 4 slots in which we want to put two cookies or two dividers. This is like the first case where we have 4 items, from which we want to choose two items. The answer is $C(4,2) = \frac{4!}{(4-2)! \cdot 2!} = 6$ ways to satisfy that order of 8 cookies.

This author writes problems for the American Mathematics Contests and some of the problems are based on counting techniques. Counting techniques offer a rich and varied supply of challenging problems that are not always easy to solve. Counting was our first step in mathematics and we continue to pursue counting today; for example, we still don't know if the number of twin primes is finite or infinite. Twin primes are two prime numbers that differ by 2.

Odds

The odds for an event is the probability of the event divided by the probability of the complement of the event. That is, if $P(E)$ is the probability of the event and $P(\bar{E})=1-P(E)$ is the probability of the complement of *E*, then for the event, $odds=\dfrac{P(E)}{1-P(E)}$.

Example 8.

If the probability of Longshot winning a horse race is $\dfrac{1}{17}$, then what are the odds for Longshot winning?

Solution: The complement event of Longshot winning the race is Longshot losing the race, and the probability of Longshot losing the race is $1-\dfrac{1}{17}=\dfrac{16}{17}$. Therefore the odds for Longshot to win the horse race is $\dfrac{\frac{1}{17}}{\frac{16}{17}}=\dfrac{1}{16}$. However, in horse racing the odds are always given as against the horse winning. Thus out of 17 races Longshot will lose 16 and win 1 race, so the odds for Longshot are 16:1. This is the same as $odds = P(lose)/P(win)=\dfrac{16}{17}/\dfrac{1}{17}=\dfrac{16}{1}=16:1$.

Expected Value

The expected value of a probability distribution is the sum of the values times the probabilities of the values. In symbols we write $E(x)=\sum x\cdot p(x)$, where *E*(*x*) is the expected value. Since $p(x)$ is the number of ways *x* can occur divided by the total number of outcomes, then the expected value is really the weighted mean or weighted average of all the values. The expected value is the mean of the probability distribution. Another way to think of the expected value is the mean of a weighted probability distribution, where the sum of all the weights equals 1, that is, $\sum p(x)=1$.

Example 9.

Find the expected value for the toss of two dice.

Solution. Open the data set chpt5example9.sav and find the weighted mean using *x* as the variable with the weights *prob* (Use Data → Weight Cases, weight cases by *prob*). Then find the mean of *x*. The expected value (mean) is 7, which is the same mean obtained when we found the weighted mean using *freq* as the weight (Use Data → Weight Cases, weight cases by *freq*). This is because $\sum p(x) = \sum \frac{freq}{\sum freq}$, so the probability $p(x)$ has the same weight as *freq*.

Example 10.

A business associate must choose among three options. Option A has a 10% chance of resulting in a $250,000 gain but otherwise will result in a $10,000 loss. Option B has a 50% chance of gaining $40,000 and a 50% chance of losing $2,000. Finally Option C has a 5% chance of gaining $800,000 but otherwise will result in a loss of $20,000. Which option has the best chance of a positive gain?

Solution. Open chpt5example10.sav. To find the expected value for each option, we need to weight the cases individually. Thus in Option A we weight cases with Pa (Plan A) and the expected value is $16,000. For Option B we weight cases with Pb (Plan B) and the expected value is $19,000. Finally for Option C we weight cases with Pc (Plan C) and the expected value is $21,000. We see that Option C has the highest expected value but also the highest possible loss. Determining the number of standard deviations the loss is from the mean might indicate how likely a loss will occur in these three scenarios. The formula for the variance of a probability distribution is $\text{var}(x) = \sigma_x^2 = \sum (x_i - \mu_x)^2 p_i$, where $p_i = \frac{f_i}{\sum f_i}$, but SPSS cannot calculate the variance with p_i as the weights; they must be changed to f_i. So we add three more columns where we multiply each of the probabilities by 100 to remove the decimals (use Transform → Compute variable, call the target variable *fa* and form the equation in Numeric Expression Pa*100, repeat for the other two variables Pb and Pc) and find the standard deviations for each option with *fa*, *fb*, and *fc* as the weights; see **Figure 8a**. Option A has a SD of $78,390, Option B has a SD of $21,110, and Option C has a SD of $179,600; see **Figure 8b**. We see from these values that while Option C has the highest positive mean, it also has the highest standard deviation, which means it has greater variability. Option B has the second-highest positive mean and the least variability, therefore Option B might be a better choice. Option B also has the lowest coefficient of variation; see Chapter 4.

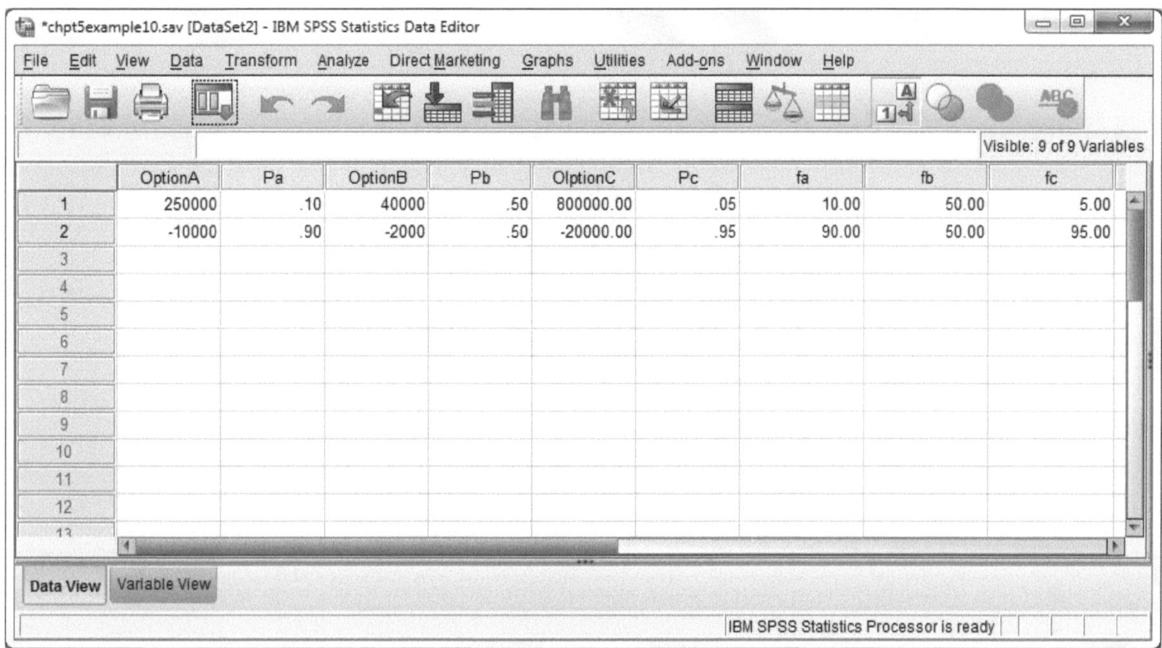

	OptionA	Pa	OptionB	Pb	OlptionC	Pc	fa	fb	fc
1	250000	.10	40000	.50	800000.00	.05	10.00	50.00	5.00
2	-10000	.90	-2000	.50	-20000.00	.95	90.00	50.00	95.00

Figure 8

Statistics

Option A

N	Valid	100
	Missing	0
Mean		16000.00
Std. Deviation		78392.950

Statistics

Option B

N	Valid	100
	Missing	0
Mean		19000.00
Std. Deviation		21105.794

Statistics

Option C

N	Valid	100
	Missing	0
Mean		21000.0000
Std. Deviation		179615.18912

Figure 8b

Mean and Standard Deviation for a Binomial Distribution

The expected value of a binomial distribution with n trials and probability of a success p is $E(x)=0\binom{n}{0}p^0q^n+1\binom{n}{1}p^1q^{n-1}+2\binom{n}{2}p^2q^{n-2}+\cdots+x\binom{n}{x}p^xq^{n-x}+\cdots+n\binom{n}{n}p^nq^0$. Using sigma notation we have $E(x)=\sum_{x=0}^{n}x\binom{n}{x}p^xq^{n-x}$. Using factorial notation for the binomial coefficients and doing some math, we have $E(x)=\sum_{x=1}^{n}x\frac{n!}{x!(n-x)!}p^xq^{n-x}=\sum_{x=1}^{n}\frac{n!}{(x-1)!(n-x)!}p^xq^{n-x}$

$$=np\sum_{x=1}^{n}\frac{(n-1)!}{(x-1)!(n-x)!}p^{x-1}q^{n-x}=np\sum_{x=1}^{n}\binom{n-1}{x-1}p^{x-1}q^{n-x}$$

Letting $k = x - 1$ (called reindexing the sum), we obtain $E(x)=np\sum_{k=0}^{n-1}\binom{n-1}{k}p^kq^{(n-1)-k}=np\left[p+q\right]^{n-1}=np$.

Thus the mean of the binomial distribution is np. Using $E\left(x^2\right)=\sum_{x=1}^{n}x^2\binom{n}{x}p^xq^{n-x}=n^2p^2-np^2+np$, (see Problem 34b) we can show $\text{var}(x)=E\left(x^2\right)-\left[E(x)\right]^2=-np^2+np=np(1-p)=npq$. So the standard deviation for a binomial distribution is $\sigma=\sqrt{npq}$.

Geometric Distribution

The geometric distribution gives the probability of the first success after r trials. That is, the first $r - 1$ trials are failures and the rth trial is a success. The formula is $P(r)=pq^{r-1}$ where p is the probability of a success and q is the probability of a failure. The mean is $\mu=\frac{1}{p}$ and the standard deviation is $\sigma=\frac{\sqrt{q}}{p}$; see Problem 32.

Example 11.

A key chain has 6 keys, only one of which will open a door. If keys are tried at random, then find the probabilities that the door will open on the first try, second try, third try,..., sixth try.

Solution: This is a geometric distribution because we don't care how often we open the door (as we would in a binomial distribution). All we care about is the probability of how many tries before we open the door once. The value for p is 1/6 since we have 6 keys and only one opens the door. So there's a probability of $1/6 \approx .1667$ the door will open on the first try and 5/6 it won't open. If we pick the wrong key, then we put the key back and try again. Unfortunately for us, we don't discard the key, but put it back on the key chain and randomly select a key again. Conceivably we could be trying keys forever. The probability we get the right key the second time is $5/6*1/6 = 5/36 \approx .1389$, the third time is 5/6*5/6*1/6 = .1157 and so on. SPSS will do the procedure for us. Open a new data editor and put the numbers 1 through 6 in the first column and call it x, and change the measure to scale. Use Transform → Compute

Variable, call the target variable p, click on PDF & Noncentral PDF under Function Group, and double click PDF.Geom. Move x into the first parameter and 1/6 as the second parameter and click ok. The data editor should be the same as in **Figure 9**; change the decimals for p to 4.

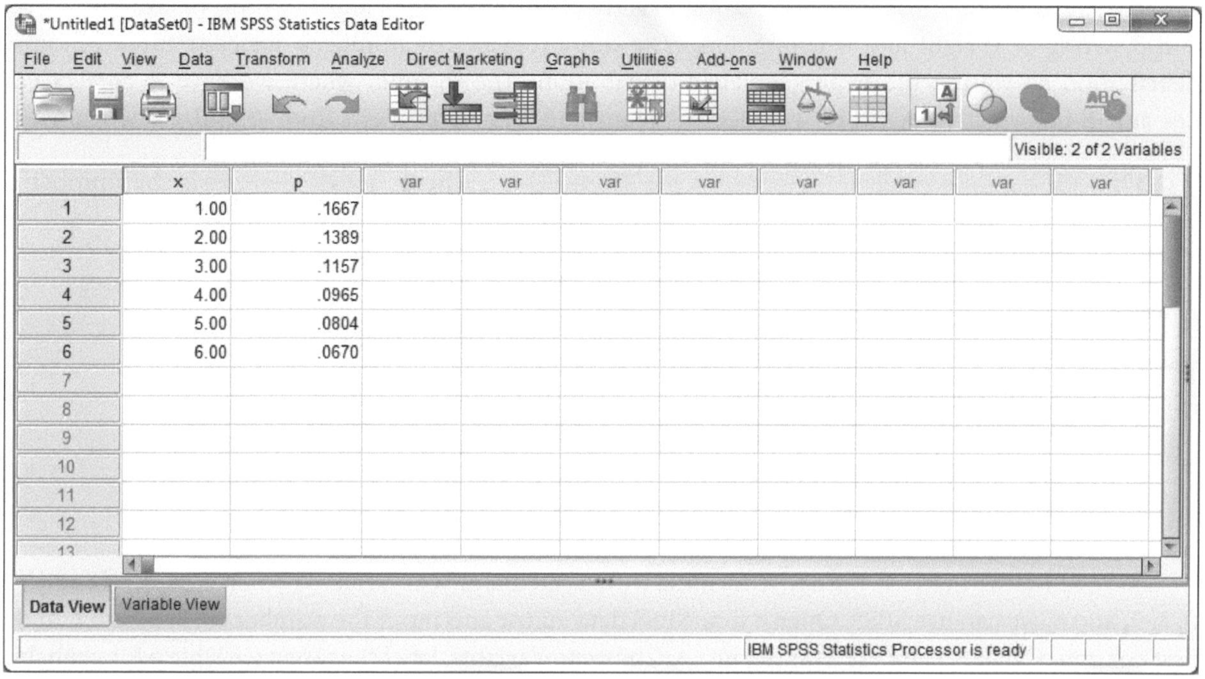

*Untitled1 [DataSet0] - IBM SPSS Statistics Data Editor

	x	p
1	1.00	.1667
2	2.00	.1389
3	3.00	.1157
4	4.00	.0965
5	5.00	.0804
6	6.00	.0670

Figure 9

The probabilities for each value 1 through 6 are under the p column. Thus the probability that the key will open the door on the first try is .1667, the probability the key will open the door on the second try is .1389, the third try is .1157, fourth try is .0965, fifth try is .0804, and the sixth try is .0670.

These probabilities don't add up to 1 because the number of tries is infinite. The infinite sum of the probabilities will be 1. However, at some point, the probability of a try will be so small it'll practically be 0; hence we can expect the door will open in a finite number of tries. Actually if $n = 49$, the cumulative probability (add 49 to x and change PDF to CDF under numeric expression in the Compute Variable window) is equal to .9999. This tells us the probability the door will be opened within 49 tries is 99.99%, which is practically 100%.

Hypergeometric Distribution

A hypergeometric distribution will consider successes and failures in r trials. This is the same as a binomial distribution, except the trials are not independent. That is, the trials are performed without replacement. Consider a lottery drawing where six numbers are chosen from 54 numbers and numbers are not replaced. The probability of having all 6 numbers is $1/C(54, 6)$, where $C(54, 6)$ is the binomial coefficient; that is, the number of combinations of 6 numbers from 54 numbers or $C(54,6) = \frac{54!}{48!6!} = 25,827,165$

Thus the probability of getting all 6 winning numbers is $\frac{1}{25{,}827{,}165} \approx .0000000387189$. Now suppose we want the probability of getting exactly 5 numbers right and 1 wrong or exactly 3 right and 3 wrong or 0 right and 6 wrong. Since we don't replace the numbers, we can't use a binomial distribution so we use the hypergeometric distribution. The formula where we have A correct values and B incorrect values after we select n values and want x of them from A and $n - x$ from B is $P(x) = \frac{A!}{(A-x)!x!} \cdot \frac{B!}{(B-n+x)!(n-x)!} \div \frac{(A+B)!}{(A+B-n)!n!}$, where $\frac{(A+B)!}{(A+B-n)!n!}$ is the number of ways we can select n numbers from $A + B$ numbers. SPSS will also do the calculations for us.

Example 12.

Using 54 numbers as the Lotto, find the probability of

a) exactly 5 of the winning numbers
b) exactly 3 of the winning numbers
c) none of the winning numbers
d) all of the winning numbers

Solution: Rather than use the above formula with a calculator with values $A = 6$, $B = 48$, $n = 6$, and $x =$ 0, 3, 5, and 6, we can use SPSS. Open a new SPSS data editor and input the numbers 0, 3, 5, and 6 in the first column. Call it x. Next use Transform → Compute Variable, label the target variable p for probability, click on PDF & Noncentral PDF in the functions group, and under Functions and Special Variables click on Pdf.Hyper. Pdf.Hyper calls for 4 parameters. The first parameter is the x variable; that is, we're looking for the probabilities for those particular x values. The second parameter is the total number of numbers, in this case 54 numbers. The third parameter asks for the number of correct numbers that are drawn, the size of A, and the fourth parameter asks for how many numbers we will draw from the total, our sample size, which is n. Both A and n are 6. In a Lotto drawing some kind of machine picks A correct numbers and we pick a sample of n numbers. In our numeric expression window we should have PDF. HYPER(x,54,6,6). Click ok and increase the number of decimals in the p variable to 10. You will also have to increase the width of the variable. Increase the width to 11 and the number of decimals to 10; see **Figure 10a**. The data view should be the same as in **Figure 10b**.

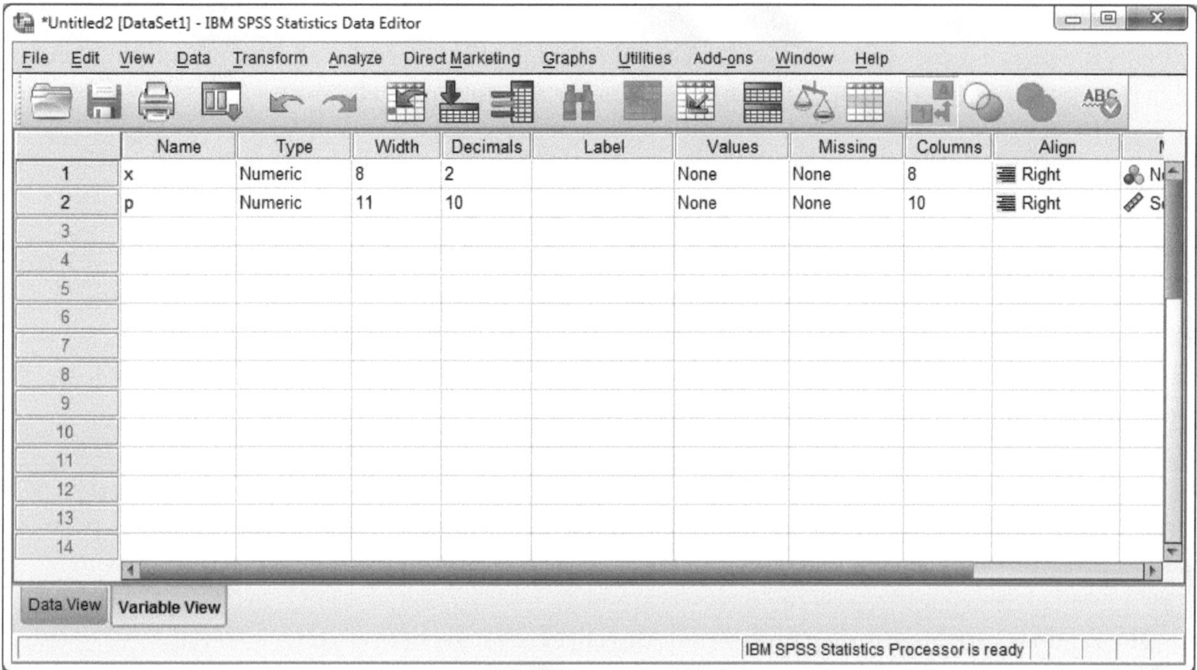

Figure 10a

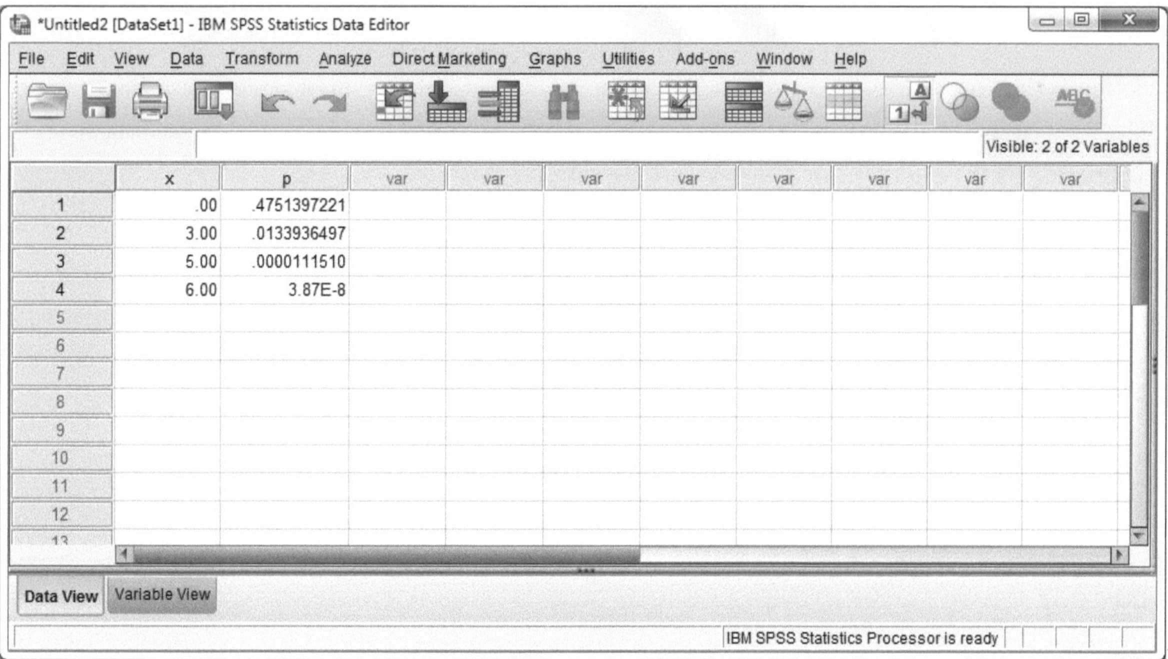

Figure 10b

The value 3.87E-8 is scientific notation and represents the number 0.0000000378 in standard notation. (The decimal point was moved 8 places to the left.) The formulas for the mean and standard deviation of the hypergeometric distribution are left for another math class.

Poisson Distribution

Our last distribution to discuss in this chapter is the Poisson distribution. As you may surmise from seeing all the PDFs listed in the Compute Variable window, there are many more PDFs listed than those that we've covered. We're trying to cover only those PDFs that are relevant to an Introductory Statistics class. We hope the interested reader will continue to pursue the PDFs we didn't cover and learn the power of the computer while working statistical exercises.

The Poisson distribution is a discrete probability distribution that applies when a certain frequency is associated to a certain interval, such as the number of groundhogs in an acre, the number of visitors entering a museum during a ten-minute interval, or the number of typing mistakes per page of a manuscript. The following are the requirements for a Poisson distribution.

1. The random variable x is the number of occurrences of an event over some interval.
2. The occurrences must be random.
3. The occurrences must be independent of each other.
4. The occurrences must be uniformly distributed over the interval being used.

The mean of a Poisson distribution is the number of occurrences over the interval and the standard deviation is the square root of the mean. The mean is $\mu = \frac{f}{I}$, where f is the frequency, the number of occurrences, I is the interval, and the standard deviation is $\sigma = \sqrt{\mu} = \sqrt{\frac{f}{I}}$.

Example 14.

Certain cultures are superstitious about groundhogs coming out of their burrows on a particular winter day. Suppose one square mile has 1,600 groundhogs living in burrows. What's the probability that an acre selected at random will have 4 groundhogs living in burrows?

Solution: The formula for the Poisson distribution is $P(x) = \frac{\mu^x \cdot e^{-\mu}}{x!}$, where e is the Euler number and is approximately 2.71828. In order to use SPSS, we need to find the mean μ, which is the frequency per interval. Since a square mile has 640 acres and there are 1,600 groundhogs per square mile, there are $\frac{1600}{640} = 2.5$ groundhogs per acre. In this problem the intervals are acres so the mean value per acre is 2.5. Using a calculator we find $P(4) = \frac{2.5^4 \cdot e^{-2.5}}{4!} = .133602$. Using SPSS we open a new data editor, put 4 in the first column, call it x, and change the measure to scale. Then use Transform → Compute

Variable. We'll call the target variable p or any pertinent name, put Pdf.Poisson into the numeric expression and the two parameters x and 2.5 inside the parentheses. The result should look like **Figure 11a**. Click OK and the data editor will have the Poisson probability in the column headed by p; see **Figure 11b**.

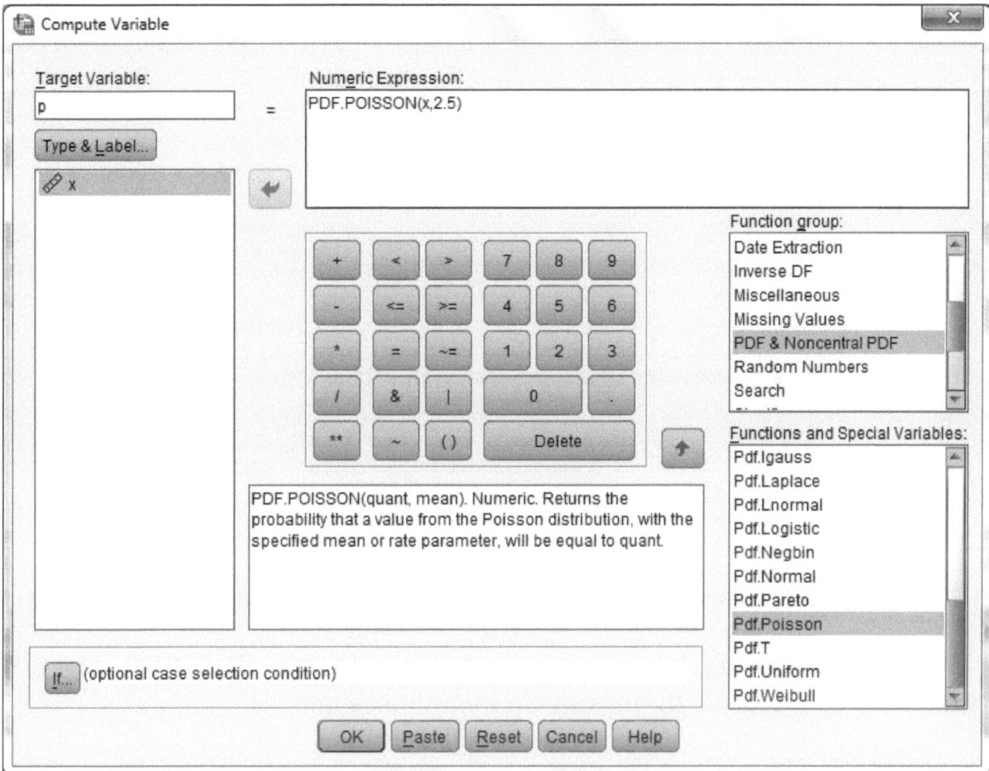

Figure 11a

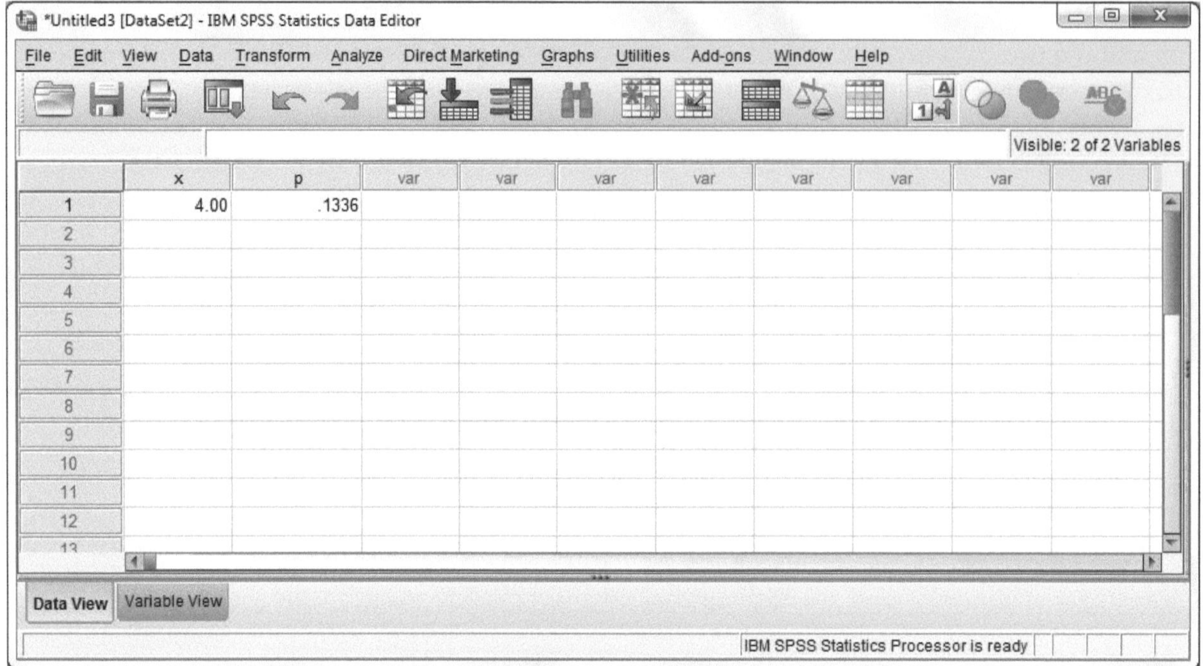

Figure 11b

Example 15.

The Getty Museum has a limit of 1,400 visitors per day when its business day is 8 hours long. What is the probability that 40 people will visit the museum in a ten-minute interval?

Solution: Open a new SPSS data editor and put 40 in the first column, call it *x*. Use Transform → Compute Variable, put *p* in the target variable, and move Pdf.Poisson into the Numeric expression. Move *x* into the first parameter and 1400/48 into the second parameter. Why 1400/48? The interval of concern is a ten-minute interval, so we need to find the average number of visitors in a ten-minute interval. Since the limit is 1,400 visitors (we'll assume the limit is reached every day) and there are 6 ten-minute intervals in each hour of an 8-hour day, there are 6*8 = 48 ten-minute intervals in an 8-hour day. The average number of persons visiting the museum in a ten-minute interval is therefore 1400/48. Click OK and the probability, *p*, under the *p* column in the data editor is .0104. Increase the number of decimals to 4; see **Figure 12**.

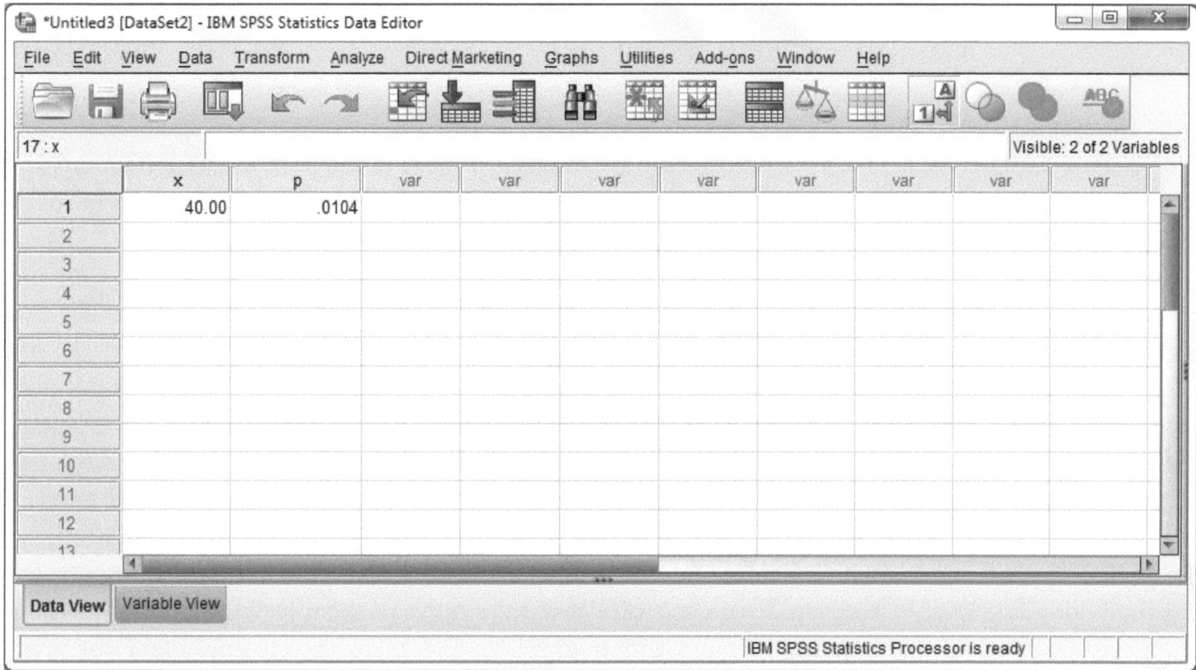

Figure 12

Example 16.

Assuming a person usually makes about 3 typing errors per line (some errors are self-correcting) and there are about 40 lines per page, then what is the probability the person will make 100 typing errors while typing a page?

Solution: The interval of concern is a page of typing with an average of 120 errors per page (counting the self-correcting errors). Using the same procedure as above (100 for x and 120 for μ), we get $P(x = 100) = .0068,$ which is the percentage of getting exactly 100 errors while typing a page.

Tree Diagrams

One last useful technique to calculate a given probability is to draw a tree diagram. A tree diagram consists of nodes (vertices) and branches (lines). At each node the tree can branch out in many directions depending on how many outcomes are possible. Trees can be used to find the probabilities of independent events and dependent events. SPSS doesn't have a multinomial distribution function, such as the probability of three independent variables in n trials; therefore, a tree can help diagram the solution. Two concepts we should note about trees:

1. Each time a tree branches from a node, the sum total of all the probabilities on all the branches is 1, and

2. The sum total of all the probabilities on the terminal nodes is 1.

Example 17.

Workplace accidents are categorized in three groups: minor, moderate, and severe. The probability that a given accident is minor is 0.5, that it is moderate is 0.4, and that it is severe is 0.1. Two accidents occur independently in one month. Find the probability of all possible outcomes.

Solution: The tree branches out as in **Figure 13**. There are 9 terminal nodes, 3 choices for each of the types of injury. The first terminal node (top node) is the probability of minor accident twice, the fifth terminal node (middle node) is for moderate twice, and the ninth terminal node (bottom node) is for severe twice, hence the terminal probabilities for minor, moderate, and severe injuries are the squares of the probabilities for minor, moderate, and severe injuries. Thus the probability there are two minor accidents in one month is .25, two moderate accidents is .16, and two severe accidents is .01. The second terminal node from the top is the outcome of minor accident followed by moderate accident; hence, the probability is 0.5*0.4 = 0.2. The next node is minor followed by severe, so the probability for that terminal node is 0.5*0.1 = 0.05, etc. If we don't care about the order of the injuries, then the outcomes that have one minor and one moderate injury add up to 2*0.2 = 0.4. Likewise one minor and one severe add up to 2*0.05 = 0.1, and one moderate with one severe add up to 2*0.04 = 0.08. Notice that .25 + .16 + .01 + .4 + .1 + .08 = 1.

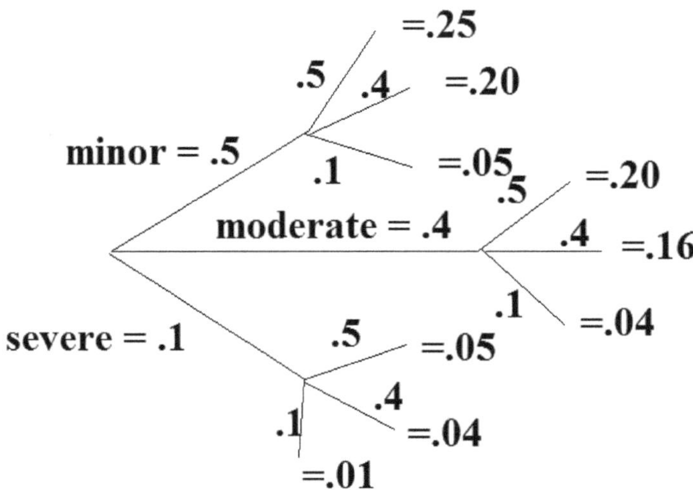

Figure 13

Example 18.

A box contains 10 balls, 3 of which are red, 2 are green, and 5 are blue. Two balls are randomly selected without replacement. Find the probabilities of all possible outcomes.

Solution: Again we'll have a tree with 9 terminal nodes, 3 choices for the first ball and 3 choices for the second ball. However, the probabilities change depending on which branch of the tree we follow; see **Figure 14**. The top terminal node represents a red ball followed by a red ball. The probability for the first red ball is 3/10. Since one ball was removed and that ball was red, the probability for the second red ball is 2/9. So the probability two red balls are drawn from the box is $\frac{3}{10}\cdot\frac{2}{9}=\frac{6}{90}$ (Normally we leave terminal probabilities in non-reduced form: it's easier to check the mathematics.) The next terminal node represents red followed by green with probability $\frac{3}{10}\cdot\frac{2}{9}=\frac{6}{90}$ The third terminal node represents red followed by blue with probability $\frac{3}{10}\cdot\frac{5}{9}=\frac{15}{90}$, and so on. Again note the sum total of all the terminal probabilities is 1.

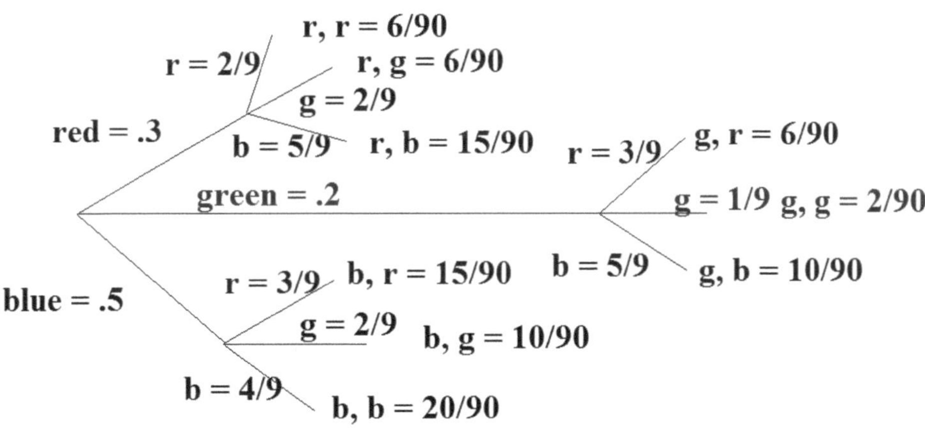

Figure 14

Although exact probabilities are useful in and of themselves, what statisticians are mostly concerned with are the cumulative probabilities, which is what we'll discuss in the next chapters of this text.

Let's review what we've learned in this chapter.

1. Probability of an event—the number of ways the event can occur divided by the total number of possible outcomes.
2. Bernoulli Trial—only two possible outcomes, success or failure with probability p or q.
3. Binomial Experiment—a total of n Bernoulli trials, each independent with the same probabilities.

4. Conditional probability, Independent events, Complementary events, Odds, Mutually disjoint events, Inclusion–Exclusion Formula
5. Bayes' Theorem—conditional probability for several disjoint sets
6. Expected Value—the mean of a probability distribution
7. Mean and Standard Deviation of Binomial Distribution
8. Geometric Distribution
9. Hypergeometric Distribution
10. Poisson Distribution
11. Trees, multinomial probabilities

Class Work 5

A classic example of the Poisson distribution is the number of deaths caused by horse kicks of men in the Prussian Army between 1875 and 1894. In 1898 it was reported that in the 20 years between 1875 and 1894, there were 196 deaths among the members of 14 corps in the Prussian Army due to horse kicks. Find the mean value per corps-year and find the probability of 0, 1, 2, 3, and 4 deaths per corps-year. Also find how close the Poisson distribution is to the actual distribution of 0 deaths in 144 corps-years, 1 death in 91 corps-years, 2 deaths in 32 corps-years, 3 deaths in 11 corps-years, and 4 deaths in 2 corps-years. A corps-year is the number of corps times the number of years, hence 14 * 20 = 280, the total number of corps-years.

Homework 5

1. Find the probability one black jack is drawn from a deck of 52 cards when two cards are drawn at random without replacement.

2. Find the probability the sum of two dice tossed at random is 10.

3. a) Find the probability a month chosen at random will have 31 days in the month. b) Find the probability a month chosen at random will have 28 or more days in the month. c) Find the probability a month chosen at random will have 32 days in the month.

4. If the smooth gene in a pea is the dominant gene and the wrinkled gene is the recessive gene and a pea must inherit the wrinkled gene from both parents in order for the pea to have the wrinkled trait, then assuming both parents are smooth and each carries a wrinkled gene, what is the probability the child pea will be wrinkled?

5. Find the probability at least one 6 is obtained in 4 throws of a fair die. (Or use SPSS to find the cumulative probability of 3 failures out of 4 tosses.)

6. Find the probability of at least one double six in 24 throws of two fair dice. If you were a gambling person, which game is the better game to bet on: At least one 6 in 4 throws of a fair die or at least one double 6 in 24 throws of two fair dice? (Some people consider this problem to be the birth of modern probability theory.) (Hint: Find the probability of zero 6's and subtract from 1.) (Or use SPSS to find the cumulative probability of 23 failures out of 24 tosses.)

7. Since cumulative probabilities are discussed in the next chapters, we don't want to introduce them here. However, we can find the cumulative probability by subtracting a few probability values from 1, as we did in Example 3. If the probability of a success is 10% and the number of trials is 15, then find the probability there will be at least two successes in the fifteen trials.

8. If the probability of a success is 40% and the number of trials is 11, then find the probability there will be at least two successes in the eleven trials.

9. Using the data set Student.sav, find the probability the student had 4 accidents in the last 2 years given the student considers himself to be an average driver.

10. Using the data set Student.sav, find the probability the student rates herself above average given she had 0 accidents in the last 2 years.

11. Using the data set Colleges.sav, find the probability a school is a private school given that it's in Alaska.

12. Using the data set Colleges.sav, find the probability a school is in Massachusetts given that it's a public school.

13. Are the events "student had 4 accidents in the last 2 years" and "student rates himself as an average driver" independent? See Problem 9.

14. Are the events "school in Massachusetts" and "public school" independent events? See Problem 12.

15. If the two events A and B are independent and $P\left(A|B\right)=.3$, then find $P(A)$.

16. A system has two components placed in series so that the system fails if either of the two components fails. The second component is twice as likely to fail as the first. If the 2 components operate independently, and if the probability the entire system will pass is .72, then find the probability that the second component will pass.

17. Ten cards are drawn from a deck of cards with replacement. What is the probability of not drawing an ace from any of those 10 draws? What is the minimum number of cards that need to be drawn so that the probability of getting an ace is larger than the probability of not getting an ace?

18. Let $P(A \cup B) = .8, P(A) = .6$ and $P(B) = .5$. Then $P(\overline{A} \cup \overline{B}) = ?$

19. The odds that Fat Chance will lose the horse race are 20 to 1. What is the probability Fat Chance will win the horse race? Remember in horse racing odds are backwards.

20. A drawer contains 3 red socks, 4 blue socks, and 5 green socks. Two socks are drawn at random without replacement. What is the probability both socks are the same color?

21. What is the probability it takes exactly 4 throws of a fair die to get one 6?

22. What is the probability it takes 12 throws of a fair die to get exactly two 6's?

23. What is the probability it takes 10 throws of two fair dice to get exactly eight 7's?

A) $\frac{35^2}{36^9}$ B) $\frac{5^2}{6^8}$ C) $\frac{5^2}{6^{10}}$ D) $\frac{35^2}{36^{10}}$ E) $\frac{45(5)^2}{6^{10}}$

24. Show that two events that do not have zero probability and are mutually exclusive cannot be independent events.

25. The inclusion–exclusion formula can be generalized to more than 2 events (or sets). For three events the inclusion–exclusion formula is
$$P(A \cup B \cup C) = P(A) + P(B) + P(C) - P(A \cap B) - P(A \cap C) - P(B \cap C) + P(A \cap B \cap C)$$

Suppose A, B, and C are all independent events with $P(A) = .2, P(B) = .3, P(C) = .4$ andevery combination of two events satisfies the product rule for the probability of the intersection. Find $P(A \cup B \cup C)$.

26. Let A be the subset "a computer chip is defective" and let B be divided into two subsets called Factory 1, B_1, and Factory 2, B_2. Also suppose 60% of the chips are made in Factory 1 and 40% are made in Factory 2. Let the conditional probabilities a chip is defective given it's made in Factory 1 be 35% and the chip is defective given it's made in Factory 2 be 25%. Use Bayes' Theorem to find the probability that a chip came from Factory 1 given it's defective. chpt5problem26.sav

27. When using SPSS to work Bayes' Theorem, we need to make a contingency table and weight all cases with the frequency. Such an example is given in the following table. Use Bayes' Theorem (and SPSS) to find the probability that the attacker was a stranger given the crime was assault.

Attacker/Crime	Homicide	Robbery	Assault
Stranger	12	379	727

Acquaintance/Relative	39	106	642
Unknown	18	20	57

chpt5problem27.sav

28. Using the above table, find the probability the crime was robbery given the perpetrator was an acquaintance. chpt5problem27.sav

29. Using the above table, find the probability the perpetrator was an unknown given the crime was a homicide. chpt5problem27.sav

30. Given that $\sigma^2 = \sum\left[(x-\mu)^2 \cdot P(x)\right]$, show that $\sigma^2 = \left[\sum x^2 \cdot P(x)\right] - \mu^2$.

31. Prove the infinite sum of the geometric distribution is 1.

32. Show that the mean and standard deviation for a geometric distribution is mean, $\mu = \dfrac{1}{p}$ and standard deviation, $\sigma = \dfrac{\sqrt{q}}{p}$ Use $\sum_{x=1}^{\infty} xr^x = \dfrac{r}{(1-r)^2}$ and $\sum_{x=1}^{\infty} x^2 r^x = \dfrac{r(r+1)}{(1-r)^3}$, where $|r|<1$. (See problems 38, & 39)

33. Find p to the nearest percent such that the probability is 50%; there will be 0 or 1 success out of 15 trials where p is the probability of a success.

34. Show that in a binomial distribution, the expected value $E(x) = np$.

35. Show that in a binomial distribution, $E(x^2) = \sum_{x=1}^{n} x^2 \binom{n}{x} p^x q^{n-x} = n^2p^2 - np^2 + np$.

36. Show that in a binomial distribution, the standard deviation $\sigma = \sqrt{npq}$.

37. In later sections we'll think of probability as area. That is, the probability of an event can be represented by the area of a region divided by the entire area of all the output. Use area to answer the following question. Let x have a uniform distribution on the interval (1, 3). What is the probability that the sum of 2 independent observations of x is greater than 5?

A) $\dfrac{1}{18}$ B) $\dfrac{1}{8}$ C) $\dfrac{1}{4}$ D) $\dfrac{1}{2}$ E) $\dfrac{5}{8}$

38. Prove $\sum_{x=1}^{\infty} xr^x = \dfrac{r}{(1-r)^2}$ where $|r|<1$.

39. Prove $\sum_{x=1}^{\infty} x^2 r^x = \dfrac{r(r+1)}{(1-r)^3}$ where $|r|<1$.

40. In analyzing hits by V-1 buzz bombs in World War II, South London was subdivided into 576 regions, each with an area of 0.25 km^2. A total of 535 bombs hit the combined area of 576 regions. If a region is randomly selected, find the probability it was hit exactly twice. Use the Poisson distribution.

41. For a recent year there were 1,400 terrorist attacks worldwide. Use the Poisson distribution to find the probability there will be 7 terrorist attacks worldwide in one day of that year.

42. From the article "A Classroom Note on the Poisson Distribution: A Model for Homicidal Deaths in Richmond, VA for 1991" *Mathematics and Computer Education*, Winston A. Richards states there were 116 homicide deaths. Using one day as the interval and the Poisson distribution, find the probability that on a randomly selected day, the number of homicide deaths is 0, 1, 2, 3, 4. Compare the calculated probabilities with the actual results of 268 days, no homicides; 79 days, 1 homicide; 17 days, 2 homicides; 1 day, 3 homicides; and no days with more than 3 homicides. In a future chapter (Chapter 8) we'll test to see how accurate the Poisson distribution is to the actual distribution.

43. Verify the following odds for the California Lotto. (Pick 5 numbers from 1 to 56 and one mega number from 1 to 46.) Use the Hypergeometric Distribution.

Hits	Odds
All 5 of 5 and Mega	175,711,536
All 5 of 5	3,904,701
Any 4 of 5 and Mega	689,065
Any 4 of 5	15,313
Any 3 of 5 and Mega	13,781
Any 3 of 5	306
Any 2 of 5 and Mega	844
Any 1 of 5 and Mega	141
None of 5, Only Mega	75

(**The Pascal Fermat Correspondence**) The problem that initiated this correspondence was the problem of the division of the stakes (points). Our abbreviated version is to suppose two teams (players) equally matched are playing a series where the winner takes all the stakes (points, money) that each side puts up to play the game. Now suppose the series has to be canceled at some point. The question is how should the money (stakes) be divided? To be specific, suppose two players *A* and *B* each put in $1,000 to play a game 7 times and the best 4 out of 7 wins all the money, $2,000. After 3 games, suppose player *A* is ahead 2 to 1. That is, player *A* won 2 games and player *B* won 1 game. Now suppose some catastrophe causes the series to be terminated and neither player can finish the series. How should the money be divided? Four games are left to completely determine the outcome, so we need to consider what the possibilities are for either player to win the series. This is a binomial problem where *A* indicates *A* wins and *B* indicates *B* wins. Expand $(A+B)^4$ to get $A^4+4A^3B+6A^2B^2+4AB^3+B^4$. Now *A* needs two more wins to win the series and *B* needs three more wins. *A* can get 2 or more wins in the first, second, and third terms of the above expansion and *B* can get three or more wins in the fourth and fifth terms of the above expansion. Hence there are 11 ways (1 + 4 + 6) A can win and only 5 ways (4 + 1) B can win. Hence the stakes should be divided 11:5, or A should get \$1,375 and B should get \$625.

44. How should the stakes be divided if the series is best 5 out of 9 games and A has won 3 games while B won 1 game?

Just for fun: One of the problems I (Steven) wrote for the American Invitational Mathematics Examination dated Tuesday, March 23, 2004 was #8. Define a *regular n-pointed star* to be the union of n line segments $\overline{P_1P_2}, \overline{P_2P_3}, \ldots, \overline{P_nP_1}$ such that the points $P_1, P_2, \ldots, P_n$ are coplanar and no three of them are colinear, each of the n line segments intersects at least one of the other line segments other than an endpoint, all of the angles at $P_1, P_2, \ldots, P_n$ are congruent, all of the line segments $\overline{P_1P_2}, \overline{P_2P_3}, \ldots, \overline{P_nP_1}$ are congruent, and the path $P_1P_2 \ldots P_nP_1$ turns counterclockwise at an angle of less than $180°$ at each vertex.

There are no regular 3-pointed, 4-pointed, or 6-pointed stars. All regular 5-pointed stars are similar, but there are two non-similar 7-pointed stars; see **Figure 15**. How many non-similar regular 1,000-pointed stars are there?

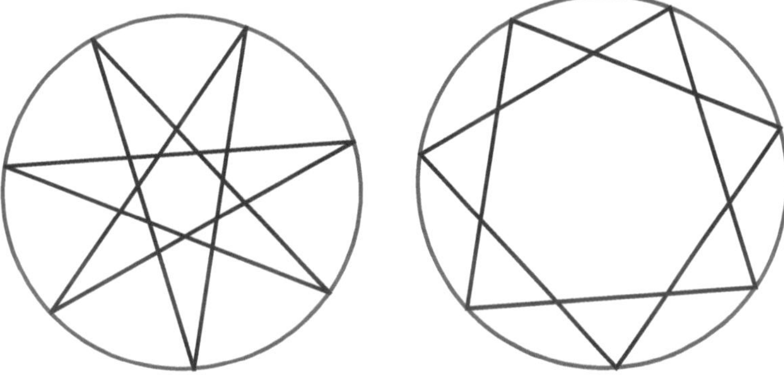

Figure 15

References

(1) Space Shuttle—Wikipedia
(2) Colleges; StatLib—Datasets Archive
(3) Binomial coefficient—Wikipedia

Chapter 6

The Normal Distribution and the Central Limit Theorem

Objectives

After completing this chapter, a student should:

✓ Know the three results of the Central Limit Theorem
✓ Know what is defined as a Bell-Shaped curve
✓ Understand the Standard Normal Distribution
✓ Find the probability of an event in a Normal Distribution
✓ Find a boundary value of an event in a Normal Distribution

As we saw in the sleep example (Chapter 2, Example 3), not all natural phenomena are normally distributed. But that's okay because normally statisticians are interested in the distribution of sample means, not individual values. The Central Limit Theorem (CLT for short) guarantees us the distribution of sample means for samples with size larger than 30 will have a normal distribution regardless of the distribution of the original population. The Central Limit Theorem has three parts and is given below; we'll leave the proof for an advanced class.

Central Limit Theorem

If the sample size is large, n greater than 30, then

1. The distribution of the sample means is normal, regardless of the distribution of the original sample.
2. The mean of the sample means is equal to the population mean.

3. The standard deviation of the sample means, called the standard error of the mean, is the sample standard deviation divided by the square root of n, the sample size. (Technically the standard error of the mean is the population standard deviation, σ, divided by the square root of the sample size, n, but for most cases if σ is not known and n is large, then s is used in place of σ.)

Example 1.

(Optional) Create 5 samples of random numbers from 1 to 10 of size 36, chpt6example1.sav. Find the five sample means and show that the mean of the sample means is approximately 5.5, the population sample mean. Also show the mean of the sample standard deviations is approximately $\sqrt{8.25} \approx 2.87228$, the population standard deviation, and the standard deviations of the sample means is approximately $\frac{\sqrt{8.25}}{6} \approx .478714$, the standard error of the mean.

Solution: Open a new data editor and enter the numbers 1 through 36 into the first column or use chpt6example1.sav. Label that variable "number"; see **Figure 1a**. We're letting n equal 36 so that the square root will be an integer, an easier number to work with. We'll create 5 samples of random variables picked from the numbers 1 through 10. Use Transform → Compute Variable, call the first Target Variable random1. In the Function group click on Arithmetic and scroll down in Functions and Special Variables until you find Trunc (truncate chops off the decimal part of a random number, so that the random number will be an integer between 1 and 10). Double click on that. Next go back to Function group, scroll down to Random Numbers, click and double click on Rv.Uniform. In the Numeric Expression put 1 for the first question mark and 11 for the second question mark. The window should look like **Figure 1b.** Hit OK. SPSS has generated 36 numbers chosen randomly from 1 to 10. This is like putting the numbers 1 through 10 into a hat, drawing one number at a time, writing it down, putting the number back, and repeating for 36 draws. The population mean is 5.5 and the population standard deviation is $\sqrt{8.25}$; these were obtained using mathematics for the theoretical distribution of a discrete population. We'll perform this experiment 5 times to get 5 different means and 5 different standard deviations. In target variable, change the variable to random2, hit OK and repeat until there are 5 samples. (The numbers may be different as shown in Figure 1a if the random generator is changed; however, the goal of this example is to verify the CLT and that will not change no matter what random numbers are selected.)

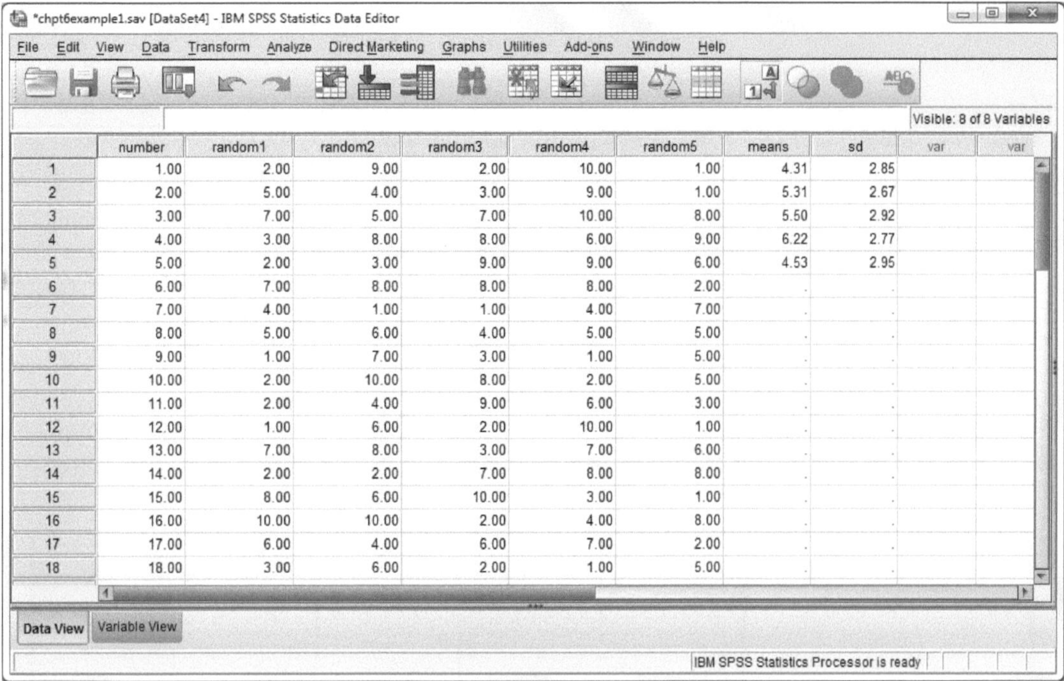

	number	random1	random2	random3	random4	random5	means	sd	var	var
1	1.00	2.00	9.00	2.00	10.00	1.00	4.31	2.85		
2	2.00	5.00	4.00	3.00	9.00	1.00	5.31	2.67		
3	3.00	7.00	5.00	7.00	10.00	8.00	5.50	2.92		
4	4.00	3.00	8.00	8.00	6.00	9.00	6.22	2.77		
5	5.00	2.00	3.00	9.00	9.00	6.00	4.53	2.95		
6	6.00	7.00	8.00	8.00	8.00	2.00	.	.		
7	7.00	4.00	1.00	1.00	4.00	7.00	.	.		
8	8.00	5.00	6.00	4.00	5.00	5.00	.	.		
9	9.00	1.00	7.00	3.00	1.00	5.00	.	.		
10	10.00	2.00	10.00	8.00	2.00	5.00	.	.		
11	11.00	2.00	4.00	9.00	6.00	3.00	.	.		
12	12.00	1.00	6.00	2.00	10.00	1.00	.	.		
13	13.00	7.00	8.00	3.00	7.00	6.00	.	.		
14	14.00	2.00	2.00	7.00	8.00	8.00	.	.		
15	15.00	8.00	6.00	10.00	3.00	1.00	.	.		
16	16.00	10.00	10.00	2.00	4.00	8.00	.	.		
17	17.00	6.00	4.00	6.00	7.00	2.00	.	.		
18	18.00	3.00	6.00	2.00	1.00	5.00	.	.		

Figure 1a

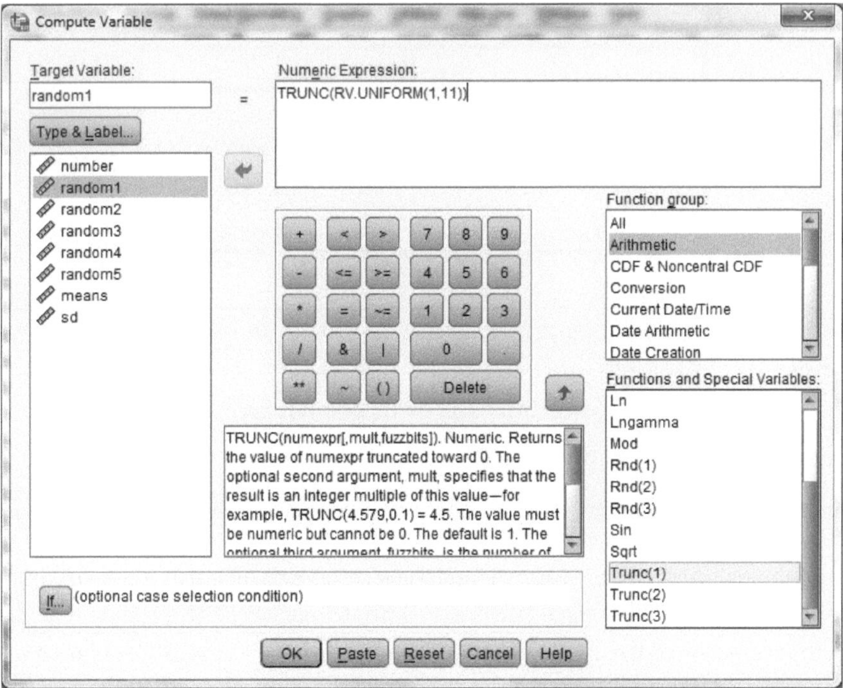

Figure 1b

Now we need to find the mean and standard deviations for the each of the samples. Use Analyze → Descriptive Statistics → Frequencies, highlight and move all the variables *random1* to *random5* into the Variable(s) window, click on Statistics tab and check Mean and Standard Deviation, continue, (deselect Display frequency tables) then hit OK. There should be 5 means and 5 standard deviations; see **Figure 1c**. Your values will be different if the random generator is changed each time the random generator is first started. Put the 5 means into another column called means and the 5 standard deviations into a column called sd, see **Figure 1a**. Now find the mean and standard deviation of the variables Means and sd. The mean of the means should be close to 5.5 (theoretical mean) and the standard deviation of the means (theoretical standard error of the mean) from Example 1 should be close to .478714. (It may not because we're dealing with a small sample size.) The mean of the standard deviations should be close to 2.87228; see **Figure 1d**. This experiment can be repeated several times until you're convinced about the validity of the central limit theorem.

Statistics

		random1	random2	random3	random4	random5
N	Valid	36	36	36	36	36
	Missing	0	0	0	0	0
Mean		4.3056	5.3056	5.5000	6.2222	4.5278
Std. Deviation		2.84675	2.67068	2.92282	2.76830	2.95186

Figure 1c

Statistics

		means	sd
N	Valid	5	5
	Missing	31	31
Mean		5.1722	2.8321
Std. Deviation		.77363	.11502

Figure 1d

Standard Error of the Mean

The standard error of the mean is always calculated when the Explore function is used. The formula is $s.e. = \frac{s}{\sqrt{n}}$, where s is the standard deviation and n is the sample size. Use Analyze → Descriptive Statistics → Frequencies, move all 5 random variables into variables list, click on Statistics, and check the S. E. Mean box. Hit OK. The average of those standard errors of the mean will be much closer to the theoretical standard error of the mean since the average of the standard deviations was closer to the theoretical standard deviation; see **Figure 1e**. (The standard deviation of the sample means was not a good estimate of the standard error of the mean because the sample size of the means was 5.)

Statistics

		random1	random2	random3	random4	random5
N	Valid	36	36	36	36	36
	Missing	0	0	0	0	0
Std. Error of Mean		.47446	.44511	.48714	.46138	.49198

Figure 1e

The Bell-Shaped Curve

The normal distribution is a bell-shaped curve as we saw in Chapter 2. The mean, median, and mode will all be in the center of the curve. As the x values go out to infinity on each side, the curve will approach the x-axis asymptotically, which means the curve will get infinitely close to the x-axis, but never pass through the x-axis, never touch the x-axis and always descend toward the x-axis, see **Figure 2a**. **Figure 2a** is a standard normal curve with mean = 0 and one standard deviation = 1. The normal distribution curve also has symmetry about the mean, if you could fold the curve at the vertical line drawn through the mean, the two branches would coincide. As the x values move away from the mean, the curve is always descending but it changes concavity. At the mean the curve is concave downward, which means the tangent line to any point on the curve is above the curve and toward the ends the curve is concave upward; the tangent lines are below the curve. The place that the curve changes concavity, called an inflection point, is the location of one standard deviation from the mean, as can be seen in **Figure 2a**. Since the total area under the normal distribution curve is 1 square unit (this fact can be shown in multi-variable calculus), the normal distribution curve is a probability distribution function. Thus we can consider the area between two x values under the curve as the probability of an event occurring between those two x values.

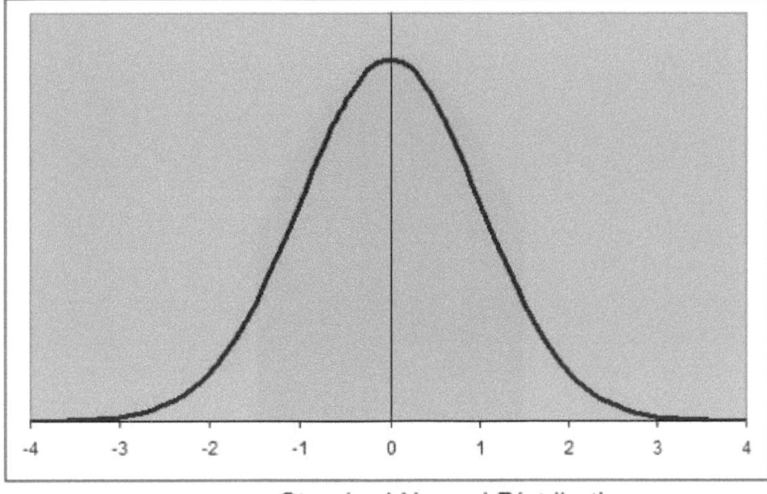

Standard Normal Distribution

Figure 2a

Let's list the characteristics of the normal distribution curve we have discussed so far.
Characteristics of a normal distribution:

i. It is a bell shaped curve
ii. The mean = the median = the mode
iii. The curve has symmetry about the mean
iv. The horizontal axis is the horizontal asymptote
v. The two tails keep descending but never touch or cross the horizontal axis.
vi. The curve changes concavity at one standard deviation
- the mound is where the curve is concave down.
- the tangent line is above the curve at the mound.
- the tangent line changes from above the curve to below the curve at the inflection point, one standard deviation.
- the tangent line is below the curve at the tails of the normal distribution.
- the curve is concave up at the tails.

vii. The total area under the curve is one square unit. This makes the normal distribution a probability distribution.

Empirical Rule, 68-95-99 Rule

The area between one standard deviation to the right of the mean and one standard deviation to the left of the mean is .6827 square units or 68.27% (68%) of the total area; see **Figure 2b**. The area between two standard deviations to the right of the mean and two standard deviations to the left of the mean is .9545 square units or 95.45% (95%) of the total area; see **Figure 2c**. The area between three standard deviations to the right of the mean and three standard deviations to the left of the mean is .9973 square units or 99.73% (99%) of the total area; see **Figure 2d**. This corresponds to .6827 square units between $\mu-\sigma$ and $\mu+\sigma$, .9545 square units between $\mu-2\sigma$ and $\mu+2\sigma$, and .9973 square units between $\mu-3\sigma$ and $\mu+3\sigma$.

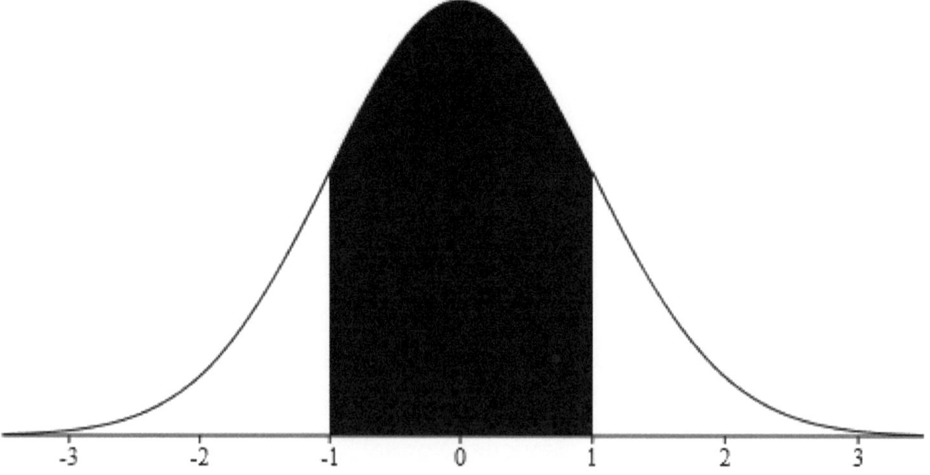

Figure 2b

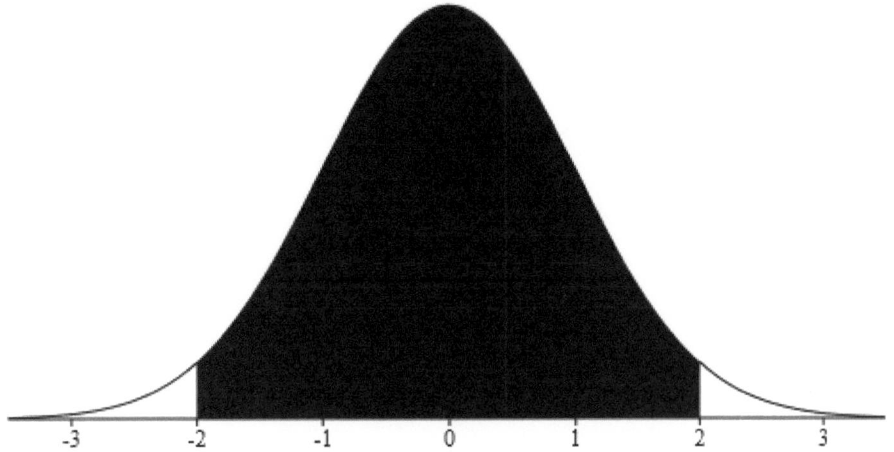

Figure 2c

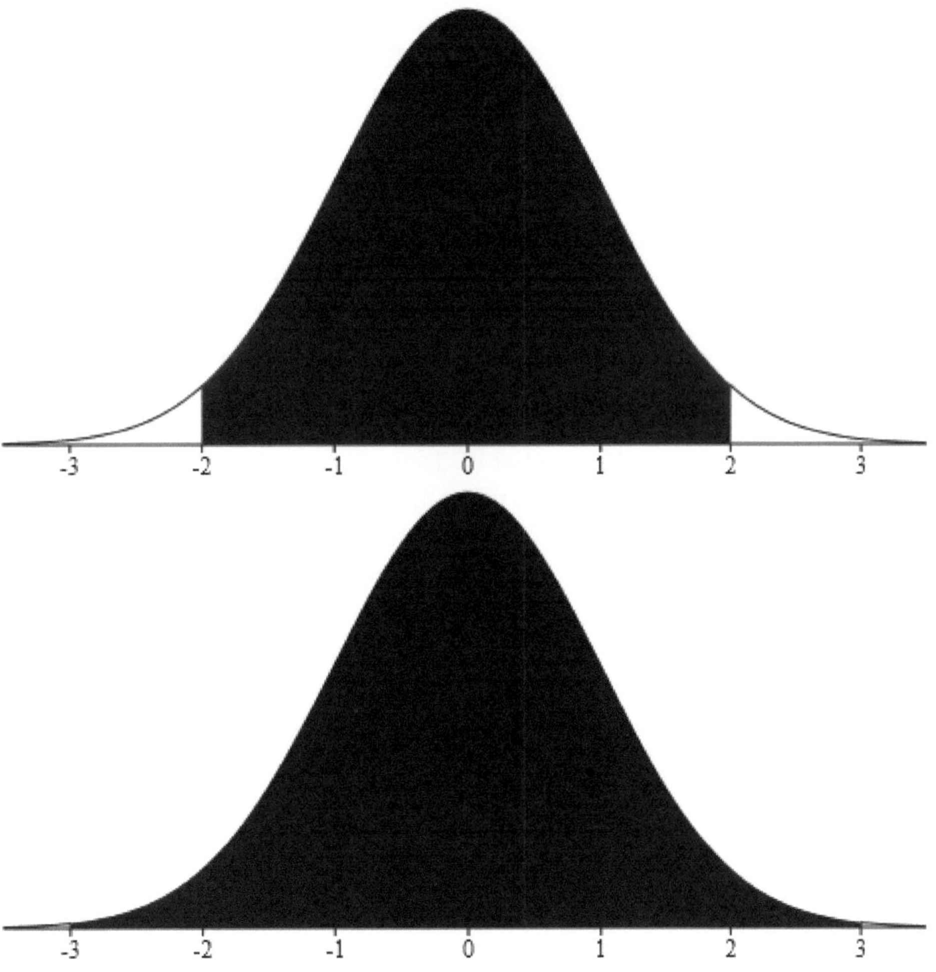

Figure 2d

These characteristics can be summarized as

i. The probability a data value lies between 1 and negative 1 standard deviations is 68.27% or .6827 square units.
ii. The probability a data values lies between 2 and negative 2 standard deviations is 95.45% or .9545 square units
iii. The probability a data value lies between 3 and negative 3 standard deviations is 99.73% or .9973 square units.

Example 2.

IQ scores are forced to be normally distributed with a mean of 100 and a standard deviation of 15. (1) Find the probability a person selected at random will have an IQ score between 85 and 115.

Solution: The empirical rule that says 68.27% of all data values lie within 1 standard deviation of the mean and since 85 is one standard deviation less than 100 and 115 is one standard deviation greater than 100, then by the empirical rule the probability a person selected at random will have an IQ between 85 and 115 is 68%. But we'll use SPSS to find the answer.

Open a new data editor. Put 85, 100, and 115 into the first column, and call the variable *x*. Use Transform → Compute Variable and put any letter into the Target Variable. (I use *p* for probability.) In the Function group click on CDF & Noncentral CDF. Double click on CDF.Normal in the Functions and Special Variables window. CDF.NORMAL will appear in the Numeric Expression; see **Figure 3a**. In the parenthesis after CDF.NORMAL are 3 question marks; this means the function is asking for three parameters. The parameters are explained in the window next to the Functions and Special Variables window. The first question mark asks for the Quantitative values for which the function will be evaluated. The second question mark asks for the mean, and the last question mark asks for the standard deviation. Highlight *x* in the variable window and move it into the numeric expression; see **Figure 3b**. Move the cursor behind the next question mark, backspace to eliminate it, and put in 100. Do the same for the last question mark and replace with 15. Hit OK. Normally we like 4 decimal places for our areas under the normal curve so increase the number of decimals to 4 via Variable View. Your data editor should look like **Figure 3c**. Recall that 100 was the mean, the total area under the normal distribution curve is 1 square unit, and the curve is symmetric on both sides of the mean, so it stands to reason the mean should have a value of .5000, which it does. The area under the *p* column gives the cumulative area up to the given *x* value under the normal distribution curve; see **Figure 3d**. Thus we can find areas between the *x* values by subtracting. The area under the normal distribution curve up to and including 115 is .8413. The area up to and including 85 is .1587. The difference between these two areas is .6826, which is what the empirical rule already told us; see **Figure 3e.** (The difference between .6826 and .6827 is rounding. SPSS rounds decimals to more than 9 decimal places, but we only round to four decimal places because that's the way it's always been done.)

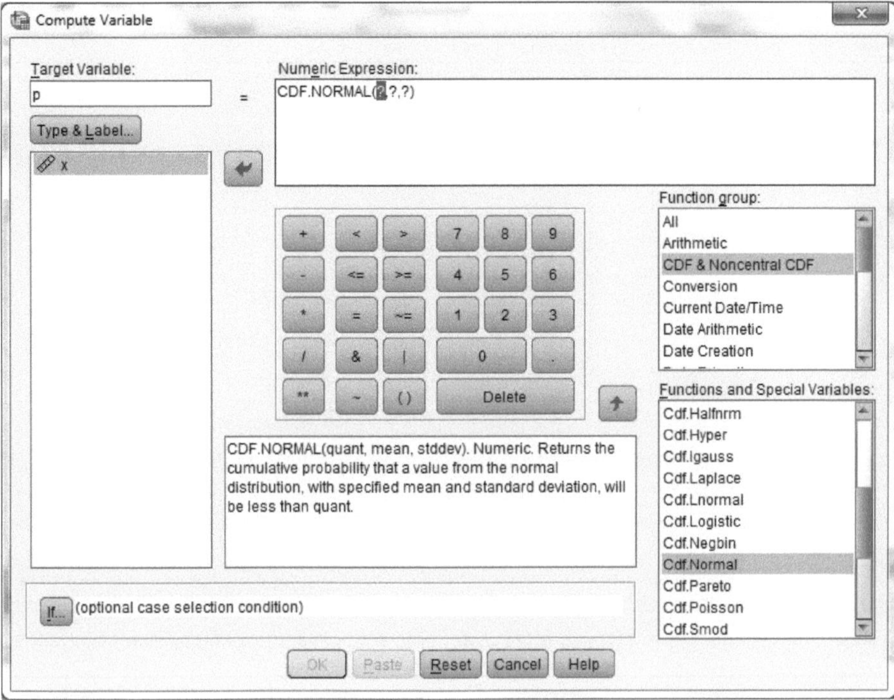

Figure 3a

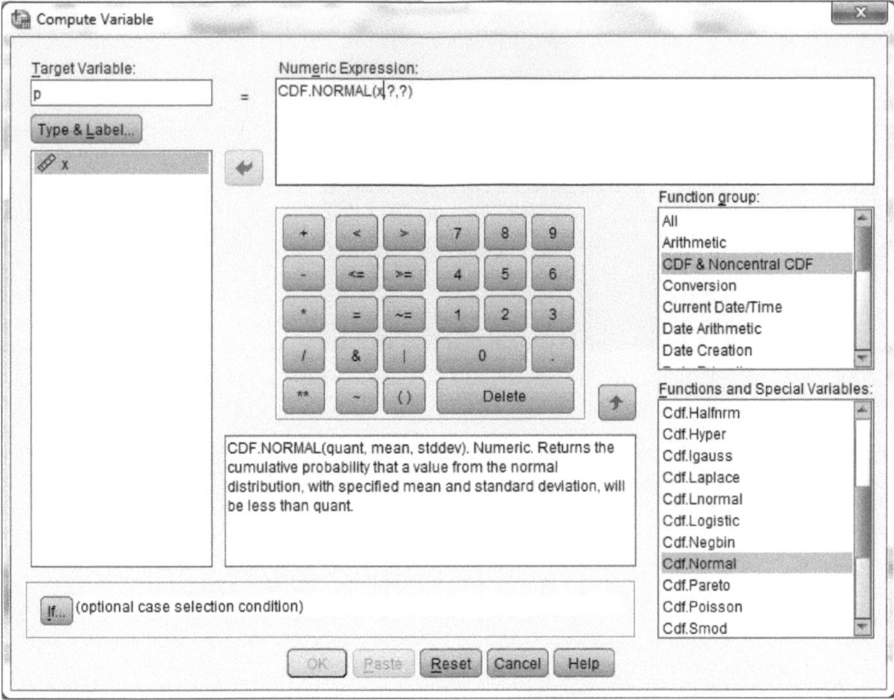

Figure 3b

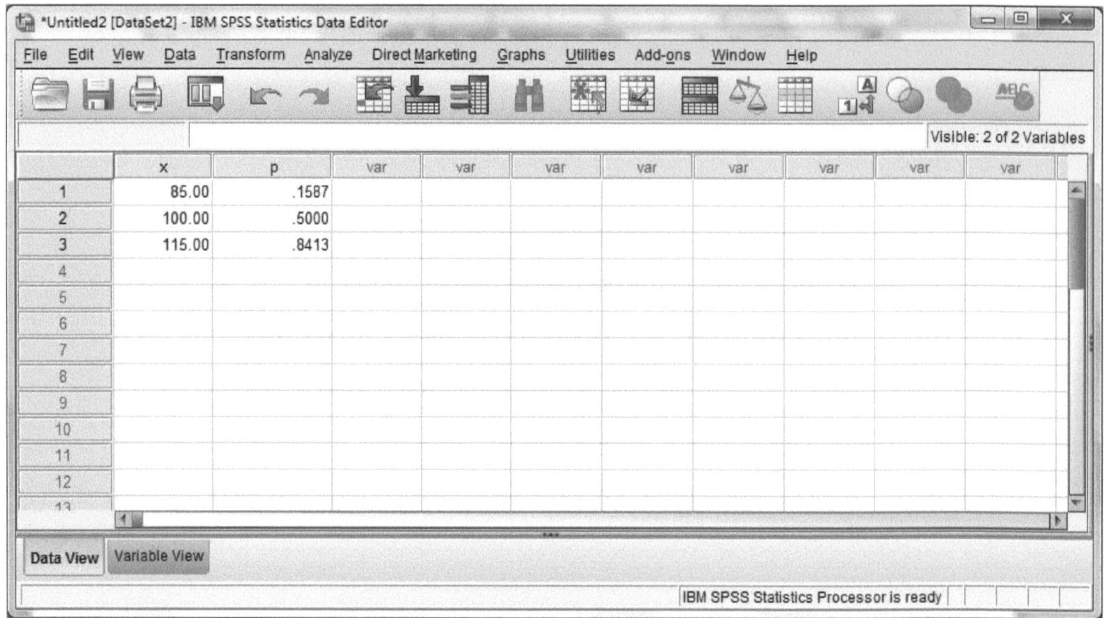

Figure 3c

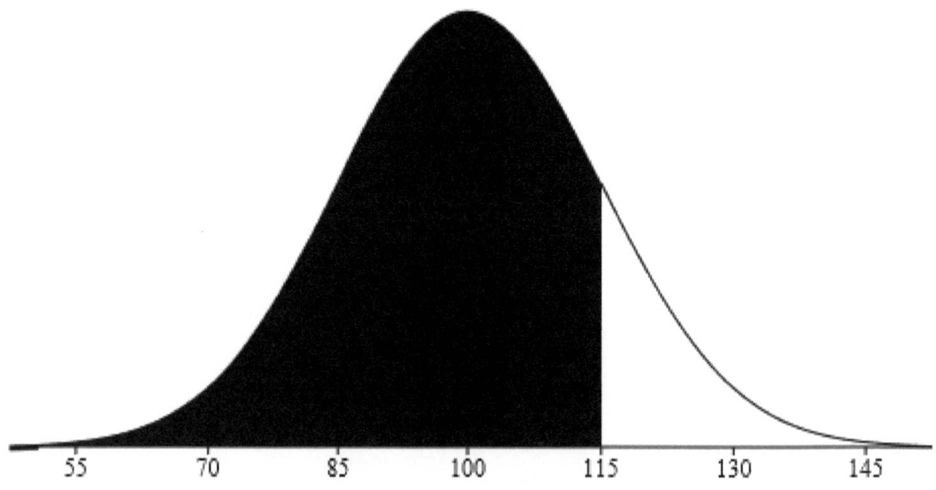

Figure 3d

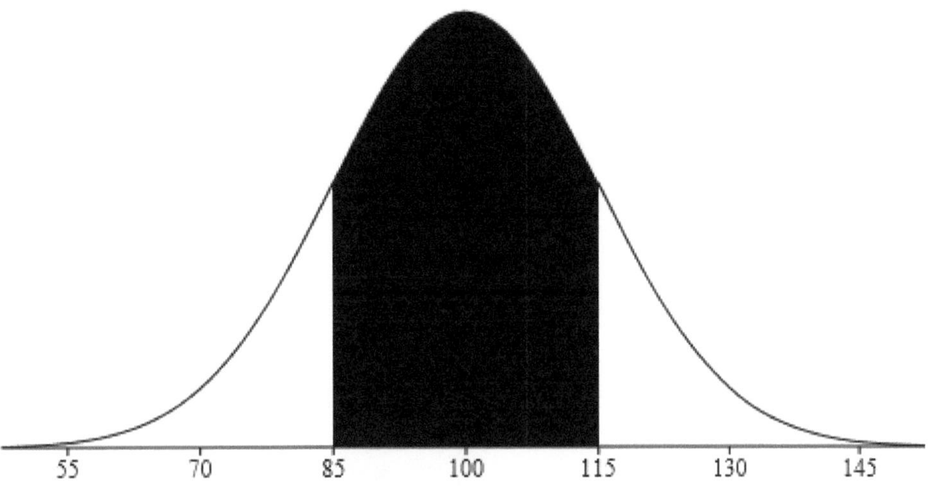

Figure 3e

Normal Distribution Tables and the Standardizing Curve (Tables? We Don't Need No Stinking Tables.)

In the 19th century mathematicians calculated by hand the area under the standard normal curve from 0 to a z-value, Area = $\int_0^z \frac{1}{\sqrt{2\pi}} e^{-\frac{x^2}{2}} dx$, using a power series and methods developed by Leonhard Euler, 1707–1783. The equation of the curve is $f(z) = \frac{1}{\sqrt{2\pi}} e^{-\frac{z^2}{2}}$, where z is the independent variable. We use z for the standard normal curve to distinguish it from the x values in a normal distribution. They calculated only the right half of the standard normal distribution curve because by symmetry the left half had the same area as the right half. Statisticians could then find the probability under any normal curve by using the standardizing formula $z = \frac{x - \mu}{\sigma}$. They still had to carry around with them a copy of the tables wherever they went. Today we need to carry around only a computer, which can fit in our pockets; hence we no longer need tables. We can quickly find any area, hence probability, under any normal distribution.

Example 3.

In the standard normal probability distribution function find the probability that a z value will fall between 0 and 2.93.

Solution: We need to find the area between the two z values given, 0 and 2.93; see **Figure 4a**. We will use CDF.NORMAL, the cumulative distribution function of the normal distribution curve. Open a new SPSS data editor window. Put 0 and 2.93 into the first column, call it z. (We don't really need to find the area for z = 0 because we already know the area to the left of 0 is .5000.) Use Transform → Compute variable, name the target variable *area* or *prob* (area and probability are synonymous when referring to a normal distribution curve), and put CDF.NORMAL into the numeric expression, (you can also type in the function rather than point and click), and use *z* for the first parameter, 0 for the second, and 1 for the third. Hit OK.

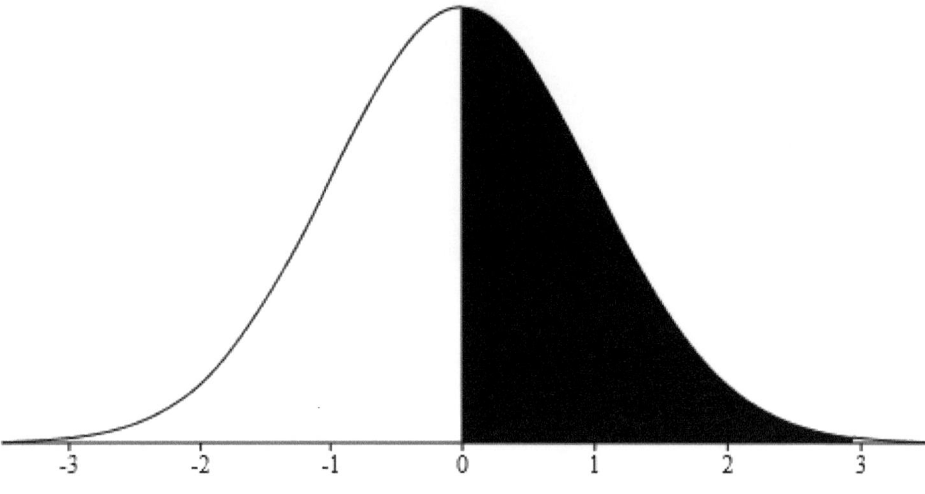

Figure 4a

Be sure to change the decimal positions to 4. The data editor should be as in **Figure 4b**. To find the area between the two *z* values we subtract .5000 from .9983 to get .4983.

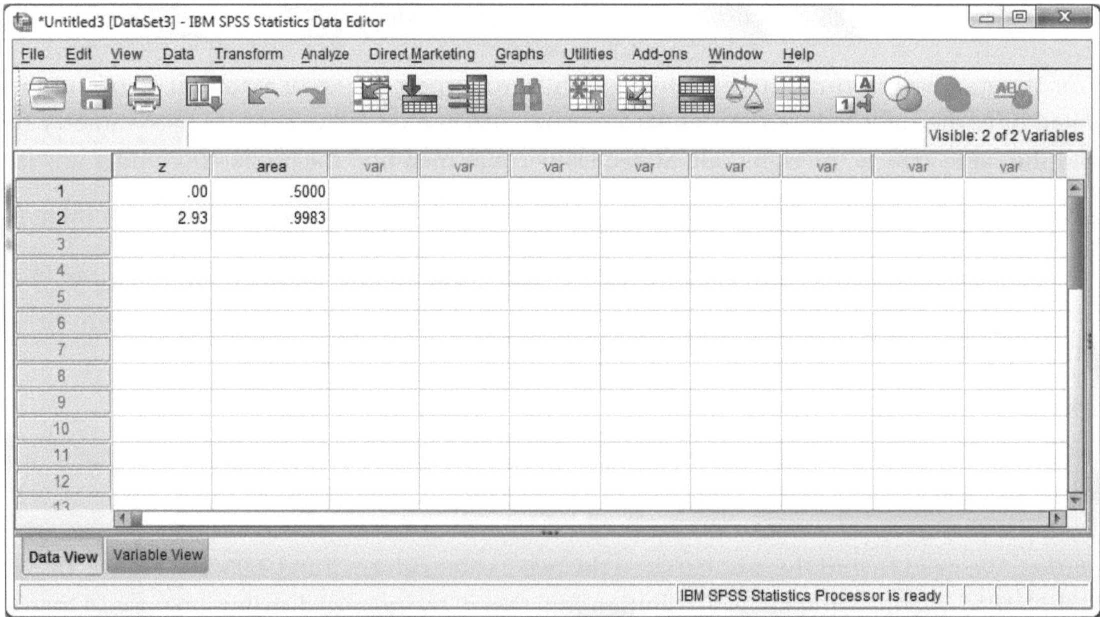

Figure 4b

If we're asked to find the probability above a specific value, then we find the cumulative probability of that value and subtract it from 1. The following example will illustrate this concept.

Example 4.

Using the standard normal distribution:

4a. Find the area to the left of -0.57.
4b. Find the area to the right of -0.57.
4c. Find the area to the right of 0.57.
4d. Find the area to the left of 0.57.

Solution:

4a. CDF.NORMAL gives the area to the left of a given z (or x) value. To find the area to the left of -0.57, which is .2843, follow the steps as in example 3, and add -0.57 in the z column; see **Figure 5a.**

4b. To find the area to the right of -0.57, we subtract the given area obtained in part 4a from 1, since the total area under the curve is 1 square unit Thus the area to the right of -0.57 is 1 – 0.2843 = .7157; see **Figure 5b.**

4c. Put 0.57 in for z and SPSS will return .7157 for the cumulative area, the area to the left of 0.57; see **Figure 5a**. Hence the area to the right of 0.57 is 1 – .7157 = .2843.

4d. The CDF.NORMAL gives the area to the left of a given z, so the value .7157 is the area to the left of 0.57; see **Figure 5c.**

Note the symmetry of the z values and their areas. The area to the left of a positive z value is the area to the right of the negative z value and the area to the left of a negative z value is the area to the right of the positive z value.

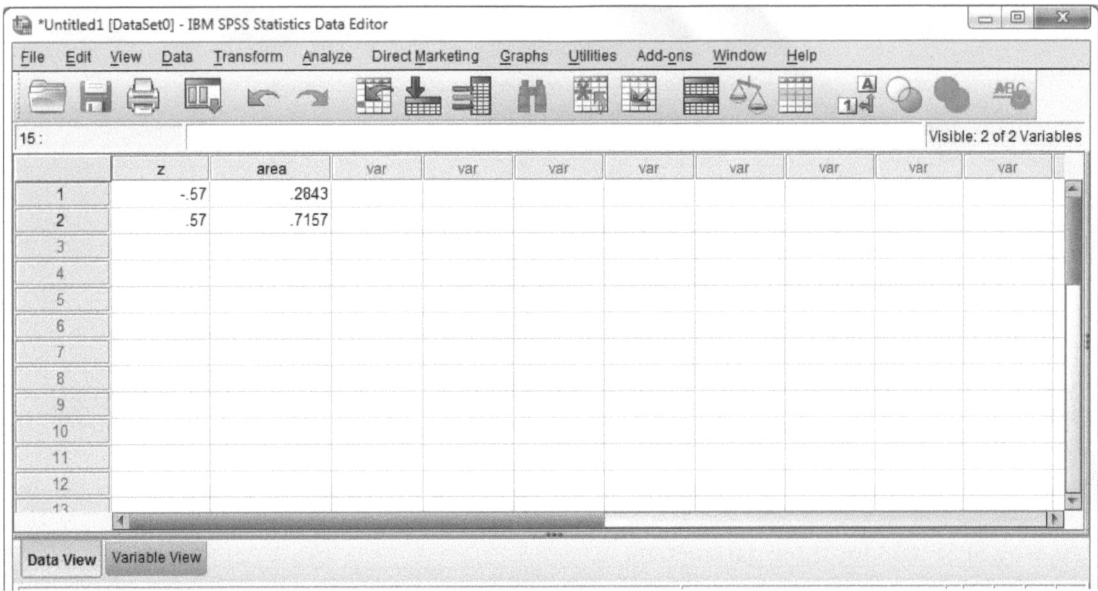

Figure 5a

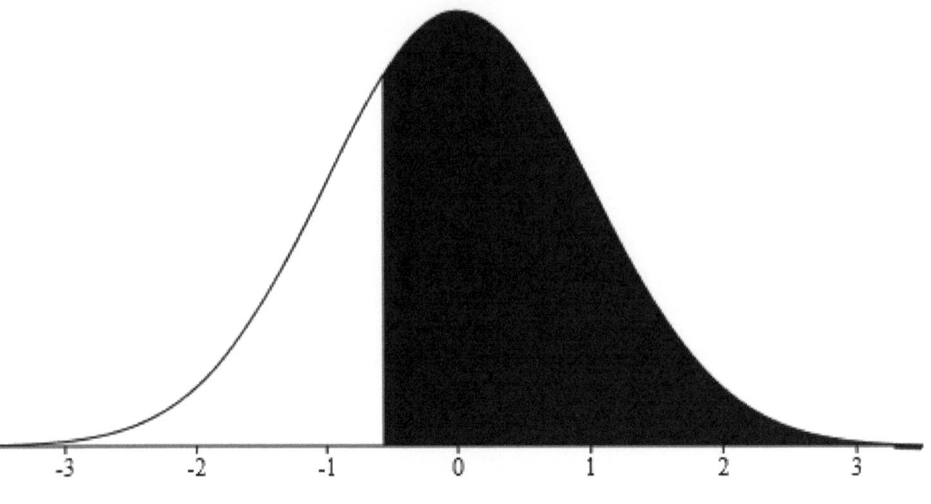

Figure 5b

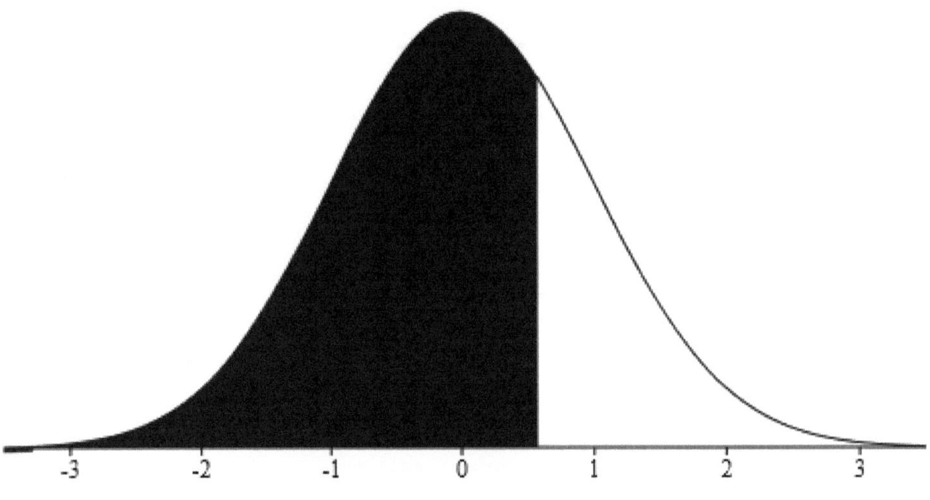

Figure 5c

Using SPSS to Find Normal Probabilities

The probability density function for the normal distribution in SPSS is $f(x)=\frac{1}{\sigma\sqrt{2\pi}}e^{-\frac{(x-\mu)^2}{2\sigma^2}}$. To find the cumulative distribution, SPSS evaluates $\int_{-\infty}^{x} f(t)dt$. Note there are three variables; x, μ, and σ, which is why SPSS asks for three parameters.

Example 5.

The average height of adult females is 63.8 inches with a standard deviation of 2.5 inches. (2)

5a. Find the probability that an adult female selected at random will be between 65 and 69 inches.
5b. Find the probability that an adult female selected at random will be less than 60 inches.
5c. Find the probability that an adult female selected at random will be greater than 60 inches.
5d. Find the probability that an adult female selected at random will be between 62 and 69 inches.

Solution: Open a new SPSS data editor and input the numbers 60, 62, 65, 69 into the first column, call it femhts, or open chpt6example5.sav. Use Transform → Compute Variable, call the target variable *area* and put CDF.NORMAL in the numeric expression. The three parameters will be femhts 63.8 and 2.5. (Recall what these parameters represent.) Hit OK. Remember to change the decimals for the variable *area* to 4.

The data editor is shown in **Figure 7a** with the corresponding areas in **figures 7b–7e**.

5a. To find the probability that an adult female selected at random will be between 65 and 69 inches, we subtract the area corresponding to 65, .6844, from the area corresponding to 69, .9812, which is .2968, hence the probability that an adult female selected at random will be between 65 and 69 inches is 29.68%; **Figure 7b**. (We move the decimal point two places to the right when we change from a decimal number to a percent; recall percent means per hundred and the second decimal position is the hundredths position.)

5b. The area under the normal distribution curve up to and including 60 inches is .0643, thus the probability that an adult female selected at random will be less than 60 inches is 6.43%; **Figure 7c**. (The normal distribution curve is a continuous curve so we can find only probability of a range of values. The curve does not have an area for a single value. Thus the probability an adult female selected at random is exactly 60 inches is 0, we cannot find the area under the curve for the value 60. So when we talk about a range of values we can always assume the endpoints are included.)

5c. Since the probability that an adult female selected at random will be less than 60 inches is 6.43%, then the probability that an adult female selected at random will be greater than 60 inches is 100% – 6.43% = 93.57%. (1 – .0643 = .9357); **Figure 7d**.

5d. Subtracting the area for a height less than 62, .2358, from the area for a height less than 69, .9812, yields .7454. The probability is 74.54%; **Figure 7e**.

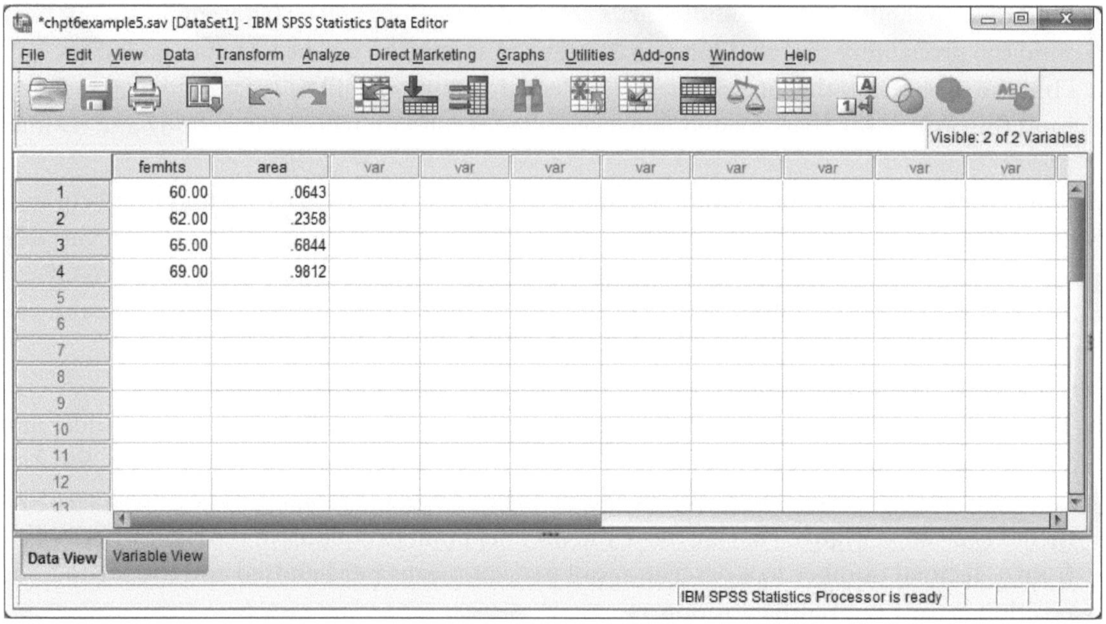

*chpt6example5.sav [DataSet1] - IBM SPSS Statistics Data Editor
File Edit View Data Transform Analyze Direct Marketing Graphs Utilities Add-ons Window Help
Visible: 2 of 2 Variables
femhts area var var var var var var var var
1 60.00 .0643
2 62.00 .2358
3 65.00 .6844
4 69.00 .9812
Data View Variable View
IBM SPSS Statistics Processor is ready

Figure 7a

56.3
58.8
61.3
63.8
66.3
68.8
71.3

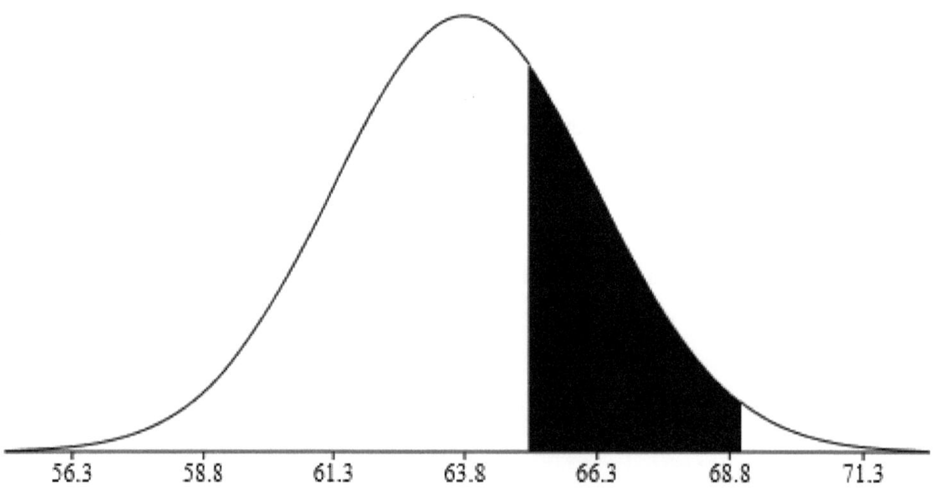

Figure 7b

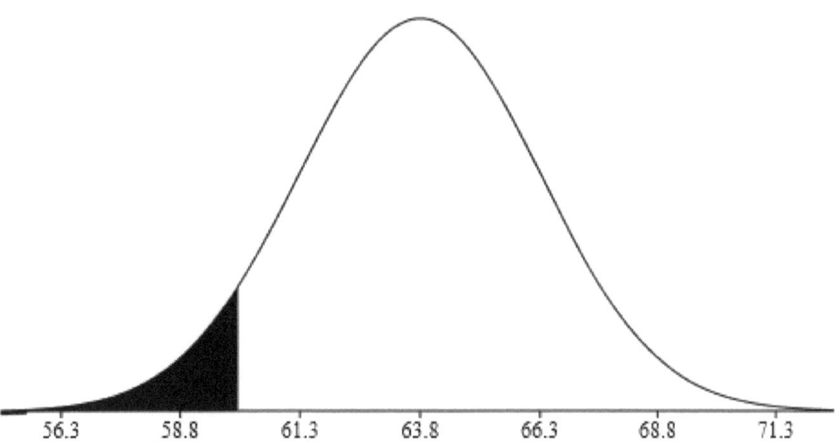

Figure 7c

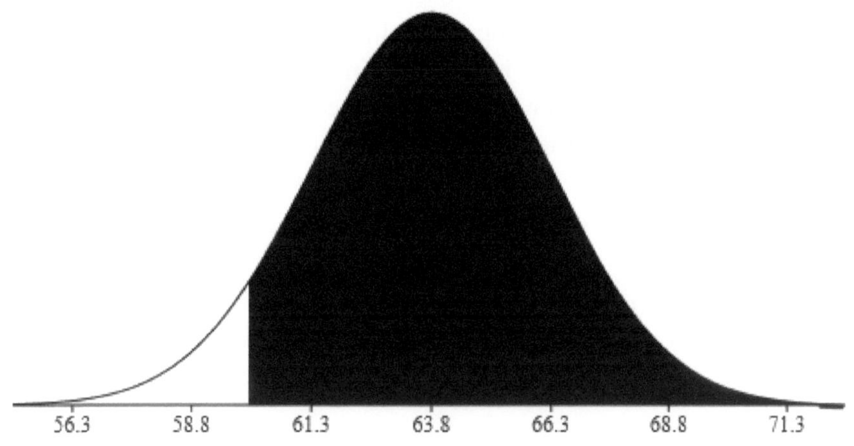

Figure 7d

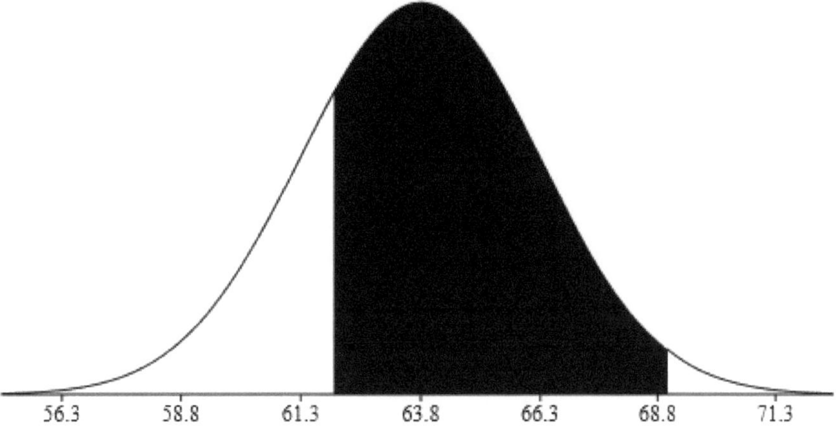

Figure 7e

The standardizing formula also enables us to find a raw score given a percentile or probability for an event. What we need to do is find the corresponding z score for a given probability with a known mean and standard deviation and then solve for x, the raw score. The equation is $x = z\sigma + \mu$. To find the z-score that corresponds to a given probability, we use the inverse normal distribution function.

The Inverse Normal Distribution Function

Example 6a.

Use the inverse normal distribution function to find critical z-scores for the areas below the z-score being .75, .80, .90, .95, .99, .995, .9995.

Solution: To find the z score for these areas, first open a new SPSS data editor and put .75, .80, .90, .95, .99, .995, .9995 in the first column, call it p. Use Transform → Compute Variable, call the target variable z, and scroll down the functions group to the Inverse DF function and click on it. Now scroll down the Functions and Special Variables to Idf.Normal and double click on that. The function calls for 3 parameters. Move p into the first position, put 0 as the mean, second position, and 1 as the standard deviation, third position. Click OK and the critical z value will appear in the z column of the data editor, **see Figure 8a**. For convenience sake we've listed some critical z scores in Table 2, where the area below the z score is the cumulative distribution value.

	p	z
1	.7500	.6745
2	.8000	.8416
3	.9000	1.2816
4	.9500	1.6449
5	.9750	1.9600
6	.9900	2.3263
7	.9950	2.5758
8	.9995	3.2905

Figure 8a

Table 2: Some critical z scores, from Figure 8a.

Area below z score	.7500	.8000	.9000	.9500	.9750	.9900	.9950	.9995
Critical z scores	.6745	.8416	1.2816	1.6449	1.960	2.3263	2.5758	3.291

Example 6b.

SAT scores also fit a normal distribution. Suppose the Math SAT scores have a mean of 520 and a standard deviation of 100 points and the scores run from 200 to 800. What raw score would be the cutoff point for the 90th percentile?

Since we want the 90th percentile, the critical z score is 1.2816. The equation we need to solve is $x = z\sigma + \mu$ for x. Substituting we get $x = 1.2816 \times 100 + 520$ or $x = 648.16$. Thus a score of 648 or above puts a person in the 90th percentile for the year pertaining to this mean and standard deviation. Each year the mean and standard deviation of SAT scores change, but they hover close to the same values, approximately 500. For the year 2011 the mean for critical reading was 497 with a standard deviation of 114, the mean for math was 514 with a standard deviation of 117, and the mean for writing was 489 with a standard deviation of 113.

SPSS can also find the critical x value and thereby we can skip using the formula $x = z\sigma + \mu$. Use Transform → Compute Variable, call the target variable x, and scroll down the functions group to the Inverse DF function and click on it. Now scroll down the Functions and Special Variables to Idf.Normal and double click on that. The function calls for 3 parameters. Move p into the first position, 520 into the second position, the mean, and 100 into the third position, the standard deviation. Hit OK and the data editor will have the 90th percentile score, which is 648.16; see **Figure 8b**.

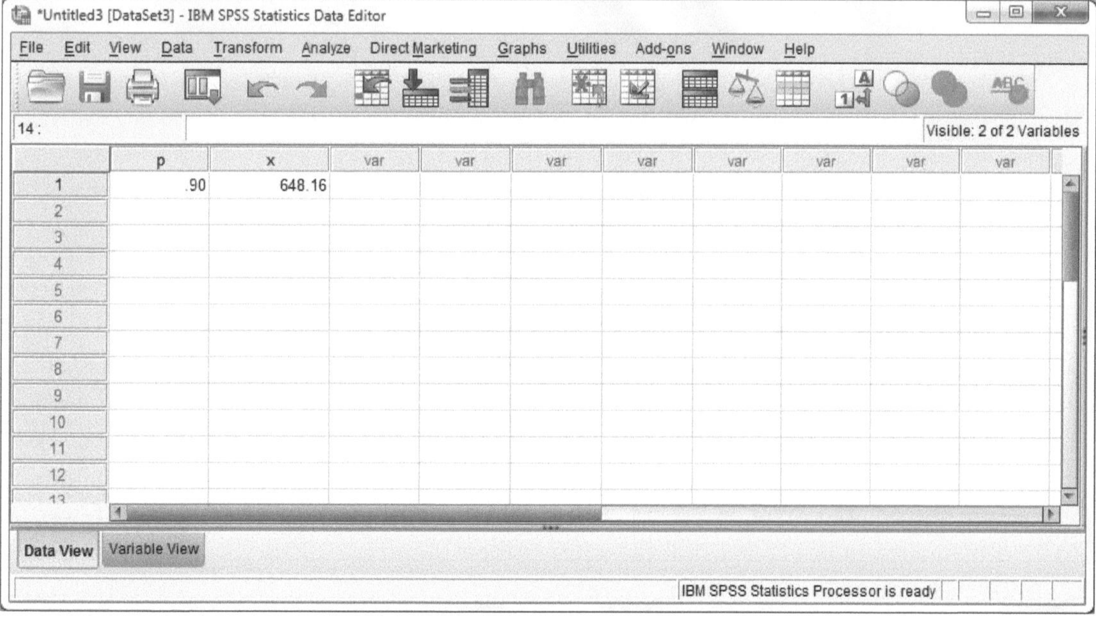

Figure 8b

Using the Normal Distribution to Find the Probability of a Percentage or Proportion

We can also use the normal distribution to find the probability of a proportion. A proportion or a ratio is basically a fraction, such as $\frac{x}{n}$, where x is less than or equal to n. The mean is the proportion $p = \frac{x}{n}$, and the standard deviation is $\sqrt{\frac{pq}{n}}$, where $q = 1 - p$, and n is the sample size. The standardizing formula is $z = \frac{\hat{p} - p}{\sqrt{\frac{pq}{n}}}$, where $\hat{p} = \frac{r}{n}$ and r is an integer value between 0 and n. P hat, $\hat{p}$, is the obtained value and p is the assumed value. When n is small and our proportion is based on an integral number, then we need to adjust $\frac{x}{n}$ with a continuity correction of $\frac{0.5}{n}$. Whether we add or subtract $\frac{0.5}{n}$ depends on whether we're trying to find the area to the left of $\hat{p}$ or the area to the right of $\hat{p}$. If our proportion is based on a percentage, then we don't need a continuity correction since percentages are already continuous.

Example 7.

Suppose a manufacturer claims 3 out of 5 consumers prefer the manufacturer's brand. Now suppose a county fair held a blind test comparing the manufacturer's brand against many other brands and only 63 out of 117 consumers preferred the manufacturer's brand. Assuming the manufacturer is correct, what is the probability that a proportion less than or equal to 63 out of 117 will occur?

Solution: In this example $p = \frac{3}{5} = .6 \quad q = \frac{2}{5} = .4 \quad n = 117$. Since our question deals with the proportion being less than or equal to 63 out of 117, we will add $\frac{.5}{117}$ to $\frac{63}{117}$. We want to find the area under the normal distribution curve less than $\frac{63}{117} + \frac{.5}{117} = \frac{63.5}{117} \approx .542735$. Open a new SPSS data editor and put .542735 into the first column, and call it *propor*. (Instead of calculating the proportion with a calculator we can let SPSS do that for us. Put 63.5 into a column, call it *x* and use Transform → Compute Variable. Call the target variable *propor* and in the numeric expression put the formula *x*/117, hit OK.) Use Transform → Compute Variable, call the target variable *area*, and put CDF.NORMAL in the numeric expression. The three parameters will be *propor*, .6, and SQRT(.6*.4/117). (You don't need to change the standard deviation into a decimal because SPSS can work with arithmetic expressions in the function CDF.NORMAL. The value for *area* should be .1030. This says that if the manufacturer's claim is correct, that is 3 out of 5 consumers prefer their brand, then the probability that fewer than or equal to 63 consumers out of 117 in a blind test prefer their brand is 10.30%.

Using the Normal Distribution to Approximate a Binomial Distribution

The normal distribution, which is a continuous distribution, can also be used to approximate a binomial distribution, which is a discrete distribution provided the mean is greater than 5. The mean of a binomial distribution is *np* and to apply the normal distribution to the binomial distribution, the mean must be greater than 5 (assuming *p* is less than *q*, otherwise *nq* must be greater than 5). The standardizing formula is $z = \frac{r - np}{\sqrt{npq}}$, where *r* is a whole number between 0 and *n*. However, since the normal distribution is continuous, then to find the probability of a single *r* value we need to adjust the boundary values for the normal distribution to accept a range of values that include *r*. We find the probability of the interval (*r* - .5, *r* + .5).

Example 8a.

In a binomial distribution the mean is *np* and the standard deviation is $\sqrt{npq}$, where $q = 1 - p$. If the probability of catching an albacore on any given day is .1 and you fish for albacore for 75 days straight, then what's the probability you'll catch exactly 7 albacore?

Solution: In this problem $r = 7$ and since the mean is $.1(75) = 7.5 > 5$, then we can use the normal distribution to approximate the binomial distribution. Open a new SPSS data editor. To adjust for the continuous distribution we'll use half a point below 7, which is 6.5 and half a point above 7, which is 7.5 for our boundary values. Put 6.5 and 7.5 into the first column, call it *x*. Use Transform → Compute Variable, label the target variable *prob*, and put CDF.NORMAL in the numeric expression. The parameters will be *x*, 7.5, and SQRT(75*.1*.9). Hit OK, be sure to increase the decimals to 4 for the variable *prob*. The two areas should be .3502 for the value 6.5 and .5000 for the value 7.5, which makes sense since 7.5 is the mean. The difference is .1498, which is the probability of getting exactly 7 albacore. That is, the probability of catching exactly 7 albacore is 14.98% or ≈ 15%.

The function group PDF & Noncentral PDF is the Probability Distribution Function (CDF is Cumulative Distribution Function) and we can use this group to find the probability of a single value.

Example 8b.

Use the function group PDF & Noncentral PDF to find the probability of catching exactly 7 albacore.

Solution: Open a new SPSS data editor, put 7 in the first column, and call it *x*. Use Transform → Compute Variable, call the target variable *prob*, and click on PDF & Noncentral PDF. Scroll down until you find Pdf.Binom in the functions and Special Variables window. Double click on Pdf.Binom. Put *x*, 75, and .1 in for the three parameters. Click OK and be sure to increase the decimals to 4 places. The probability is .1535, which is the probability of getting exactly 7 albacore. As we can see these two probabilities are identical when rounded to two decimal places.

Conditions that Allow a Normal Approximate to a Binomial Distribution

The sample size *n* needs to be large and both *np* and *nq* need to be greater than 5. We can show the binomial distribution will appear normal if *np* is greater than 5 in the following example.

Example 9.

Draw a histogram for the binomial distribution with $n = 15$ and $p = .5$.

Solution: Open a new SPSS data editor. Put the numbers 0 through 15 in the first column, call it x. Use Transform → Compute Variable, call the target variable p and place PDF.BINOM(x,15,.5) in the numeric expression, hit OK. The data editor should have the probabilities for the values 0 through 15 in the p variable (column). Increase the decimals and notice the symmetry of the values; see **Figure 9a**. The symmetry exists because the probabilities for p and q are the same and the binomial coefficients are symmetric. To draw our histogram we need to weight the variable x with p. Use Data → Weight cases, click Weight cases by and move p into the Frequency Variable window, hit OK. Next use Graphs → Legacy Dialogs → Histogram, put x in the variable window, hit OK. SPSS doesn't draw a very good histogram so we need to change the number of bins. Right click (or double click) on the graph, open the SPSS chart object, which gives a chart editor, click on Options → Un-bin element, then again Options → Bin element. In the Properties window click on Custom and Interval width, then change the interval width to 1 and anchor the first interval at 0. Click Apply. Close the windows and the histogram should look like **Figure 9b**. This is a normal distribution. Lastly we can add the normal distribution curve to the graph, but we need to create a new variable. Use Transform → Compute Variable, call the target variable f, and put the variable p in the numeric expression window and multiply by 10,000 (p*10000). Go to Data → Weight Cases and weight case by f. Then Graphs → Legacy Dialogs → Histogram, put x in the variable window, and also check Display normal curve. Hit OK and the graph should be like **Figure 9c**.

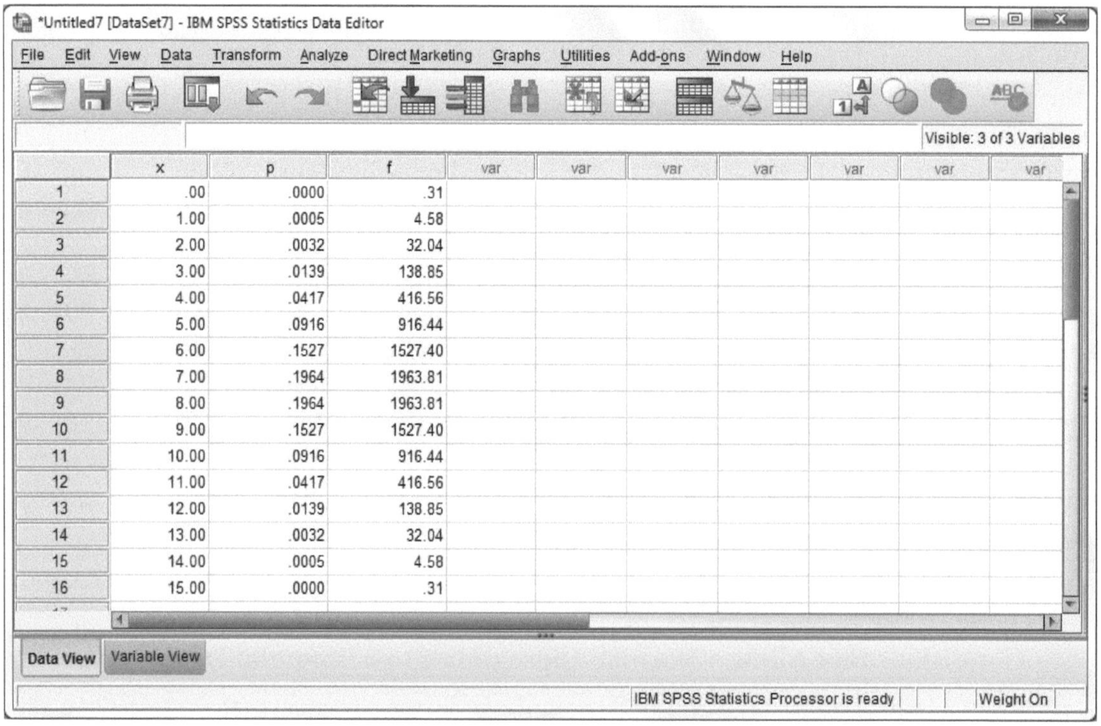

*Untitled7 [DataSet7] - IBM SPSS Statistics Data Editor

File Edit View Data Transform Analyze Direct Marketing Graphs Utilities Add-ons Window Help

Visible: 3 of 3 Variables

	x	p	f	var	var	var	var	var	var	var
1	.00	.0000	.31							
2	1.00	.0005	4.58							
3	2.00	.0032	32.04							
4	3.00	.0139	138.85							
5	4.00	.0417	416.56							
6	5.00	.0916	916.44							
7	6.00	.1527	1527.40							
8	7.00	.1964	1963.81							
9	8.00	.1964	1963.81							
10	9.00	.1527	1527.40							
11	10.00	.0916	916.44							
12	11.00	.0417	416.56							
13	12.00	.0139	138.85							
14	13.00	.0032	32.04							
15	14.00	.0005	4.58							
16	15.00	.0000	.31							

Data View Variable View

IBM SPSS Statistics Processor is ready Weight On

Figure 9a

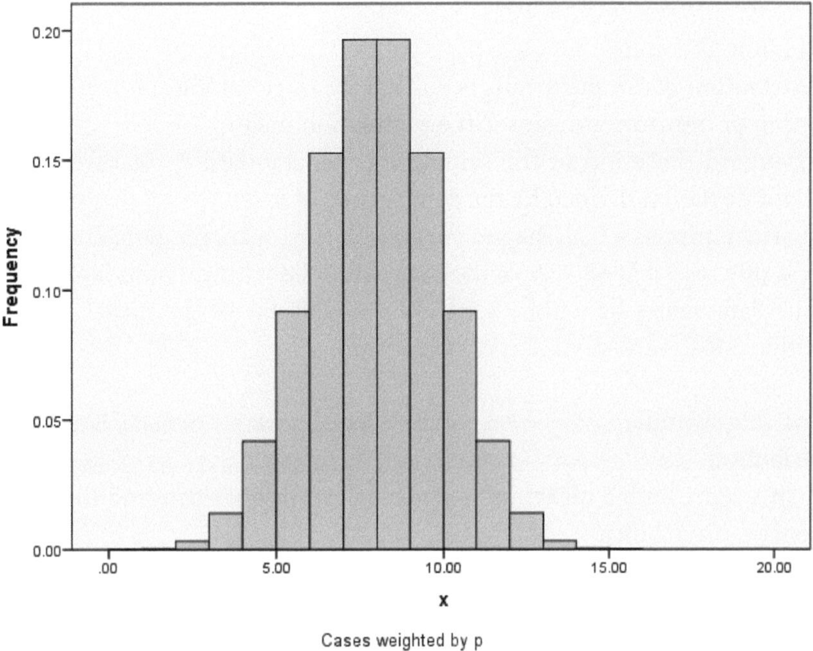

Figure 9b

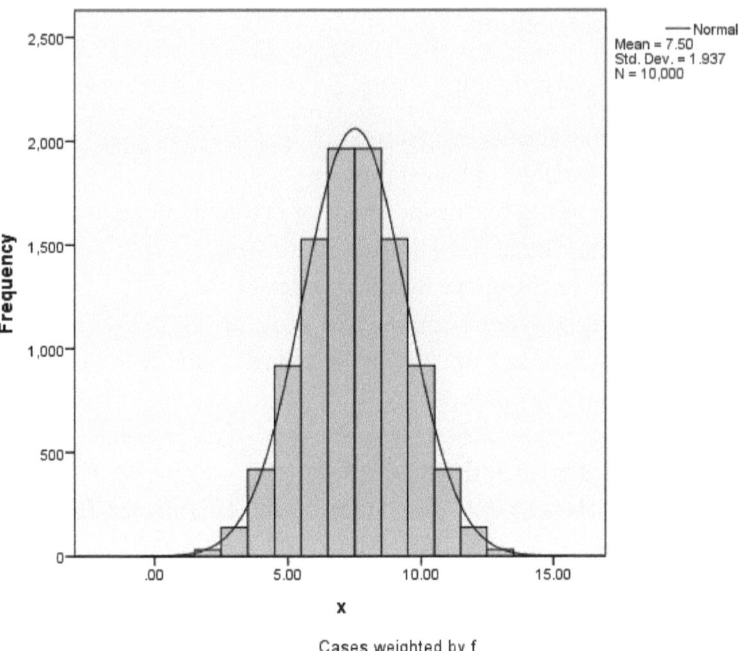

Figure 9c

Let's review what we learned in this chapter.

1. The Central Limit Theorem
   i.) The distribution of sample means is normal for large samples, $n > 30$.
   ii.) The mean of the sample means is the population mean.
   iii.) The standard deviation of the sample means, standard error of the mean, is the sample standard deviation divided by the square root of n.
2. The normal distribution is a bell-shaped curve with the mean, median, and mode all in the center.
3. The empirical rule says that 68.27% of the data values lie within 1 standard deviation of the mean, 95.45% of the data values lie within 2 standard deviations of the mean, and 99.73% of the data values lie within 3 standard deviations of the mean.
4. The standardizing formula $z = \dfrac{x - \mu}{\sigma}$ converts an arbitrary normal distribution into a standard normal distribution.
   i.) Given an x score with a mean and standard deviation we can find the corresponding z score and hence a probability.
   ii.) Given a probability we can find a corresponding z score, and with a known mean and standard deviation we can find the corresponding x score.
5. Using the normal distribution to find probabilities for proportions with the mean and standard deviation given as $\mu = p, \quad \sigma = \sqrt{\dfrac{pq}{n}}$, where $q = 1 - p$.
6. The normal approximation to a binomial distribution with mean and standard deviation given as $\mu = np, \quad \sigma = \sqrt{npq}$, where $q = 1 - p$.
7. Characteristics of a normal distribution:
   i.) It is a bell-shaped curve
   ii.) The mean = the median = the mode
   iii.) The curve has symmetry about the mean
   iv.) The horizontal axis is the horizontal asymptote
   v.) The two tails keep descending but never touch or cross the horizontal axis.
   vi.) The curve changes concavity at one standard deviation
      - the mound is where the curve is concave down.
      - the tangent line is above the curve at the mound.
      - the tangent line changes from above the curve to below the curve at the inflection
      - point, one standard deviation.
      - the tangent line is below the curve at the tails of the normal distribution.
      - the curve is concave up at the tails.
   vii.) The total area under the curve is one square unit. This makes the normal distribution a probability distribution.
   viii.) The probability a data value lies between 1 and negative 1 standard deviations is 68.27% or .6827 square units.
   xi.) The probability a data values lies between 2 and negative 2 standard deviations is 95.45% or .9545 square units.

x.) The probability a data value lies between 3 and negative 3 standard deviations is 99.73% or .9973 square units.

Class Work 6

Part A

Use the empirical rule (Chapter 6, page 6) to find the boundary values for 68%, 95%, and 99% of the data values in a normal distribution with mean 16 and standard deviation 2.5.

Part B

Since IQ is forced to be a normal distribution, let's find the probabilities of some IQ ranges using the normal distribution curve. Use a mean of 100 and a standard deviation of 15.

A.) Find the probability a person selected at random has an IQ score greater than 140.
B.) Find the probability a person selected at random has an IQ score between 115 and 130.
C.) Find the probability a person selected at random has an IQ score less than 110.
D.) Find the probability a person selected at random has an IQ score greater than 80.
E.) Find the probability a person selected at random has an IQ score between 80 and 110.
F.) Find the IQ score for a person who scores in the 90th percentile.
G.) Find the range of IQ scores that contain the middle 50% of IQ scores.

Homework 6

1. If a population has a mean of 6 and a standard deviation of 3.282 and a random sample of size 36 is selected from the population, then what should be the values of the sample mean, sample standard deviation, and standard error of the mean?

2. If a population has a mean of 27 and a standard deviation of 8.61 and a random sample of size 49 is selected from the population, then what should be the values of the sample mean, sample standard deviation, and standard error of the mean?

3. If a population has a mean of 102 and a standard deviation of 18.8 and a random sample of size 64 is selected from the population, then what should be the values of the sample mean, sample standard deviation, and standard error of the mean?

4. If a population has a mean of 20.61 and a standard deviation of 29.16 and a random sample of size 81 is selected from the population, then what should be the values of the sample mean, sample standard deviation, and standard error of the mean?

5. If a normal distribution has a population mean of 20 and a population standard deviation of 6, then use the empirical rule to find the boundary values that contain 68.27% of the data values.

6. If a normal distribution has a population mean of 20 and a population standard deviation of 6, then use the empirical rule to find the boundary values that contain 95.45% of the data values.

7. If a normal distribution has a population mean of 20 and a population standard deviation of 6, then use the empirical rule to find the boundary values that contain 99.73% of the data values.

8. If a normal distribution has a population mean of 20 and a population standard deviation of 6, then find the z scores that correspond to x scores of 17, 18, 24, and 32.

9. If a normal distribution has a population mean of 102 and a population standard deviation of 18.8, then find the z scores that correspond to x scores of 58.76, 95.608, 121.176, and 145.24.

10. If a normal distribution has a population mean of 20 and a population standard deviation of 6, then find the x scores that correspond to z scores of -0.5, -0.333, 0.667, and 2.

11. If a normal distribution has a population mean of 102 and a population standard deviation of 18.8, then find the x scores that correspond to z scores of -2.3, -0.34, 1.02, and 2.3.

For the following problems use IDF.NORMAL

12. Find the z score that corresponds to a cumulative probability of .80.

13. Find the z score that corresponds to a probability above the z score of .15.

14. Find the z scores that correspond to a probability between the z scores of .37.

15. Find the z scores that correspond to a probability outside the z scores of .37.

16. If a normal distribution has a mean of 102 and a standard deviation of 18.8, then find the x value such that 20% of the data values are above that x value.

17. SAT scores also fit a normal distribution. Assume the mean for the verbal section is 538 with a standard deviation of 70 and Bob scores in the 90th percentile. What is Bob's raw score?

18. A granite countertop company insists 90% of all homeowners prefer granite countertops in their kitchen. Assuming the company is correct, find the probability that more than 95% out of 100 homeowners prefer granite countertops in their kitchen.

19. A CD manufacturer claims only 5% of the CDs manufactured are defective. In a shipment of 200 CDs, find the probability that no more than 4% are defective.

20. Some scientists claim that only 75% of the public belives global warming is a direct result of humankind interaction with the environment, burning of fossil fuels, etc. Assuming this percentage is correct, find the probability in a group of 100 people that more than 90% of them believe global warming is indeed a direct result of humankind interaction with the environment.

21. A cookie manufacturer claims only 10% of the cookies' ingredients are artificial. Assuming this percentage is correct, find the probability in a package of 60 cookies that 25% or more of the cookies' ingredients are artificial. Artificial ingredients include refined sugar and preservatives.

22. In a multiple-choice test that has 30 questions, where each question has 5 choices and the test taker is simply guessing on all questions, then what's the probability the test taker will score 19 or above?

23. In a multiple-choice test that has 30 questions, where each question has 5 choices and the test taker is simply guessing on all questions, then what's the probability the test taker will score 15 or above?

24. In a multiple-choice test that has 30 questions, where each question has 5 choices and the test taker is simply guessing on all questions, then what's the probability the test taker will score below 6?

25. In a multiple-choice test that has 30 questions, where each question has 5 choices and the test taker is simply guessing on all questions, then what's the probability the test taker will get at least one question right?

26. Show that the standardizing formula for a proportion is equivalent to the standardizing formula for a normal approximation to a binomial distribution by dividing the top and bottom of the standardizing formula of a binomial distribution by n.

27. If a normal distribution has a population mean of 20 and a population standard deviation of 6, then find the probability that a) x is between 24 and 32, b) x is less than 17, and c) x is greater than 18.

Just for fun: Fingerprint Identification

Minutiae-based algorithms compare several minutia points (ridge ending, bifurcation, and short ridge) extracted from the original image stored in a template with those extracted from a candidate fingerprint. In one method of fingerprint comparison, only 7(or more) minutiae out of 48 need to match for a person to be considered a suspect for a crime. Assuming the probability of a match for any single minutia to be 6%, find the probability that a person drawn at random is a suspect for the crime.

References

(1) Wikipedia
(2) About.com Pediatrics

Chapter 7

t-Test for Independent Groups with Significance

Objectives

After completing this chapter, a student should:

- ✓ Understand what is meant by inferential statistics
- ✓ Understand the t Distribution
- ✓ Know the eight steps to test a hypothesis
- ✓ Write the decision in a complete sentence in context of the problem
- ✓ Know what is meant by acceptance region and rejection region
- ✓ Be able to apply the normal distribution to a binomial distribution
- ✓ Know what is a type I error and a type II error
- ✓ Know what is meant by the power of the test
- ✓ Know how to compute a confidence interval and find the margin of error
- ✓ Be able to compute a sample size for a specific level of confidence

Inferential Statistics

Inferential statistics has three methods of analysis: 1) Test to discover if a p-value is significant; 2) Compare an obtained value against a critical value; and 3) Find a confidence interval for a population parameter given a sample statistic. We will deal with all three methods of analysis in this text but our main focus will be on whether or not a p-value is significant (method 1).

Recall that a normal distribution curve is bell-shaped and the Central Limit Theorem allows us to use a normal curve for the means of sample sets if the sample set has size larger than 30 data values. If

the sample set has fewer than 30 data values, then we need to use the student's t-distribution.

The derivation of the t-distribution was first published in 1908 by William Gosset (picture) while he worked at Guinness Brewery in Dublin. Gosset, with the help of Karl Pearson, wrote a paper concerning small samples and developing small-sample methods. Because the brewery did not want confidential information leaked to the public, they prevented Gosset from publishing his paper. After convincing the brewery his paper contained no information of practical use to competing brewers, he was allowed to publish under the pseudonym Student. R. A. Fisher, who is responsible for the Fisher F-distribution, referred to Gosset's work as the "Student's distribution." The name stuck so today we refer to Gosset's distribution as the student's t-distribution.

The T-Distribution

SPSS will, by default, always use the student's t-distribution regardless of the size of the data set, since it is in general more accurate. The student's t-distribution is slightly different for each value n, the size of the data sets, but they all have the same bell-shaped curve. The mean will equal the median will equal the mode = 0, the horizontal axis, the t-axis, will be the asymptote and the t value close to ± 1 will still be where the curve changes concavity (the inflection point gets closer to a t value of ± 1 as n gets large). Also the total area under the curve is 1 square area and the curve is symmetric about the mean, which has a t value of 0. See **Figure 1**.

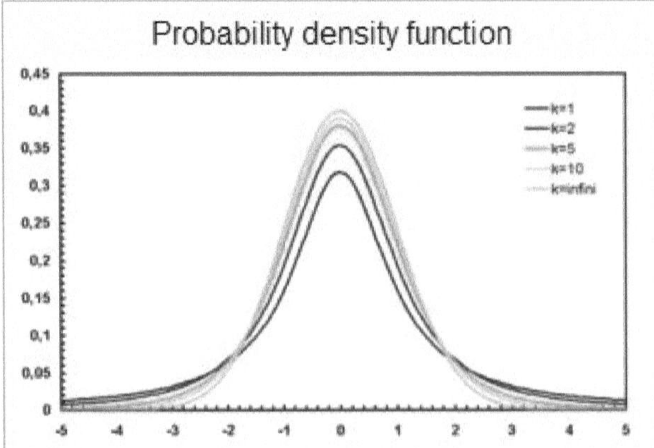

K equals the degrees of freedom

Figure 1

The probability density function is $f(t)=\frac{\Gamma((\nu+1)/2)}{\sqrt{\nu\pi}\Gamma(\nu/2)}\left(1+t^2/\nu\right)^{-(\nu+1)/2}$, where Γ is the Gamma function (the factorial function for all complex numbers). Recall in the normal distribution we had three parameters, x, μ, and σ. Here we only have two parameters, t and ν, called nu. Nu is also called the degrees of freedom and is equal to $n - 1$ (we say $df = n - 1$), note n must be larger than or equal to 2 so that the degrees of freedom will be 1 or greater. Just as for each different value of μ and σ will generate a different normal distribution curve, so will each value of ν generate a different t-distribution curve.

Example 1.

Find the cumulative probability for degrees of freedom of 2, 5, 10, 30, and 100. Use t values of 1, 1.414, 1.732, 2, 2.5, 3, and ∞ (z scores). How do the values correspond to a cumulative normal distribution (z scores)?

Solution: Open a new SPSS data editor and put the numbers 1, 1.414, 1.732, 2, 2.5, and 3 into the first column, call it t. Next use Transform → Compute Variable, call the target variable $p2$, and put CDF.T in the numerical expression. (This can be done using CDF & Noncentral CDF in Function group and Cdf.T in Functions and Special Variables.) The Cumulative Distribution Function for the t variable asks for two parameters, the t values and the degrees of freedom. We will put t in for the t values and 2 for the degrees of freedom; see **Figure 2a**. The data editor should be changed as in **Figure 2b**—be sure to increase the number of decimal positions to 4. Continue adding more variables $p5$, $p10$, $p30$, and $p100$ for the degrees of freedom 5, 10, 30 and 100. The table should look like **Figure 2c**. As a final comparison make a last variable for the Normal Cumulative distribution function (Transform → Compute Variable) and call the target variable pz. Click on CDF & Noncentral CDF and Cdf.Normal, use t for the quant., 0 for the mean, and 1 for the standard deviation, as we did in chapter 6. Notice how the probability values tend to the normal distribution as n gets larger; see **Figure 2d**. Also notice the cumulative probability is smaller for a smaller value of t and vice versa. This implies the area under the tail on the right is larger for smaller values of t, as can be seen in **Figure 1**. This important consequence will be realized later in a section called the Power of the Test. The conclusion is a larger sample size will always increase the power of the test.

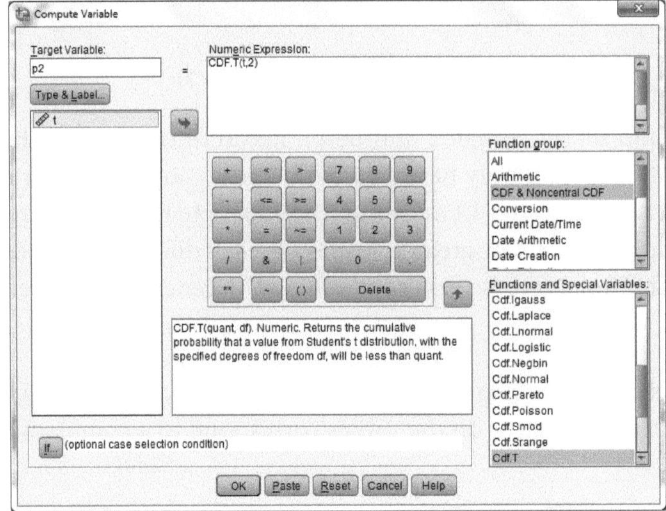

Figure 2a

*Untitled1 [DataSet0] - IBM SPSS Statistics Data Editor

File Edit View Data Transform Analyze Direct Marketing Graphs Utilities Add-ons Window Help

Weight Cases

Visible: 2 of 2 Variables

	t	p2
1	1.00	.7887
2	1.41	.8535
3	1.73	.8873
4	2.00	.9082
5	2.50	.9352
6	3.00	.9523

Data View | Variable View

Weight Cases | IBM SPSS Statistics Processor is ready

Figure 2b

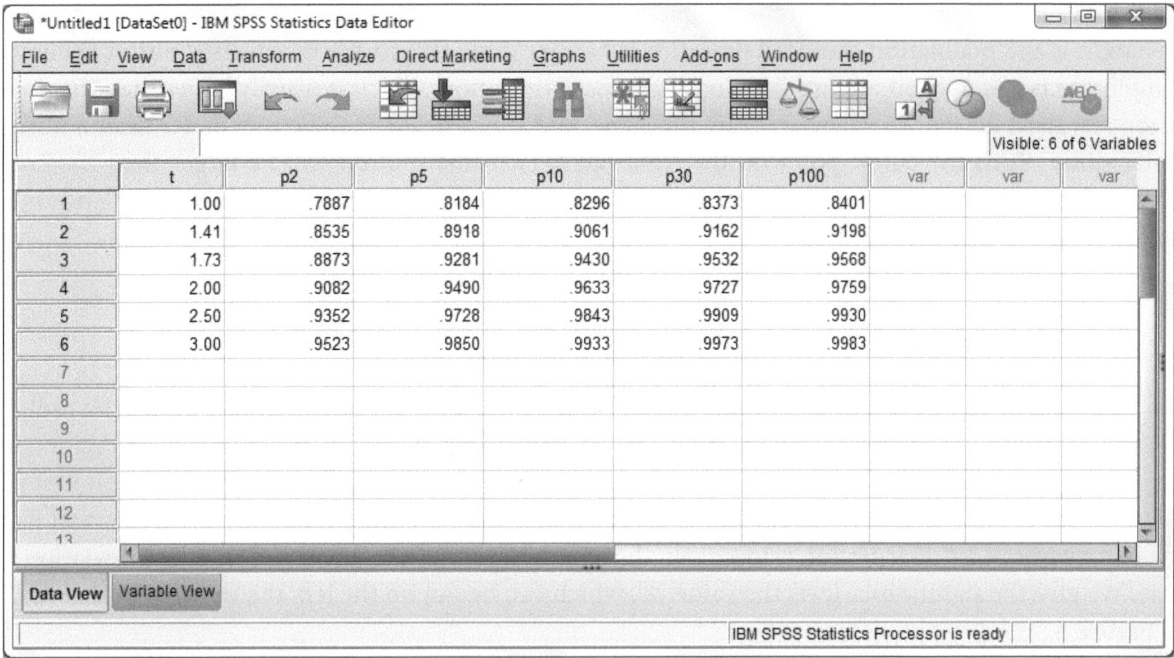

	t	p2	p5	p10	p30	p100
1	1.00	.7887	.8184	.8296	.8373	.8401
2	1.41	.8535	.8918	.9061	.9162	.9198
3	1.73	.8873	.9281	.9430	.9532	.9568
4	2.00	.9082	.9490	.9633	.9727	.9759
5	2.50	.9352	.9728	.9843	.9909	.9930
6	3.00	.9523	.9850	.9933	.9973	.9983

Figure 2c

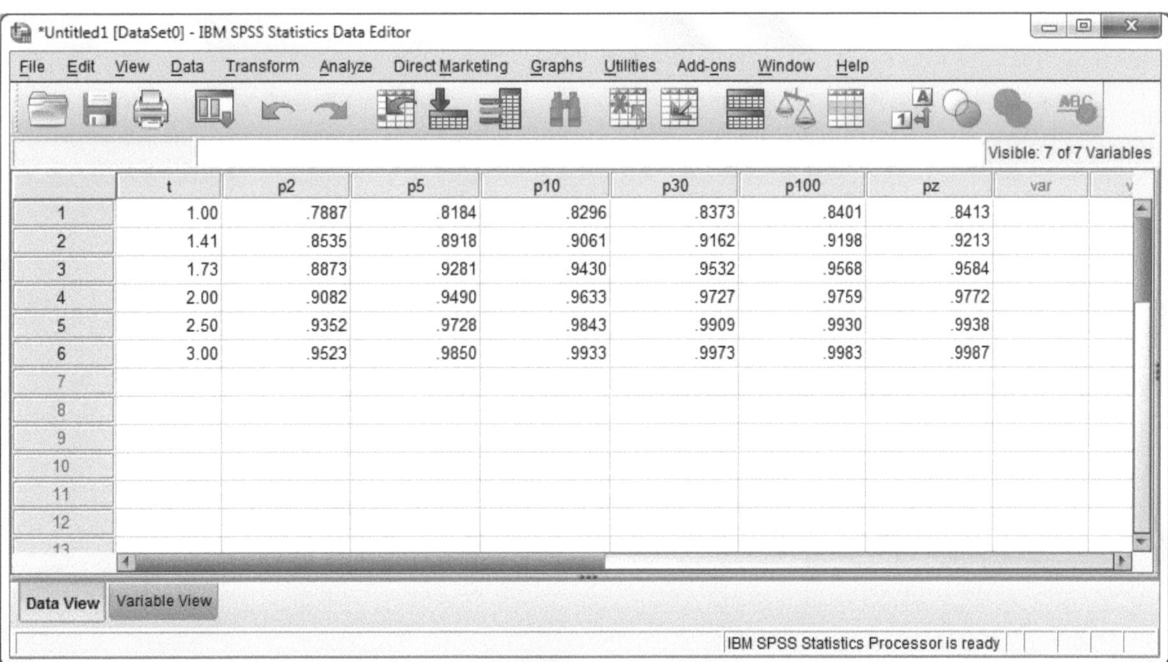

	t	p2	p5	p10	p30	p100	pz
1	1.00	.7887	.8184	.8296	.8373	.8401	.8413
2	1.41	.8535	.8918	.9061	.9162	.9198	.9213
3	1.73	.8873	.9281	.9430	.9532	.9568	.9584
4	2.00	.9082	.9490	.9633	.9727	.9759	.9772
5	2.50	.9352	.9728	.9843	.9909	.9930	.9938
6	3.00	.9523	.9850	.9933	.9973	.9983	.9987

Figure 2d

In a normal distribution, when the sample size is larger than or equal to 30, we assume the standard deviation is the population standard deviation when we draw the curve. In a *t*-distribution, when the sample size is less than 30, we use the sample standard deviation, which is usually larger than the population standard deviation, when we draw the curve and this will make the curve flatter than a normal distribution. Thus the cutoff point on the right tail for a *t*-distribution will be larger than the cutoff point on the right tail for a normal distribution. This makes the *t*-distribution more robust and thus a preferred distribution for testing.

The Significance Region

The significance region is that area under the tail ends of a *t*-distribution where we reject the null hypothesis. Our null hypothesis is always that the test value does not lie in the significance region, and our null hypothesis is always assumed to be true. To test if a statistic is significant we assume it isn't and check to see if the statistic lies in the tail ends of a *t*-distribution. The probability a statistic will lie in the tail regions of a *t*-distribution is low given the assumption that the mean and standard deviation are true. We give the significance level the value α, which can be put on the left, the right, or both sides, as in **Figure 3**.

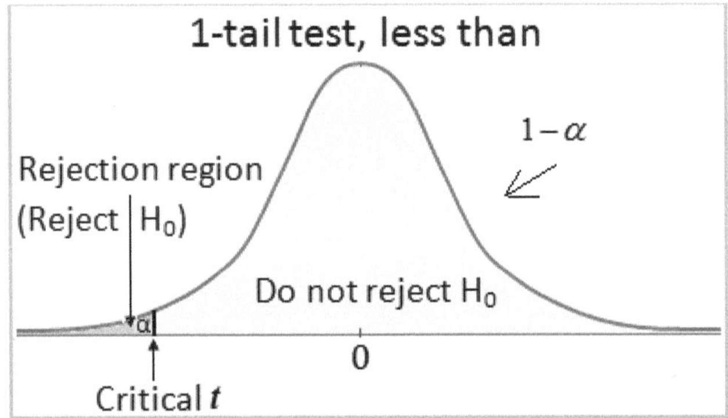

Alternative hypothesis is less than or left tailed

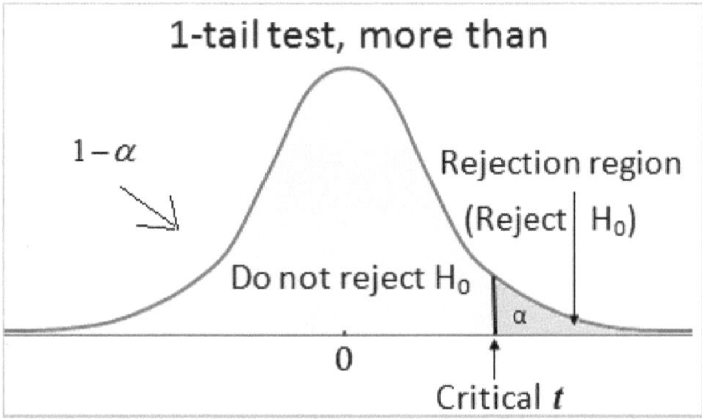

Alternative hypothesis is greater than or right tailed

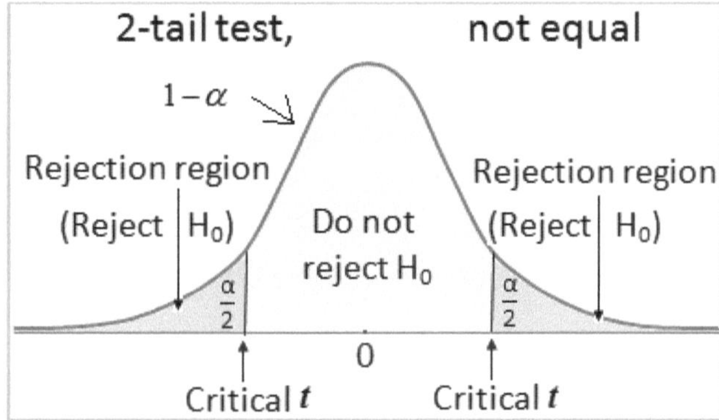

Alternative hypothesis is not equal to or two tailed

Figure 3

The null hypothesis is that the test value is not significant. The alternative hypothesis is that the test value is significant. The null hypothesis is accepted if our test value lies in the region indicated by $1-\alpha$, and we say the test value is not significant. If the test value lies in the region indicated by α, or $\frac{\alpha}{2}$ for a two-tailed test, then we say the test value is significant and we reject the null hypothesis. This does not imply the null hypothesis is false. All we are saying is the test value is significant.

If the alternative hypothesis is two-sided, then α is split in half and the halves are placed on each side of the curve. If the alternative hypothesis is positive, then α is placed on the right side and if the alternative hypothesis is negative, then α is placed on the left side.

Example 2.

Let's test that a person selected at random will have an IQ greater than 125.

Solution: The normal distribution of IQ has the mean of 100 and a standard deviation of 15. To test if a person has an IQ larger than 125 is to find the area (probability) under the normal distribution curve to the right of 125; see **Figure 4**. As we saw previously, the CDF.NORMAL function will give us the answer. Open a new data editor and put 125 into an *x* variable. Use Transform → Compute Variable, call the target variable *a*, and use the function CDF.NORMAL with *x*, 100, and 15 as the parameters, and hit OK. We will get a probability of .9522. This tells us that less than 5% (1 – .9522 = .0478) of the population will have an IQ greater than 125. Thus to pick a person at random and have their IQ greater than 125 is an unusual occurrence. We say if this case happens regularly, then it is significant, which implies the population from which our samples are being culled must have an IQ greater than 100.

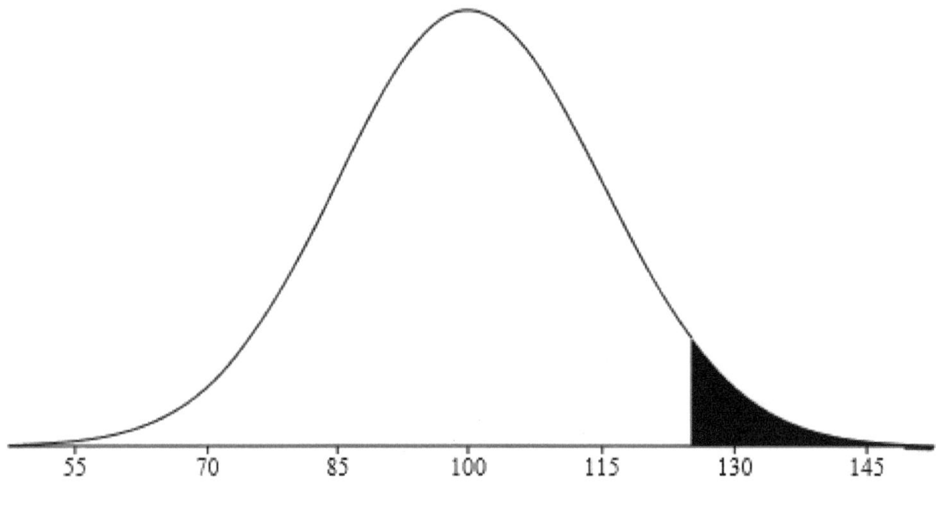

Figure 4

For example if we keep selecting kids at random from a higher-performing school and test their IQs with the result always coming back that they're higher than 125, then we can say with significant probability that the average IQ of students in that school is greater than 100. We can't say it's greater than 125 because we don't know what is the mean IQ of those kids; all we can say is it's statistically significant that the mean IQ is greater than 100. We use the phrase "statistically significant" because there's always the probability that in selecting a random sample, the sample just happens to be one of those rare occurrences where the data values all lie on one side of the mean.

The Larger the Sample Size the More Significant the Statistic

Of course if we increase our sample size, then by the third result of the CLT, the standard deviation of the sample means will get smaller, forcing the sample means toward the true population mean, thereby decreasing the margin of error for our confidence interval and thus making our statistic more significant. This means our estimate will actually be closer to the true population mean for a larger sample size; see **Figure 1**.

T-Distribution Is Used for a Small Sample

In reality we usually can't pick a huge sample; for example, in crash-testing a car, a car company can't crash-test all its cars. There would be no cars left to sell. Furthermore crash-testing a car is very expensive, so the company will crash-test only a few and base their parameters on the statistics of the results of those few cars tested. Thus we see the necessity for the student's t-distribution.

Formula for the T Distribution

The formula SPSS uses to find the t-value is $t = \frac{\overline{x} - \mu}{\frac{s}{\sqrt{n}}}$, where we assume the distribution of x is somewhat of a bell-shaped curve. If we cannot assume the distribution is somewhat bell-shaped, then we must use a sample size larger than 30, because the first part of the CLT implies the distribution of the sample means of samples of size greater than 30 is normal.

One-Sample T-Test and Assumptions

We use a one-sample t-test when we are comparing the mean of one sample against a fixed mean. We assume the sample is randomly drawn for a population that has a somewhat normal distribution.

Example 3.

A car company crash-tested 6 of its cars and recorded the time for full deployment of the airbags. The results were 0.29, 0.32, 0.33, 0.28, 0.35, 0.33 seconds for full deployment of the airbags. Government guidelines say the airbags should be fully deployed within 0.3 seconds. Do the car company airbags meet the government guidelines? (chpt7example3.sav)

Solution: The assumption is the car company's airbags do meet the government guidelines; this is called the status quo. The null hypothesis always states that the status quo is true. The alternative hypothesis is what is being tested. The test is to see if the car company's airbags do not meet the government guidelines. This is called the alternative claim. The alternative claim needs to be established beyond a certain degree of statistical certainty, otherwise we'll stick with the status quo.

Eight Steps for Hypothesis Testing

Wherever statisticians do a test we follow strict guidelines ourselves. We have written them as an eight-step process. They are as follows:

Null Hypothesis

Step 1. The Null Hypothesis is the status quo. In this step we assume that nothing has changed, things are the way they've always been, there's no need for change. This is called the benefit of the doubt. We will give the car company the benefit of the doubt and say their airbags deploy within 0.3 seconds, that is to say, the mean deployment time for all their airbags is 0.3 seconds. In mathematical terms we write $H_0 : \mu = 0.3$. Note: the null hypothesis actually includes 0.3 seconds or less, but we do not include the inequality with the null hypothesis in this text. We always refer the inequality to the alternative hypothesis.

Alterative Hypothesis

Step 2. The Alternative Hypothesis is the challenge hypothesis. Here a government employee might challenge the company's claim that their airbags do indeed meet the specifications that they deploy fully within 0.3 seconds of a crash. This is what's on trial; this is what's always assumed to be false. "The airbags deploy fully within 0.3 seconds of a crash" is assumed to be true and "the airbags do not deploy fully within 0.3 seconds" is assumed to be false. As in our criminal justice system, "the accused is assumed to be innocent" is true and "the accused is assumed to be not innocent" is false. In the criminal justice system, the prosecution is what is on trial. The prosecution needs to prove its case beyond a reasonable doubt that the accused is not innocent. In the alternative hypothesis we want to show beyond reasonable doubt the null hypothesis is false, the reasonable doubt being about 5%. Thus if we can show the mean deployment time for the airbags is in a 5% region of the tail on the right, see **Figure 5a**, then we can conclude the mean deployment time for the airbags is greater than 0.3 seconds. In mathematical terms we write $H_1 : \mu > 0.3$ or some texts might write $H_a : \mu > 0.3$, H_a for the alternative hypothesis. We will use H_1 in this text.

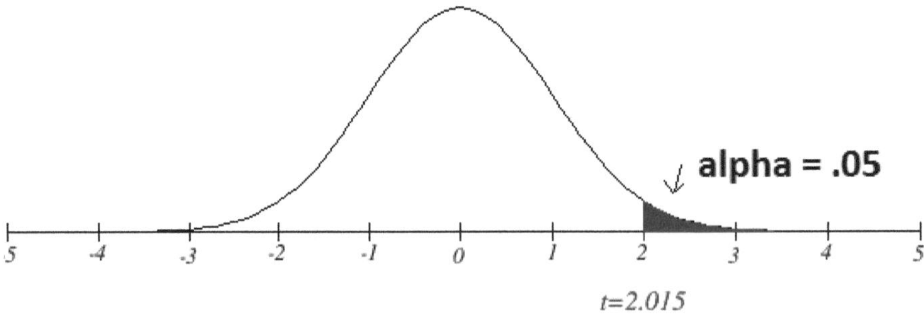

Figure 5a

As noted above some statisticians like to be all-inclusive, so they include inequality with the null hypothesis. That is, if our test value is less than 0.3, then we'll be able to handle that case. We have the cases = 0.3 and > 0.3 covered, but we didn't specify the case < 0.3 seconds. So some statisticians use $H_0 : \mu \leq 0.3$ for the null hypothesis and $H_1 : \mu > 0.3$ for the alternative hypothesis.

We won't include inequalities for the null hypothesis in this text. Our null hypothesis will be equal (=) only, $H_0 : \mu = c$, where *c* is some condition and our alternative hypothesis will be one of these three cases: $H_1 : \mu < c$, $H_1 : \mu > c$ $H_1 : \mu \neq c$; see **Figure 3**. Our decision is usually based on the *p*-value, but we can also compare obtained values against critical values.

Significance Level

Step 3. Set the significance level. The significance level must be set before the test is performed. In most applications the significance level is set at 5%, but we also allow for 1% and 10% significance. In fact we can choose any significance level we like, but we must be considerate of what we call errors, which will be explained later. Unless otherwise instructed, always assume the significance level is 5%. We use the letter α, alpha, for the significance level. In mathematical terms we write $\alpha = .05$.

Decide which Test to Use

Step 4. Decide what test we're going to use and if it's a two-tailed or one-tailed test, and list assumptions. In this case it's obvious we're going to use a one-sample *t*-test since we haven't learned any other test yet, but in later chapters we'll have to decide from a list of several possible tests. This step will be more important later. We can see we have a one-tailed test since that alternative hypothesis is >, not $\neq$. If the alternative hypothesis has $\neq$, then the test is two-tailed and the rejection region could either be in the left tail or the right tail; see **Figure 3**. If the alternative hypothesis has > or < the test is one-tailed. If the symbol > is used then it's a one-tailed test to the right, and the rejection region is in the tail on the right; on the other hand if the symbol < is used, then it's a one-tailed test to the left, and the rejection region is in the tail on the left; see **Figure 5b**.

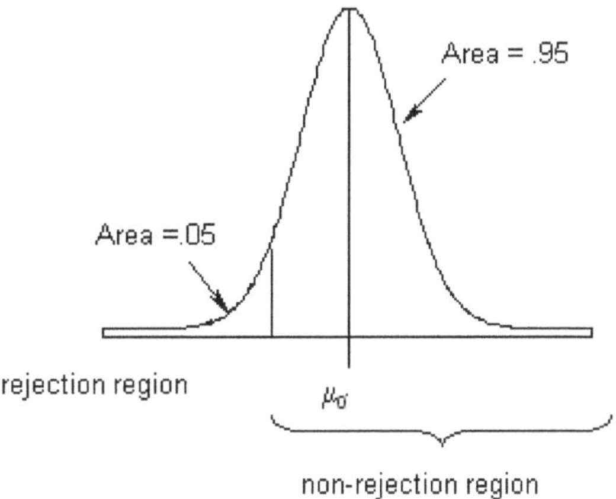

Figure 5b

Find the Obtained Test Value and the p-Value

Step 5. Perform the test. We're going to do a one-sample t-test because we have one sample and we'll test it against 0.3. Open a new SPSS data editor, and input the values 0.29, 0.32, 0.33, 0.28, 0.35, 0.33 into the first column and call it *deploy_time*, or open data set chapter7example3.sav. Use Analyze → Compare Means → One Sample T test, put *deploy_time* in the Test Variable(s) window, and change the Test Value to 0.3; see **Figure 5c**. Hit OK and there should be two output windows, as shown in **Figure 5d**. The first window gives the statistics of our sample, the sample size N is 6, the mean is .3167, the standard deviation is .02658, and the standard error of the mean (the standard deviation for sample means of this type) is .01085.

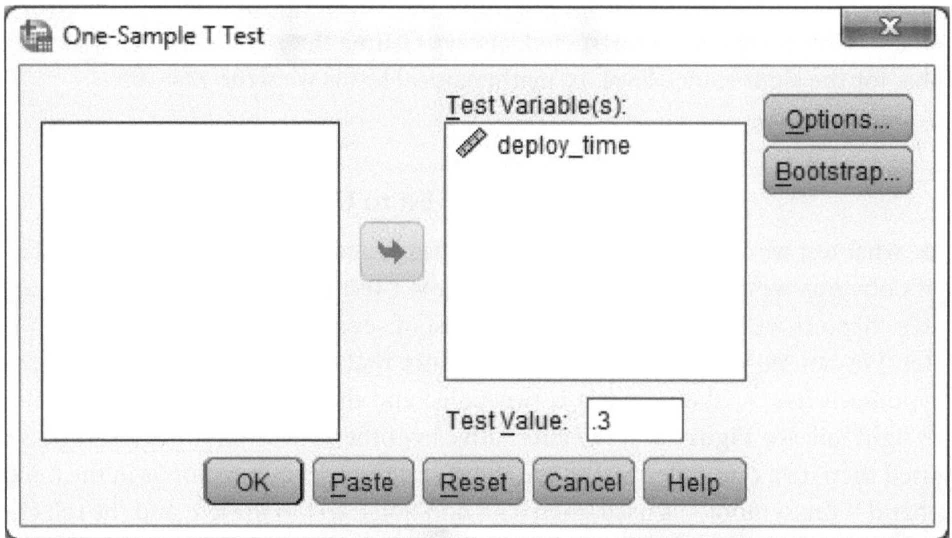

Figure 5c

One-Sample Statistics

	N	Mean	Std. Deviation	Std. Error Mean
deploy_time	6	.3167	.02658	.01085

One-Sample Test

	Test Value = .3					
	t	df	Sig. (2-tailed)	Mean Difference	95% Confidence Interval of the Difference	
					Lower	Upper
deploy_time	1.536	5	.185	.01667	-.0112	.0446

Figure 5d

(Recall the standard error of the mean is the standard deviation divided by the square root of n, the sample size. Check to see if .02658 divided by the square root of 6 is .01085. Also check if the same t-value is obtained by using the formula. It is, we checked. Any differences are due to rounding.)

The second table is what we're interested in, the One-Sample Test. This table will tell us if the data pass the test. Remember our significance level is 5%. The first value is the *t* value, which is 1.536. We can use this *t* value to test against a critical *t* value that we can find using the inverse function, which we'll do below. The next value is ν, the degrees of freedom, which is 5 for this example. Remember df, the degrees of freedom, is $n - 1$. The next value is the significance for this test. SPSS by default will always do a two-tailed test, which means it finds the probability a sample mean will be on either side of the test value at a specified distance. In this example the specified distance is .01667. That is because our sample mean was .3167 and our test value was .3, so SPSS found the probability that a sample mean from a sample of size 6 will be .0167 from the right of 0.3 and .0167 from the left of 0.3 to be .185; see **Figure 5d**. Since we're concerned with only the right side of the distribution, we'll divide .185 by two and get a significance of 0.0925 for this test. This is called the *p*-value; see **Figure 5e**. The last two columns give us a confidence interval for the difference between the true mean and 0.3 based upon this sample. We can be 95% certain the difference between the true mean and the value 0.3 is in the interval [-.0112, .0446].

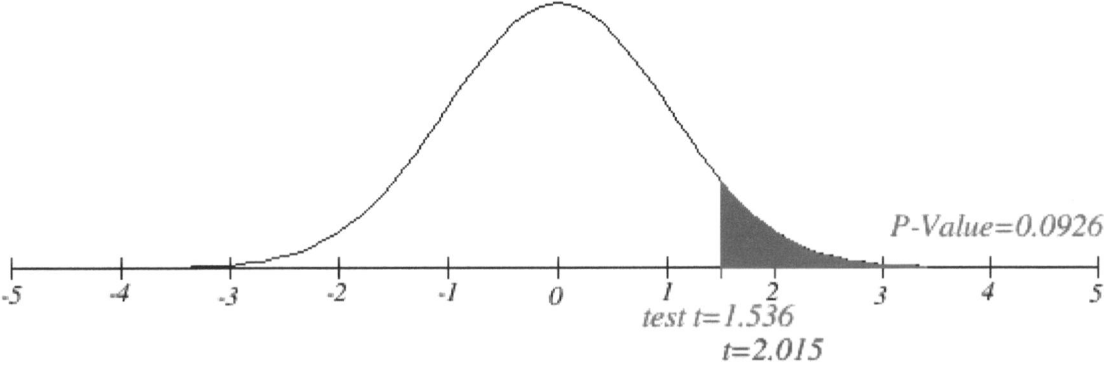

Figure 5e

Find a Critical Value

Step 6. Find a critical value for the test. To find the critical value using SPSS, we first need to find the cumulative probability up to that critical value. Since this is a one-tailed test on the right, the cumulative probability is .95 (1 – .05 = .95). Open a new SPSS data editor, put .95 in the first column, call it *p*, and use Transform → Compute Variable to create the equation *t* = IDF.T(*p*,5); see **Figure 5f**. Hit OK, $t_c = 2.015$; see **Figure 5g**. So for a one-tailed test on the right with α = .05 and *df* = 5, the critical *t* value is 2.015.

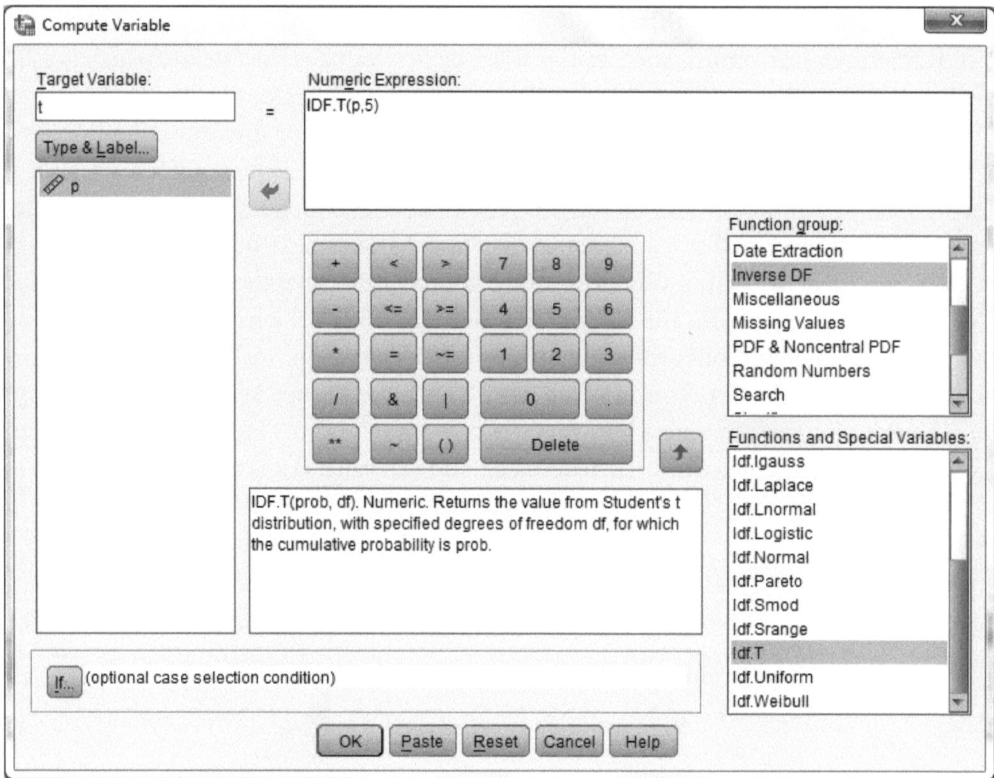

Figure 5f

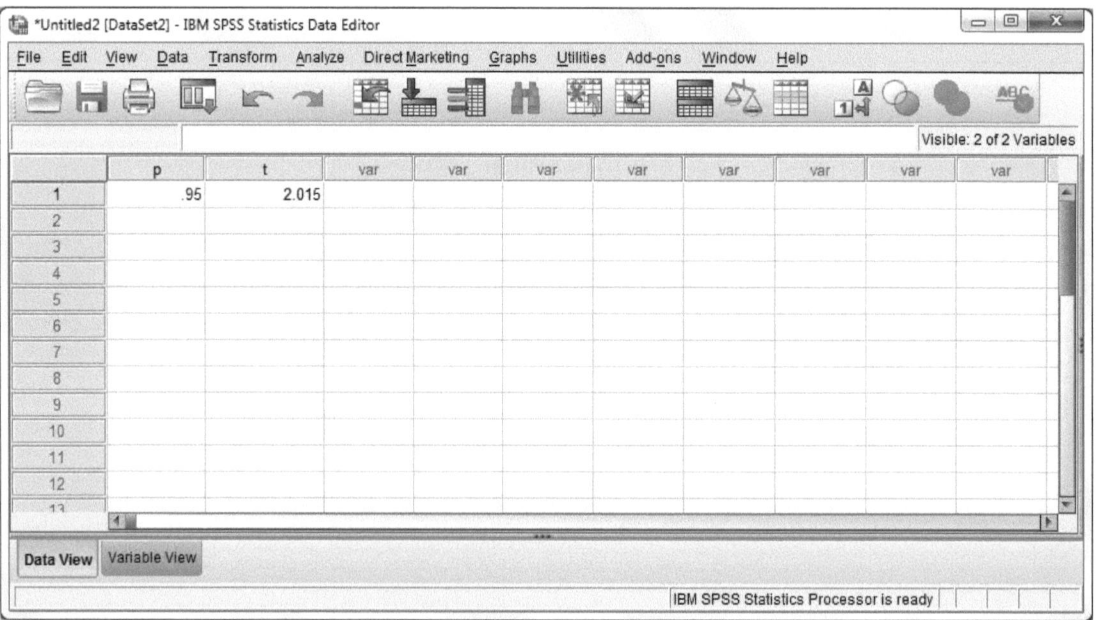

Figure 5g

Compare

Step 7. Compare. We want to compare the obtained t-value against the critical t-value and the obtained p-value against α. The obtained t-value is 1.536 and the critical t-value is 2.015. Since $1.536 < 2.015$, it follows our obtained value is not in the interval below the shaded area in **Figure 5h**. As we can see in **Figure 5e**, our obtained value is to the left of the critical value. Thus our obtained value is in the non-rejection region.

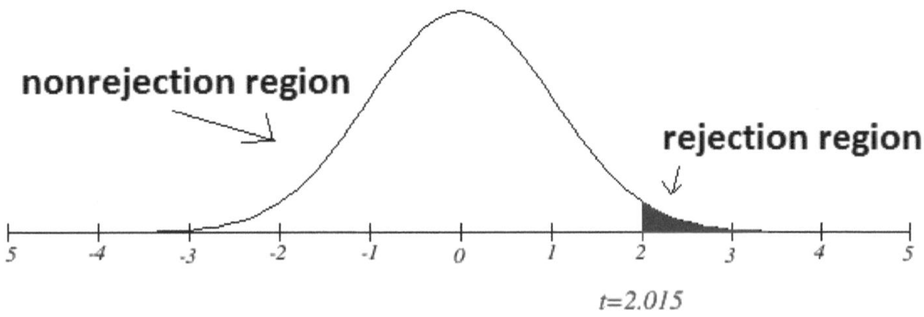

Figure 5h

We also compare our p-value against our significance level α. Our p-value is .0925, which is greater than .05; that is, $.0925 > .05$.

Make a Decision

Step 8. Decision Rule: Whenever the p-value is greater than the significance level, α, and the test is two-tailed or the obtained t-value has the same sign as the critical t-value and the p-value is greater than α then we accept the null hypothesis, that is, the status quo is correct. In the context of our example, our decision is to accept the claim that the car company's airbag deployment is within 0.3 seconds. This decision is determined because the p-value is .0925 where $\alpha = .05$. The p-value is larger than α, our significance level. Even though the mean value for the airbags to be fully deployed is 0.3167 seconds, which is greater than 0.3 seconds, it was not statistically significantly greater than 0.3 seconds, thus it was not in the range where we would reject the null hypothesis.

Alternatively if the t-value lies in the rejection region that was established by finding critical t-values, then we reject the null hypothesis. If the obtained t-value lies in the acceptance region, then we accept the null hypothesis; see **Figure 5h**. Our obtained t-value is 1.536, which is less than 2.015 (see **Figure 5e**) so our obtained t-value lies in the acceptance region. Our conclusion is to accept the null hypothesis, regardless of what really is true.

This is the heart of statistical testing: we test if a value is in the rejection or non-rejection region. We can never be absolutely certain our decision is correct, we can only be certain within a statistical percentage of being correct. There's always a possibility that we made a mistake by accepting or rejecting the null hypothesis.

Type I and Type II Errors

No one likes to make a mistake, but in statistical analysis, there's always room for error. This is because the population parameter may never be known and we can only estimate it with a sample statistic, and since the mean of samples follow a normal distribution, there's always the possibility that a sample mean could be an extreme value. For example, in determining the heights of males worldwide, it could just be that our random selector happened to select only people over 6 feet tall. Highly unlikely, but it could happen. Thus if we were to conclude the average height of males was larger than the real worldwide height of 5' 9", then we would have made a mistake. This is called a Type I error and it is represented by α. This is the probability of saying the null hypothesis is false when in reality it's true. This says we should make a change in the status quo, when in reality, no change is needed. In criminal justice a Type I error would be to convict an innocent person. A Type I error should be avoided whenever possible so we try to make the probability of a Type I error small.

But on the flip side, if we try to minimize the probability of making a Type I error, we may end up saying no change is needed when in reality change is needed. In criminal justice if we raise the bar so high so as not to convict an innocent person, we may end up setting a guilty person free. This is called a Type II error. As we saw in the airbag example above, at the 5% level no change was needed. The statistic did not fall in the 5% region. But if we set our significance level to 10%, then we could conclude a change is needed since $.0925 < .10$. Thus the car company would have to shut down production until it could retool the plant to make the airbags comply with the government guidelines of being fully deployed within 0.3 seconds, not to mention the recalls of all the cars that have already been sold that also need to be fixed to comply with the guidelines. Clearly any change usually requires more effort than to maintain the status quo, so in most cases the rule is not to change unless there's clear evidence change is necessary.

Type I and Type II errors are related in the sense that if we decrease the probability of a Type I error, then we increase the probability of making a Type II error; see **Figure 6**.

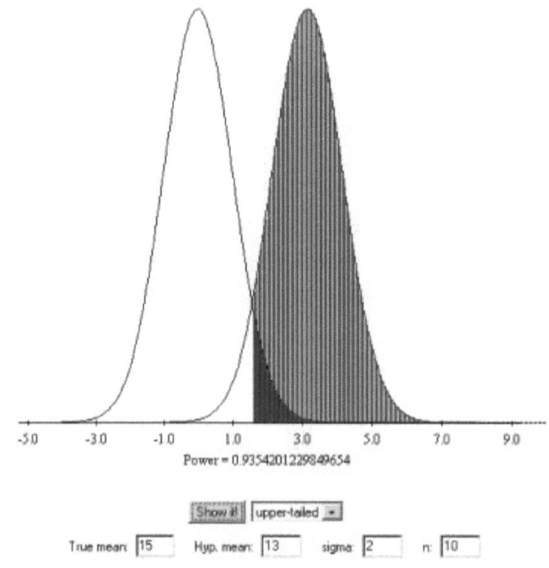

Figure 6 The curve on the left represents the null hypothesis, the curve on the right represents the alternative hypothesis. The darker area of the overlap is α and the clear area of the overlap is β. The clear area under the null hypothesis is $1 - \alpha$ and the shaded area under the alternative hypothesis is $1\text{-}\beta$.

In **Figure 6** the null hypothesis is on the left and the alternative hypothesis is on the right (the distribution curves represents the null and alternative hypotheses). The critical value is the line dividing the shaded area from the clear area. The shaded area under the left curve is α and the clear area under the right curve is β. The probability of a Type II error is assigned the letter β, beta. If we move the dividing line to the right, then α gets smaller and β gets larger. If we move the dividing line to the left, then α gets larger and β gets smaller. Thus we see if we try to reduce the probability of one type of error, then we increase the probability of the other type of error, so we need to be careful where we set out limits. This is one reason we always set α equal to 5%.

A synthesis of the types of errors is given in the following table.

Table 1

		Reality	
		Null Hypothesis is really true	Null Hypothesis is really false
Decision	Accept the Null Hypothesis	Good Decision No error is made, probability = $1-\alpha$	Bad Decision Type II error is made, probability = β
	Reject the Null Hypothesis	Bad Decision Type I error is made, probability = α	Good Decision No error is made, probability = $1-\beta$

In any test the null hypothesis has the advantage of being assumed as correct. The null hypothesis is never on trial. The alternative hypothesis is what's on trial and must be shown to be a better choice than the null hypothesis. However, we can never say the alternative hypothesis is correct—we can never make conclusions regarding the new statistic. All we can say is the null hypothesis appears to be no longer valid. Thus, in our decision, we either accept the null hypothesis or we fail to accept the null hypothesis, which is the same as "reject the null hypothesis"; we'll never say we accept the alternative hypothesis. We can only accept the alterative hypothesis after a series of many studies of similar samples give us a new mean that we can then call the new status quo. In the meantime our decision is always in terms of the null hypothesis. (Since the alternative hypothesis is what's on trial, some authors will use the terminology "fail to reject the null hypothesis," which is the same as "accept the null hypothesis." This terminology correctly implicates the alterative hypothesis as what's under review. That is, if the alternative hypothesis appears to be a better choice, then we reject the null hypothesis, whereas if there appears to be no difference between the alternative and the null hypotheses, we fail to reject the null hypothesis. We will use the terminology "accept the null hypothesis" and "reject the null hypothesis" in this text.)

Confidence Intervals

A confidence interval will give us an idea of where the new mean might lie. In particular if the old mean is included in our confidence interval, then we can't really say the new mean is any different from the old mean. Thus if our confidence interval for the sample mean includes the mean of the null hypothesis, then this is another method by which we can accept the null hypothesis.

Example 4.

Find a 90% confidence interval for the data in Example 3. Also find an 80% confidence interval.

Solution: Since the alpha level for the test in Example 3 was 5% and was a one-tailed test, then our confidence interval will have to be 90% so that we have 5% in each tail; see **Figure 7a**. SPSS by default gives a 95% confidence interval. Use Analyze → Descriptive Statistics → Explore, put deploy_time in the Dependent List, click on the Statistics tab, and change the Confidence Interval for Mean to 90%. Hit continue and OK and the output should read as in **Figure 7b**. Note the 90% confidence interval for the mean Lower Bound is .2948 and the Upper Bound is .3385. Since 0.3 is included in this range, we can say the null hypothesis is correct, and there is no need to make a change. Go back to the confidence interval and change it to read 80%, 10% in each tail. Now what is the new CI? The answer is [.3006, .3327]. Since 0.3 is not in this interval, we can conclude that if the significance level was 10%, then the null hypothesis is not correct. Finding the confidence interval is another way of testing the null hypothesis.

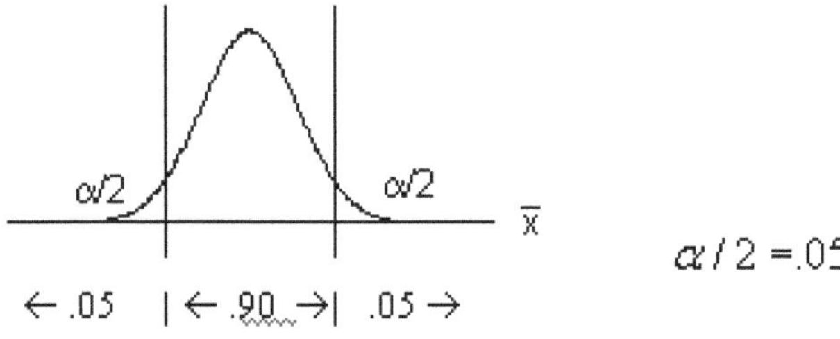

Figure 7a

Descriptives

			Statistic	Std. Error
deploy_time	Mean		.3167	.01085
	90% Confidence Interval for Mean	Lower Bound	.2948	
		Upper Bound	.3385	
	5% Trimmed Mean		.3169	
	Median		.3250	
	Variance		.001	
	Std. Deviation		.02658	
	Minimum		.28	
	Maximum		.35	
	Range		.07	
	Interquartile Range		.05	
	Skewness		-.422	.845
	Kurtosis		-1.188	1.741

Figure 7b

Power of the Test

The power of the test is found in $1-\beta$ and it tells us the probability of rejecting a false null hypothesis. In criminal justice a false null hypothesis would be the accused is guilty, and rejecting a false null hypothesis would be to convict a guilty person. If we accept a false null hypothesis we let a guilty person go free. If we reject a false null hypothesis we convict a guilty person. Just as much as we do not want to convict an innocent person we also don't want to let a guilty person go free. To prevent that from happening, we prefer the power of the test to be large. We can increase the power of the test in two ways. As we saw in **Figure 6**, moving the critical line to the left will increase α and decrease β. As β gets smaller, $1-\beta$ will get larger, and hence the power of the test will increase. Recall if β gets small, then α gets large, which implies that it is more likely a Type I error will be made. Since Type I errors are usually less desirable than Type II errors, we like to keep α small. But if we keep α small, then β is large, and the power of the test goes down. How do we increase the power of the test without adjusting α or β? The answer is to increase the sample size.

The power of the test cannot be readily determined because β cannot be readily determined until we have many samples from which to draw a new statistic. But we can increase the power of a test by increasing the sample size, which is better than trying to adjust α or β. Since a larger sample size decreases the error of the mean, then the boundary value for α will be pulled toward the mean, thereby decreasing the value of β and increasing the power of the test; see **Figures 8a, 8b, 8c**.

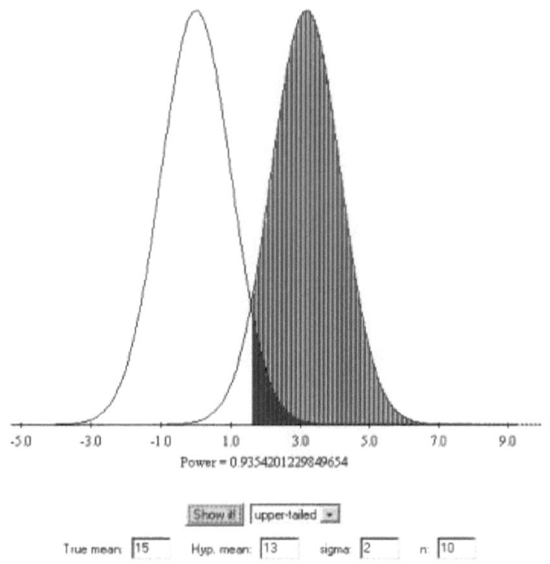

Figure 8a

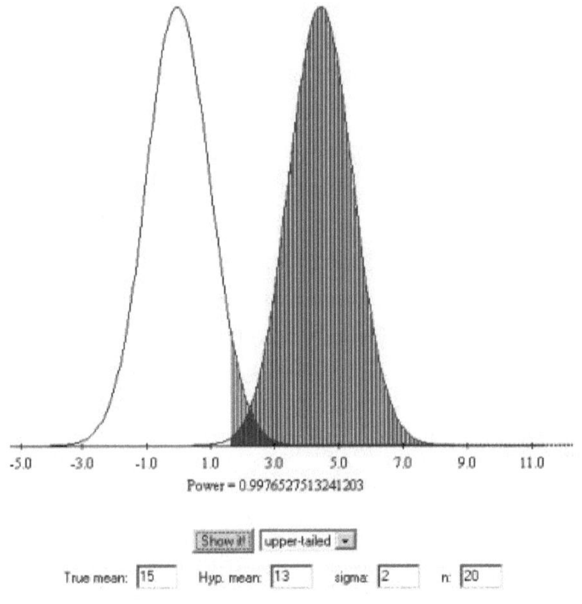

Figure 8b

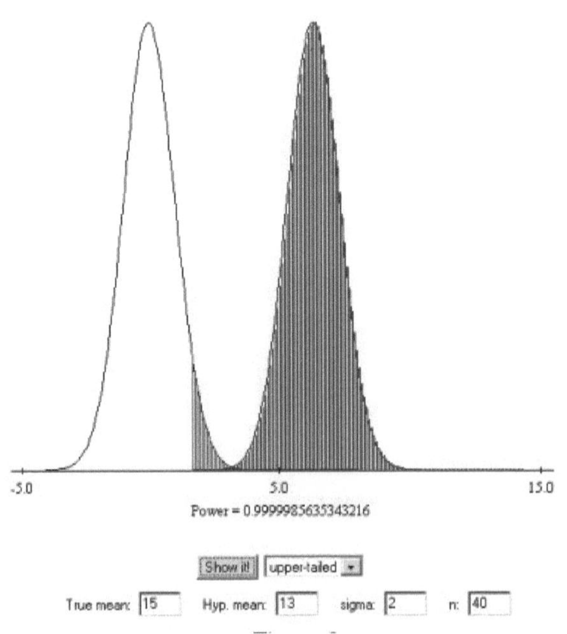

Figure 8c

In each case the curve on the left is the null hypothesis and the curve on the right is the alternative hypothesis. The shaded area under the curve on the left represents α and the shaded area under the curve on the right represents 1 - β. Notice 1 – β gets larger as n gets larger.

If we reject the null hypothesis and the power is strong (large), then we can feel good about our decision and not worry that we may have made a mistake. But if we reject the null hypothesis and the power is weak (small), then our decision may be somewhat circumspect since we may have rejected a true null hypothesis. Remember making a Type I error is considered the less desirable error. It is better to let a guilty person go free than it is to convict an innocent person.

In criminal justice the best way to convict a guilty person is to have more evidence. A subject who is brought to court on minimal evidence will most likely be set free regardless of guilt. But a subject who is brought to court with a preponderance of evidence will have a harder time establishing innocence and will most likely be convicted. This demonstrates that increasing the evidence, sample size, will increase the probability of convicting a guilty person. That is, increasing the sample size increases the power of the test, which increases the probability of rejecting a false null hypothesis.

Test a Proportion

If we want to test a proportion, we use the normal distribution, provided both *np* and *nq* are both greater than 5. The test value that we use is $z = \frac{\hat{p} - p}{\sqrt{\frac{pq}{n}}}$. In this equation $\hat{p}$ is the sample proportion and p is the population proportion, the proportion for the null hypothesis. We assume the sample proportion should equal the population proportion. The complement proportion is q, where $q = 1 - p$, and n is the sample size. The standard deviation is $\sqrt{\frac{pq}{n}}$ and this is the standard deviation for the distribution of sample proportions. As n gets very large, $\sqrt{\frac{pq}{n}}$ gets small and therefore if a difference exists between p and $\hat{p}$, then the z value will be significant. Again this reaffirms the importance of sample size.

When we use SPSS to find probability or critical z-values, we will have to use the arithmetic function SQRT in the numeric expression. We'll also use p as the population mean. Do not confuse the letter p, the population proportion, with the letter p in p-value. They represent different concepts but use the same letter.

Example 5.

Suppose a manufacturer claims 3 out of 5 consumers prefer the manufacturer's brand. Now suppose a county fair held a blind test comparing the manufacturer's brand against many other brands and only 63 out of 117 consumers preferred the manufacturer's brand. Will these results support the claim of the manufacturer?

Solution: This is the same as Example 7 in Chapter 6. We found in that example that the probability that a proportion of 63 or less out of 117 will occur, assuming 3 out of 5 prefer the manufacturer's brand, is 10.30%. This is how the test will run:

Step 1. $H_0 : p = .6$ **Null Hypothesis.** P is the true proportion of consumers who prefer the manufacturer's brand.

Step 2. $H_1: p \neq .6$ **Alternative Hypothesis.** Unless we know otherwise, we always assume the test will be two-tailed; that is, we don't know if the alternative hypothesis will be greater than or less than.

Step 3. $\alpha = .05$ **Significance level.** Set the level of significance; if no level is given, always set it to .05.

Step 4. **Determine the test statistic.** Decide what test to use. In this case we will use the normal distribution to do a test of proportion.

Step 5. **Perform the test.** Use SPSS to find the test statistic or the p-value. The cumulative probability found in Example 7 of Chapter 6 was .1030. Since that is the left tail of the normal distribution curve and this is a two-tailed test, we double that value to get .2060. (To find the area .1030, put 63.5 into a variable called x. Create a new variable called *propor* via Transform → Compute Variable, put *propor* in target variable, and put $x/117$ in the numeric expression. The area is found by using Transform → Compute Variable, call the target variable a, and put CDF.NORMAL(*propor*, .6, SQRT(.6*.4/117)) in the numeric expression.)

The test statistic is $z = \dfrac{\hat{p} - p}{\sqrt{\dfrac{pq}{n}}}$. We can find the z-value using Transform → Compute Variable. Put z in Target Variable and use IDF.NORMAL(a, 0,1) in the numeric expression. The test statistic z value is -1.264376.

Step 6. **Find a critical value(s)** to compare the test statistic against. This step can be effectively eliminated if we always find the p value. However, the critical z value from Table 2 in Chapter 6 is ± 1.96. We use $\pm$ because this is a two-tailed test.

Step 7. **Compare** the p value against the significance level, .2060 > .05. Alternatively compare the z value -1.264376 against the critical z value ± 1.96: -1.96 < -1.264376 < 1.96.

Step 8. **Decision.** Whenever the p value for a two-tailed test is greater than the significance level, we accept the null hypothesis. If the p value is less than the significance level, then we reject the null hypothesis. In this case we accept. Alternatively -1.96 < -1.264376 < +1.96, which says the test z value is in the acceptance region; see **Figure 9**.

Conclusion. Always write the conclusion in a complete sentence in context or the problem. In this example our conclusion will read as such: "There is not enough evidence to reject the claim that 60% of consumers will prefer the manufacturer's brand." In other words, the results support the claim of the manufacturer.

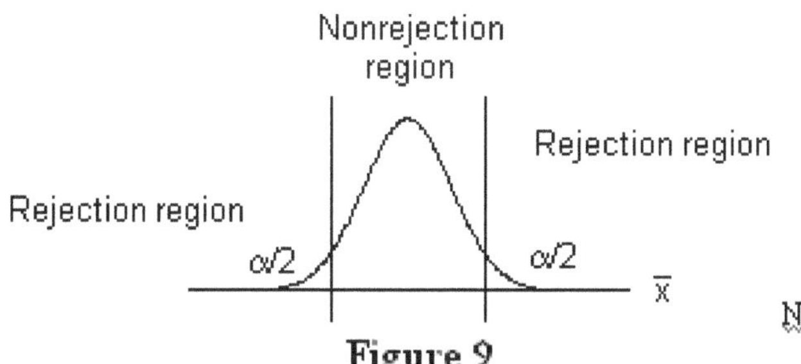

Figure 9

Nonrejection region = Acceptance region

Test a Difference of Proportions

We will next use SPSS to test difference of means and difference of proportions. The formulas for difference of proportions are $z = \dfrac{\hat{p}_1 - \hat{p}_2}{\sqrt{\dfrac{\overline{pq}}{n_1} + \dfrac{\overline{pq}}{n_2}}}$, where $\hat{p}_1 = \dfrac{r_1}{n_1}$, $\hat{p}_2 = \dfrac{r_2}{n_2}$, and $\overline{p} = \dfrac{r_1 + r_2}{n_1 + n_2}$, $\overline{q} = 1 - \overline{p}$. We can find the r_1, r_2, n_1, n_2 values by selecting cases.

Example 6.

Open the data set Colleges (1). We would like to see if the proportion of the number of applicants accepted to the number of applicants received is the same for both public and private colleges. For example, only 22% of the applicants to Stanford University in California, a private college, were accepted, while 77% of the applicants to the University of California at Riverside, a public college, were accepted. On the other hand, 42% of the applicants to the University of California at Berkeley, a public college, were accepted, whereas 90% of the applicants to Woodbury University in California, a private college, were accepted. We would like to question if the ratios of applications accepted to applications received are different for public or private colleges.

Solution: To find r_1, r_2, n_1, n_2, we need to select cases. Use Data → Select Cases, click on the "If condition is satisfied" button and then click on the "If" tab; see **Figure 10a**. Click on the nominal variable Public/Private School and move that into that window and set it equal to 1; see **Figure 10b**. We also want to eliminate the cases where the number of applicants received is not given and the number of applicants accepted is not given, since otherwise our calculations will be thrown off, so we put & appsacc > 0 & appsrec > 0 in the window also; see **Figure 10c**. Hit continue and then OK. Now in the data editor a new variable titled filter_$ will appear with the values *selected* or *not selected*, which correspond to the numbers 1 and 0 respectively. Then those schools that are private or those schools that don't have a report for the Number of Applicants accepted or Number of Applicants received are crossed off; see **Figure 10d**. Now we can find r_1 and n_1. Use Analyze → Descriptive Statistics → Frequencies, deselect display frequency tables, move the variables Number of Applicants received and Number of Applicants accepted into the variable window, click on Statistics, check Sum, and hit continue and OK. You should get the output shown in **Figure 10e**. This says the number of applicants received is 2098862 and the number of applicants accepted is 1466847 for public schools. Thus $\hat{p}_1 = \frac{1466847}{2098862} \approx .699$. Similarly we can find the proportion of the number of applicants accepted to the number of applicants received for private schools if we change pubpvt = 2 in Data → Select Cases → If Condition is Satisfied → If; see **Figure 10f.** Now we find the sum of Number of Applicants received and Number of Applicants accepted as above and we should get the output window as seen in **Figure 10g.** This says the number of applicants received is 1452635 and the number of applicants accepted is 945242 for private schools. Thus ^ $= \frac{945242}{1452635} \approx .651$. We are now ready to perform the test.

Step 1. $H_0 : p_1 - p_2 = 0$ or alternatively $H_0 : p_1 = p_2$ (We always prefer the first equality.)

Step 2. $H_1 : p_1 - p_2 \neq 0$ or alternatively $H_1 : p_1 \neq p_2$

Step 3. $\alpha = .05$

Step 4. We will use a *z* test for the difference of proportions, since the sample sizes are very large.

Step 5. Find the test value. Before we do that we need to find $\overline{p}$ and $\overline{q}$. $\overline{p} = \frac{1466847 + 945242}{2098862 + 1452635} \approx .679, \overline{q} \approx .321.$ $z = \frac{.699 - .651}{\sqrt{\frac{.679 \cdot .321}{2098862} + \frac{.691 \cdot .321}{1452635}}} \approx 95.26$, this is very large, the *p*-value is 0.

Step 6. Find the critical *z* score. Since this is a two-tailed test and the significance level is .05, then the critical *z*-score from Table 2 in Chapter 6 will correspond to the value with area to the left of .9750 because the area to the right will be .025. Or put .975 into a new variable and use the inverse function in SPSS to get the critical *z*-value; see **Figure 10h.** The critical *z* score is ± 1.96.

Step 7. Compare. Our obtained value is much greater than 1.96; see **Figure 10i.** Alternatively the *p*-value of 0 is less than .05.

Step 8. Conclusion. We will fail to accept the null hypothesis; there appears to be significant evidence that the proportion of applicants accepted to applicants received is different for public schools and private schools. (There is a higher probability of being admitted into a public school as compared to being admitted into a private school.)

Why is the *z* value so significant when the proportions, 70% compared to 65%, don't seem to be that much different? The reason is that the sample size was so huge. As the sample size gets bigger, the sample proportion becomes closer to the true proportion. Thus any differences, no matter how small, will be significant. However, as we saw above, some private schools have a higher proportion of admits than public schools and vice versa. So the question might be, Is there really a difference between the proportions of individual schools? If we treat the schools as independent of one another, we might get a very different result. This is our next example.

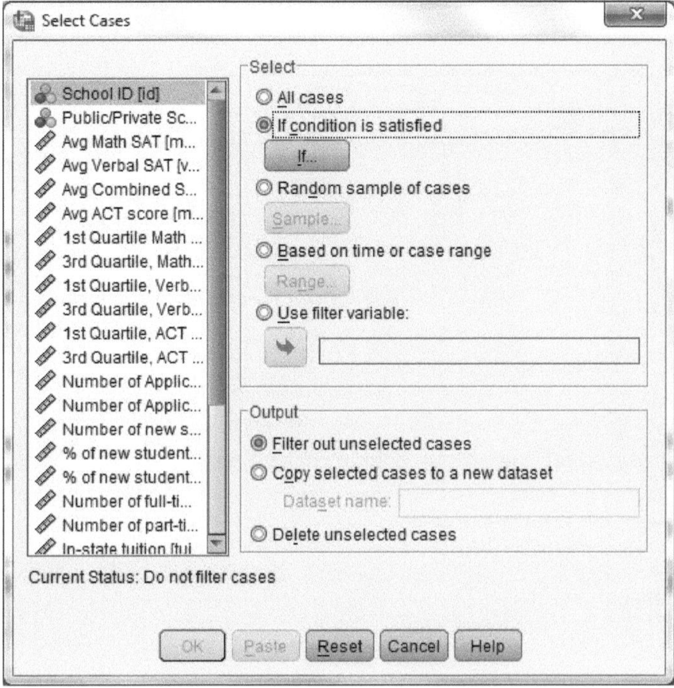

Figure 10a

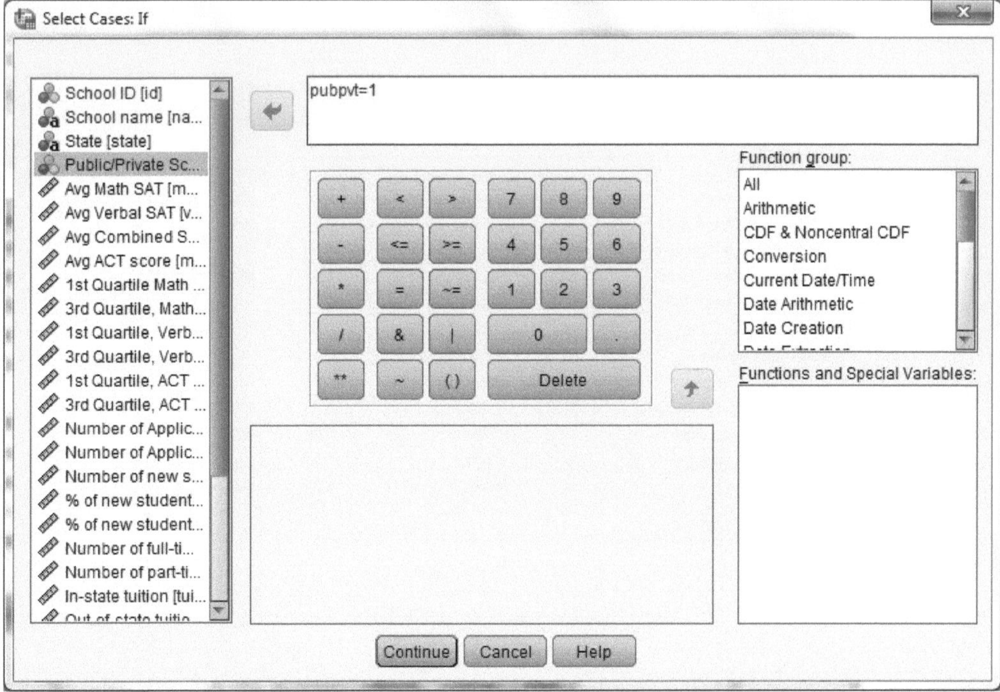

Figure 10b

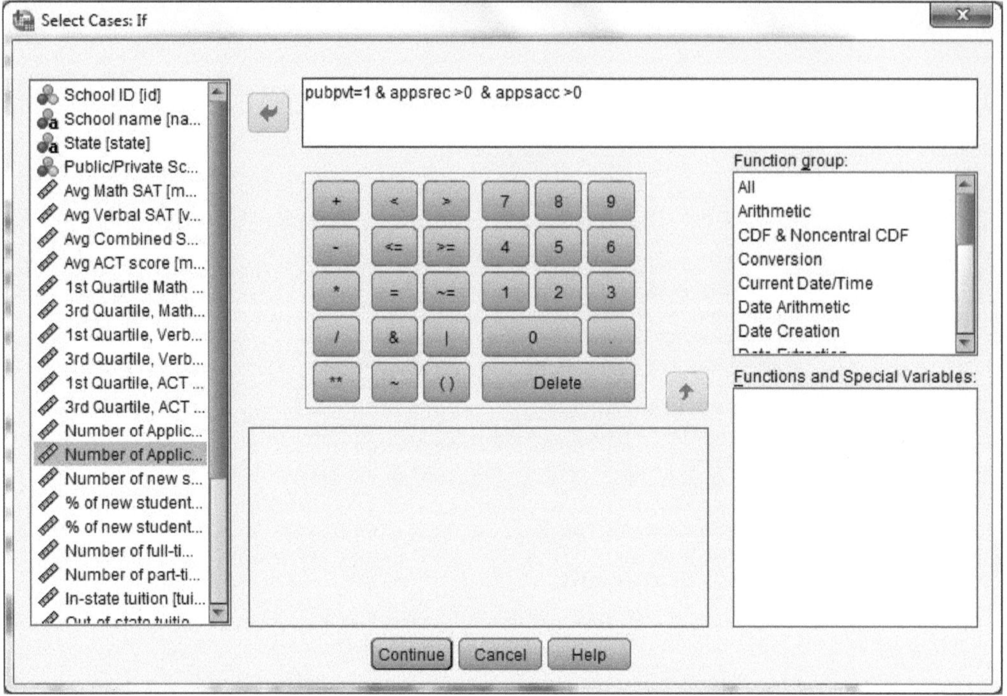

Figure 10c

*Colleges.sav [DataSet1] - IBM SPSS Statistics Data Editor

File Edit View Data Transform Analyze Direct Marketing Graphs Utilities Add-ons Window Help

Visible: 31 of 31 Variables

	id	name	state	pubpvt	mathsat	verbsat	combsat	meanact	msatq1	msatq3
1	1061	Alaska Pacific University	AK	Private	490	482	972	20	440	530
2	1063	University of Alaska at Fairba	AK	Public	499	462	961	22	.	.
3	1065	University of Alaska Southeast	AK	Public	.	.	.	.	.	.
4	11462	University of Alaska at Anchor	AK	Public	459	422	881	20	.	.
5	1002	Alabama Agri. & Mech. Univ.	AL	Public	.	.	.	17	.	.
6	1003	Faulkner University	AL	Private	.	.	.	20	.	.
7	1004	University of Montevallo	AL	Public	.	.	.	21	.	.
8	1005	Alabama State University	AL	Public	.	.	.	.	.	.
9	1009	Auburn University-Main Campus	AL	Public	575	501	1076	24	520	638
10	1012	Birmingham-Southern College	AL	Private	575	525	1100	26	470	680
11	1016	University of North Alabama	AL	Public	.	.	.	.	.	.
12	1019	Huntingdon College	AL	Private	513	446	959	23	480	570

Data View | Variable View

IBM SPSS Statistics Processor is ready | Filter On

Figure 10d

Statistics

		Number of Applications received	Number of Applications accepted
N	Valid	461	461
	Missing	0	0
Sum		2098862	1466847

Figure 10e

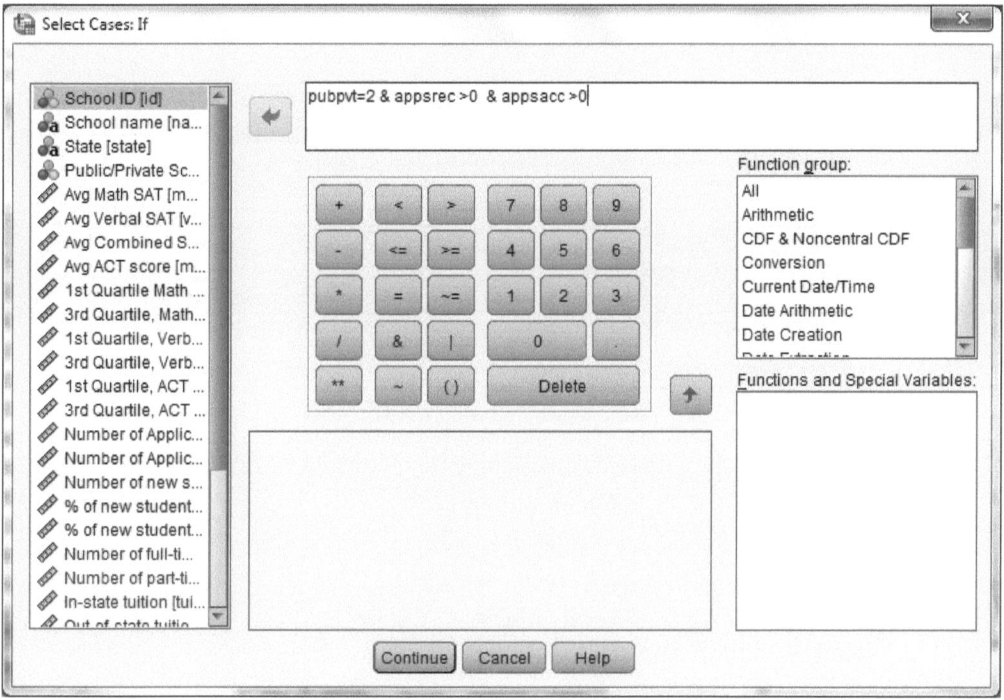

Figure 10f

Statistics

		Number of Applications received	Number of Applications accepted
N	Valid	828	828
	Missing	0	0
Sum		1452635	945242

Figure 10g

*Untitled2 [DataSet2] - IBM SPSS Statistics Data Editor

File Edit View Data Transform Analyze Direct Marketing Graphs Utilities Add-ons Window Help

Visible: 2 of 2 Variables

	a	cz	var	var	var	var	var	var	var	var
1	.975	1.96								
2										
3										
4										
5										
6										
7										
8										
9										
10										
11										
12										

Data View Variable View

IBM SPSS Statistics Processor is ready

Figure 10h

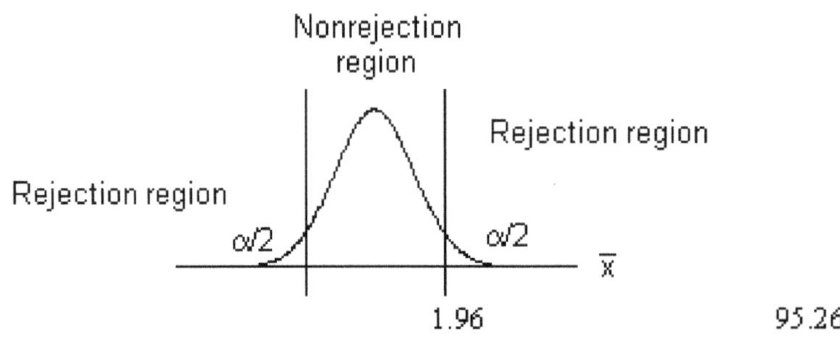

Figure 10i

Test of Two Independent Samples and Assumptions

Two sets are independent if they have no effect on each other. Such as in the public and private schools, the ratio of applicants accepted over applicants received for one public school will have no effect on the ratio of applicants accepted over applicants received for a private school. In that sense the two sets of ratios can be thought as independent of each other. As in the one-sample test we require both samples be 1) independent, 2) randomly drawn, 3) from a somewhat normal population, and 4) the population variances for both samples are equal. SPSS will compute both an equal variance test and a non-equal variance test to test if the variances are equal. As we'll see in Chapter 9, ANOVA, assuming equal variances may give a false conclusion.

Example 7.

Use a two-sample *t* test, the two samples being the public and the private schools, to see if the mean of the proportions of applicants accepted to applicants received of individual schools is different between the two samples.

Solution: If the data set Colleges is still open, then deselect cases by using Data → Select Cases, click on All cases (see **Figure 11a**), and then OK (otherwise skip this step). Now we will create a new column for proportions. Use Transform → Compute Variable, call the target variable *propor*, and put appsacc/appsrec in the Numeric Expression; see **Figure 11b**. Hit OK.

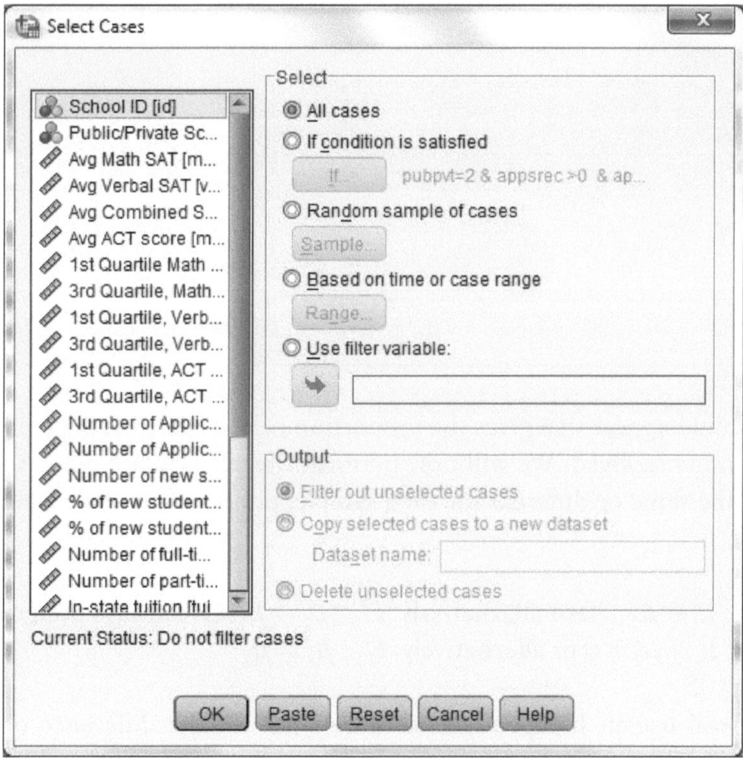

Figure 11a

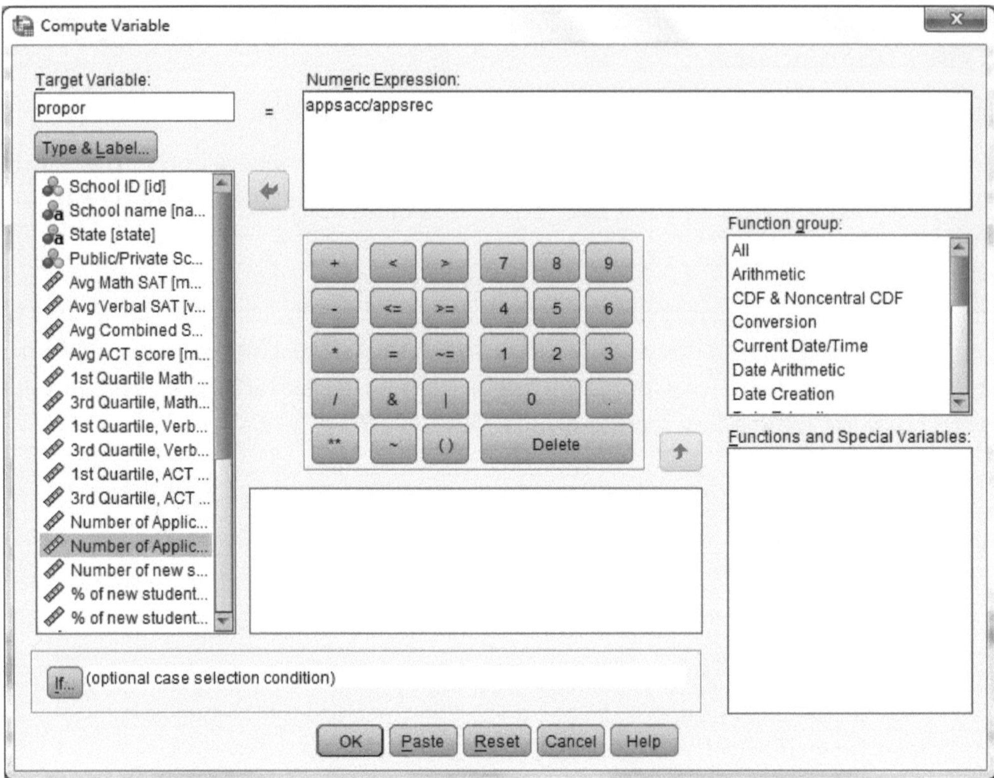

Figure 11b

A new column should appear that gives the proportion of number of applicants accepted divided by the number of applicants received. We will now perform the test. We are going to test if the means of the proportions are the same or different for each sample, public and private schools. The steps are as follows:

Step 1. $H_0 : \mu_1 - \mu_2 = 0$ or alternatively $H_0 : \mu_1 = \mu_2$ (We always prefer the first equality.)

Step 2. $H_1 : \mu_1 - \mu_2 \neq 0$ or alternatively $H_1 : \mu_1 \neq \mu_2$

Step 3. $\alpha = .05$

Step 4. We will use an Independent Samples *t*-test for the difference of means. We assume the populations are normal with equal variances and the samples are independent and randomly drawn.

Step 5. Find the obtained *t*-value and the *p*-value. Use Analyze → Compare Means → Independent-Samples T test. Put *propor* in the Test Variable window (be sure *propor* is a scale variable) and *pubpvt* in the Grouping Variable window (*pubpvt* is a nominal variable). Remember we're looking at the mean of Proportions between the two groups, public school and private school. SPSS will ask us to define the groups. Recall Private is defined as 2 and Public is defined as 1. Define the groups 1 for public, and 2 for private. See **Figure 11c**. Click on the tab Define Groups and put 1 for group 1 and 2 for group 2; see **Figure 11d**. Click continue and hit OK and this will be our output; see **Figure**

11e. Since Levene's Test for Equality of Variances has Sig .201, which is greater than .05, we can assume equality of variances. The t value is -.352, df is 1287, significance level is .725, the mean difference between the two Proportions is -.00326, and the 95% confidence interval is -.02146 to .01493.

Step 6. Since df is very large, $df = 1287$, then we can use ± 1.96 as the critical t values. These values are the critical z values that correspond to a .975 area to the left of the critical z value; see Table 2. They are also the boundary values for a 95% confidence interval for a normal distribution, which is the same as a 5% two-tailed test for a normal distribution.

Step 7. The p-value of .725 is greater than .05, or alternatively the t-value of -.352 is in the interval [-1.96, 1.96], which is the acceptance region. Also the 95% confidence interval for the difference of the sample means is [-.02146, .01493], which includes 0, implying we can be 95% confident the difference of our sample means is no different than 0.

Step 8. We accept the null hypothesis; there is no difference in the mean proportion of applicants accepted over applicants received for both public and private colleges.

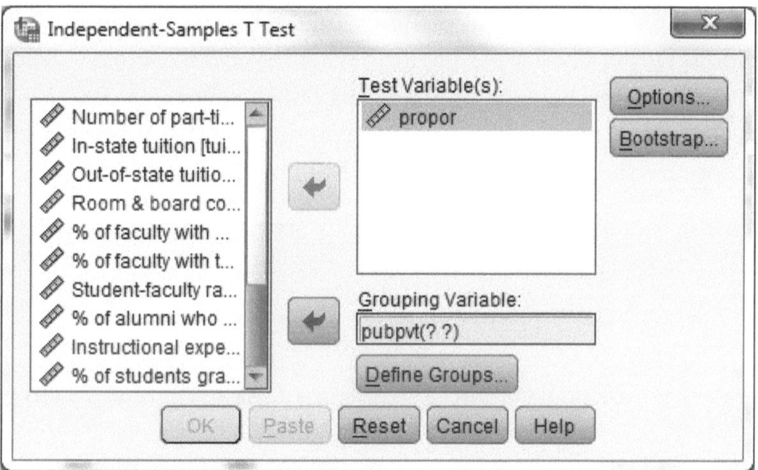

Figure 11c

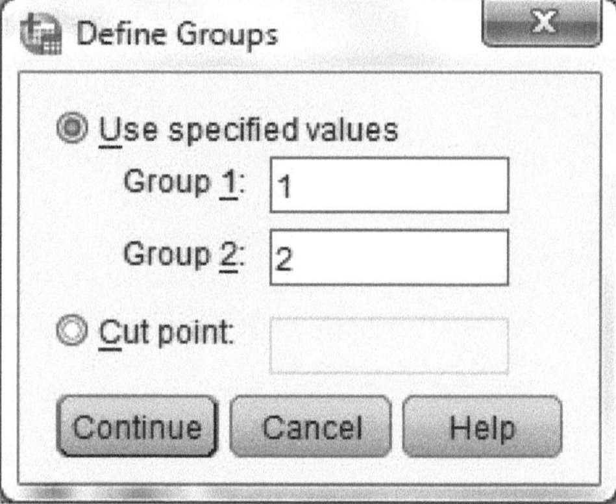

Figure 11d

Independent Samples Test

		Levene's Test for Equality of Variances		t-test for Equality of Means						
									95% Confidence Interval of the Difference	
		F	Sig.	t	df	Sig. (2-tailed)	Mean Difference	Std. Error Difference	Lower	Upper
propor	Equal variances assumed	1.638	.201	-.352	1287	.725	-.00326	.00928	-.02146	.01493
	Equal variances not assumed			-.349	930.691	.727	-.00326	.00934	-.02160	.01507

Figure 11e

The result in Example 7 is drastically different from the result in Example 6. The reason for the differences is that in Example 6, we took one proportion from all the private schools and one proportion from all the public schools, thereby wiping out all the variances of the ratios between the schools. Because the sample size is large, any small difference between proportions is meaningful. In Example 7 we took the ratios of each school and then found the mean of the ratios of the private schools and the mean of the ratios of the public schools. In this case the means of the ratios were very close for both groups, and the variances were assumed equal. By looking at the individual ratios for each school instead of the ratio of the sum of all the schools, we get a better picture as to whether or not there is a difference between being accepted into a private school and being accepted into a public school.

In Example 7 we used the ratios applicants accepted/applicants received to create a new variable and had SPSS compare the means for group 1, Public schools, to group 2, Private schools. The formula SPSS used is $t=\dfrac{\overline{x}_1-\overline{x}_2}{\sqrt{\dfrac{s_1^2}{n_1}+\dfrac{s_2^2}{n_2}}}$, if equal variances are not assumed and $t=\dfrac{\overline{x}_1-\overline{x}_2}{s\sqrt{\dfrac{1}{n_1}+\dfrac{1}{n_2}}}$, where

$s=\sqrt{\dfrac{(n_1-1)s_1^2+(n_2-1)s_2^2}{n_1+n_2-2}}$ if equal variances are assumed. As an exercise you might want to check the equation using the mean and standard deviation from the data set to make certain the *t*-value is correct. (It is, we checked.) Note: Any slight differences between the calculator results and the software (SPSS) results are due to the fact the software uses slightly different formulas for *df* and hence the *t* values and confidence intervals from SPSS are usually smaller than the values the calculator will give. The formula SPSS actually uses to compute *d.f.* is given by Satterthwaite's formula $df=\dfrac{\left(\dfrac{s_1^2}{n_1}+\dfrac{s_2^2}{n_2}\right)^2}{\dfrac{1}{n_1-1}\left(\dfrac{s_1^2}{n_1}\right)^2+\dfrac{1}{n_2-1}\left(\dfrac{s_2^2}{n_2}\right)^2}$. The calculator results are more conservative in the sense that they produce larger confidence intervals and *t* values.

t-Test for Small Samples

A *t*-test for a small sample requires the population to have somewhat of a normal distribution. We will usually make that assumption, but if we have the population data, we can always check to see if it is somewhat normal by making a histogram.

We'll do an example for a *t*-test of independent small samples and we'll check the answers using the above formulas. In case we ever want to find the standard deviations using our calculator and perhaps making a chart with the sum of *x* and the sum of x^2, here's a shortcut formula for finding the variance: $s^2 = \frac{1}{n-1}\left(\sum x^2 - n(\overline{x})^2\right)$. As an extra-credit problem, try to derive this formula from the one given in Chapter 4.

Example 8.

Recall in Chapter 1 Example 3b we were interested in the scores of preschool children after interacting with different types of toys. A test was performed giving the following results. Group A was students interacting with nondescript toys, such as stuffed animals, toy trucks, etc. Group B was given educational toys from a company that specializes in making educational toys, matching objects, working with letters, colors, etc. After a period of time, 3 weeks, the same ten-question preschool test was given to all the kids in both groups; the test was to draw a line between two pictures that had the same number of objects, identify different colors, recognize letters of the alphabet, draw polygon figures, etc, and their scores are as follows:

Group A:	7	5	9	1	6	4	7	8	9	2	4	5	3	8	9	1
Group B:	6	9	5	8	1	3	9	8	6	7	2	3	9	5	4	

We would like to test if there is a difference between the means of group A and group B. Chpt7example8.sav

Solution: Open a new SPSS data editor, or open chpt7example8. Group A has 16 data values and group B has 15 data values. Put 1 in column 1 for 16 values and 2 in the same column for 15 values. The data editor should look like **Figure 12a**. You can change the values by making 1 A and 2 B and change the measure to Nominal. In the next column put the data values associated with each of the groups, call it score, and lower the decimals to 0. The data editor should now look like **Figure 12b**. Note these figures are scrolled down to show some values from both groups.

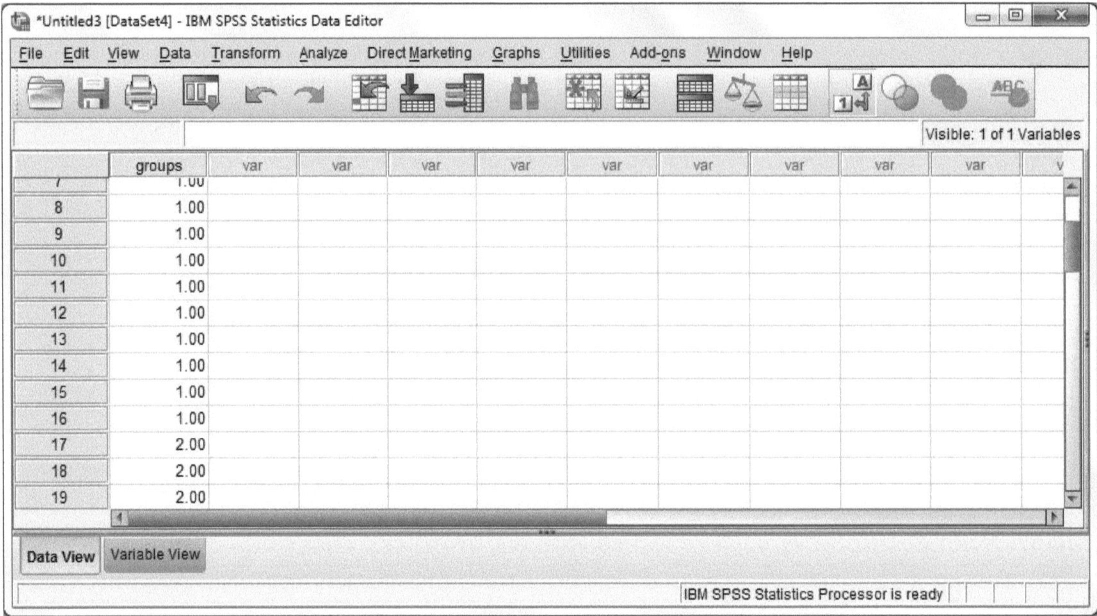

Figure 12a

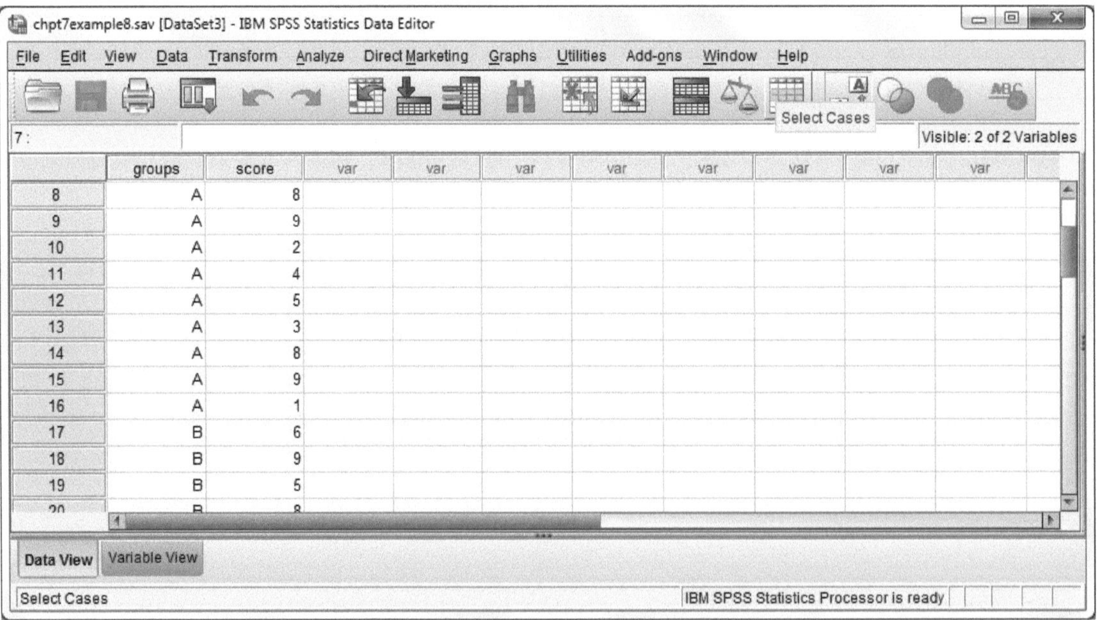

Figure 12b

Now we can perform the test. Whenever we do a test we always use the eight-step process.

Step 1. What is the null hypothesis? The status quo is it doesn't matter which toys kids play with, they are at an age where their mental development will be enhanced by imagination regardless of how their environment is constructed. They don't have the ability to decipher between academic and nonacademic interactions. Playing with one toy will have no different impact on their mental development than from playing with another toy. $H_0 : \mu_A - \mu_B = 0$

Step 2. What's the alternative hypothesis? The toy company thinks kids do pick up subtle clues from their environment. A positive academic environment will lead to a positive academic mental growth. They claim the mean score of the kids playing with their educational toys will be higher.

$H_1 : \mu_A - \mu_B < 0$ Recall if $\mu_B > \mu_A$, then $\mu_A - \mu_B < 0$.

Step 3. Set the significance level. The toy company is so sure that their educational toys will make a difference in the ten-question preschool test that they are willing to set the significance level at 1%.

$\alpha = .01$

Step 4. Choose the test. We will use a left-tail *t* test for independent samples and assume the samples are random and independently drawn from somewhat normal populations. We will also assume variances are equal.

Step 5. Obtain the test value. We'll use Analyze → Compare Means → Independent Samples T Test, put scores into the window that says Test Variable(s), and groups into the window that says Grouping Variable. Define Groups as Group 1 is 1 and Group 2 is 2. Hit Continue and OK. The output should be as in **Figure 12c**. Since the significance level for equality of variances is .755 > .05, we can assume equality of variances, thus the *t* value is -.169 and the significance, *p*-value, is .867/2 = .434. Remember this is a one-tailed test so we need to divide the sig. (2-tailed) by 2, the mean difference is -.167, and the 95% confidence interval is [-2.179, 1.846], which includes 0.

Step 6. Obtain the critical *t* value. To do this we'll put .01 (this is a left-tail test) into a new SPSS data editor, call it *p*, use Transform → Compute Variable, call the target variable *ct*, and use IDF.T(*p*,29) to obtain a *t* value for a one-tailed test on the left with a *df* of 29 and a significance level of .01. Recall $df = n_1 + n_2 - 2$, hence 16 + 15 – 2 = 29. The critical *t*-value is -2.462.

Step 7. Comparing our obtained value -.169 with -2.462, we see -.169 > -2.462, which tells us our obtained *t* value is in the acceptance region, not the rejection region; see **Figure 12d**. Alternatively the obtained *p*-value is .434, which is greater than .01. Either way we will accept the null hypothesis. Also the 95% confidence interval is [-2.179, 1.846], which contains 0, which implies we're 95% confident the difference in means is no different from 0.

Step 8. Conclusion. We will accept the null hypothesis; there is not enough evidence to suggest that these educational toys will have any effect on the mental development of preschool children. They are too young to recognize academic patterns in their world.

Independent Samples Test

		Levene's Test for Equality of Variances		t-test for Equality of Means					95% Confidence Interval of the Difference	
		F	Sig.	t	df	Sig. (2-tailed)	Mean Difference	Std. Error Difference	Lower	Upper
score	Equal variances assumed	.099	.755	-.169	29	.867	-.167	.984	-2.179	1.846
	Equal variances not assumed			-.170	28.993	.866	-.167	.982	-2.175	1.842

Figure 12c

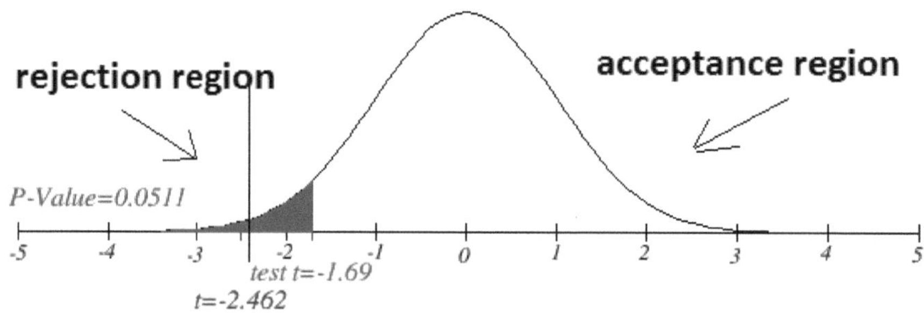

Figure 12d

Using the formulas above we have $t = \dfrac{5.5 - 5.67}{\sqrt{\dfrac{15 \cdot 2.805^2 + 14 \cdot 2.664^2}{29} \cdot \left(\dfrac{1}{16} + \dfrac{1}{15}\right)}} \approx -.172769$ for equal variances assumed and $t = \dfrac{5.5 - 5.67}{\sqrt{\dfrac{2.805^2}{16} + \dfrac{2.664^2}{15}}} \approx -.173$ for equal variance not assumed. Actually SPSS will always do a test on the variances as well as the t test. The test variable for variances is called the F ratio, after Fisher who first studied them, and the significance level is given in the second column; see **Figure 12c**. We can assume equality of variances as long as the significance level is above .05, which it was for both of the previous examples. Again note the t-values obtained by the calculator are slightly higher than the t values obtained by SPSS. This is again due to the fact SPSS uses a slightly different formula to compute *df*; see note following Example 7.

Example 9.

One important aspect of data entry is the ability to enter correct keystrokes per minute. Is there a difference between genders in the ability to enter correct keystrokes per minute? The following data, correct keystrokes per minute, were obtained:

Male:	43	56	32	45	36	48			
Female:	41	63	72	53	68	49	51	59	60

Solution: Enter the data as instructed above or open chpt7example9.sav. Perform the 8-step test process.

Step 1. $H_0 : \mu_M - \mu_F = 0$

Step 2. $H_1 : \mu_M - \mu_F \neq 0$ This is a two-tailed test because we have no idea which gender can enter more correct keystrokes.

Step 3. $\alpha = .05$

Step 4. This will be a two-tailed Independent Samples T Test. Assume the samples are random and independently drawn from somewhat normal populations, and also assume variances are equal.

Step 5. Since Levene's Test for Equality of Variances has a significance of .582, which is greater than .05, we can assume equality of variances. The output is shown in **Figure 13a**. These are the obtained values. Using the formulas and a calculator will give the same values.

Step 6. Since *df* = 13, the critical *t* value is ± 2.16. To find the critical *t* value, open a new data editor and enter .025 and .975 into the first column. (Remember we split .05 in half and put half on each side of the distribution.) Call the variable *p* for probability. Use Transform → Compute Variable, call the target variable *ct* and use IDF.T(*p*, 13). The data editor is shown in **Figure 13b**.

Step 7. The obtained *t* value is -2.845, which is less than -2.160. Alternatively the obtained *p*-value is .014, which is less than .05. Alternatively the 95% confidence interval for the difference of the means is [-24.631, -3.369], which doesn't contain 0.

Step 8. From all of the above we can conclude to not accept the null hypothesis. There is too much evidence to suggest the difference in the mean number of correct keystrokes per minute is due to something other than chance. Males and Females appear to have a different number for the number of correct keystrokes per minute.

Independent Samples Test

		Levene's Test for Equality of Variances		t-test for Equality of Means						
									95% Confidence Interval of the Difference	
		F	Sig.	t	df	Sig. (2-tailed)	Mean Difference	Std. Error Difference	Lower	Upper
tapspeed	Equal variances assumed	.319	.582	-2.845	13	.014	-14.000	4.921	-24.631	-3.369
	Equal variances not assumed			-2.927	11.865	.013	-14.000	4.784	-24.436	-3.564

Figure 13a

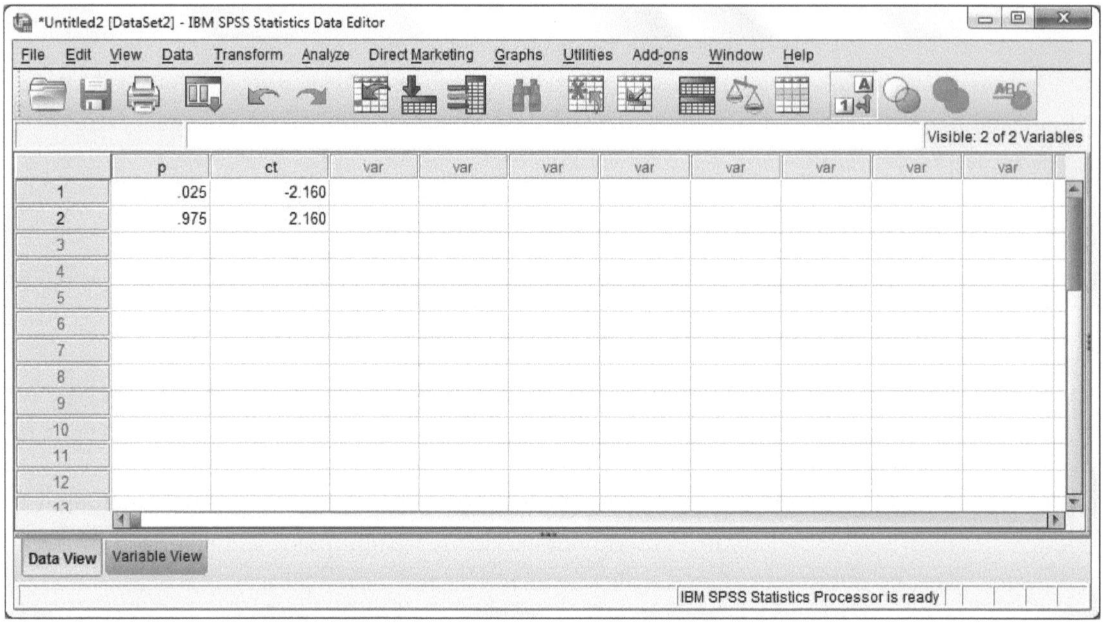

*Untitled2 [DataSet2] - IBM SPSS Statistics Data Editor

	p	ct
1	.025	-2.160
2	.975	2.160

Figure 13b

In this example we have a negative value for the difference of means so we might be tempted to say females can hit keystrokes more accurately than males, but that would be wrong. Before we do the test we must first make our choices as to what will be our null and alternative hypotheses. Since we decided in the beginning, Step 2, this would be a two-tailed test, our conclusion can be in the context of only a two-tailed test. If we originally decided to make this a one-tailed test, then our conclusion can state in which direction the results point. The point is we can conclude only what we initially stated.

Margin of Error

When we find the confidence interval, we also find what is called the margin of error. The margin of error is half the length of the confidence interval. There are different formulas for the margin of error depending on which type of test is being conducted. For a large-sample z test, the margin of error is $E = z_c \frac{\sigma}{\sqrt{n}}$, where z_c is the critical z value (see Table 2, Chapter 6), σ is the population standard deviation (which can be replaced by the sample standard deviation), and n is the sample size. A similar formula can be found for the test of a proportion: $E = z_c \sqrt{\frac{pq}{n}}$.

Calculation of Sample Size

In both these formulas we can solve for n and thus find the sample size needed to have a margin of error within a prescribed range. Solving for n in the first equation, we have $n = \left(\frac{z_c \sigma}{E}\right)^2$, and solving for n in the second equation, we have $n = pq\left(\frac{z_c}{E}\right)^2$. Since $q = 1 - p$ we can also write this formula as $n = p(1-p)\left(\frac{z_c}{E}\right)^2$. If we treat p as a variable, then this equation is similar to the equation $y = x - x^2$. From calculus (or algebra) we know this function has a maximum value of $\frac{1}{4}$, when $x = \frac{1}{2}$. If we substitute $p = \frac{1}{2}$, $z_c = 1.96$ for a 95% confidence interval, and $E = .03$, a 3% margin of error into the equation $n = p(1-p)\left(\frac{z_c}{E}\right)^2 = .25\left(\frac{z_c}{E}\right)^2$, we get n = 1067.11 ≈ 1,068. When we solve for n we always round up. This tells us the maximum size for a random sample to be 95% confident that the true proportion is within 3 percentage points is 1,068 data values. That is why whenever there is a poll listed in the media it always states the sample size as being around 1,068 and the sample statistic has a 95% confidence of being correct within 3 percentage points of the true population statistic. Therefore if one candidate is at 49% and another candidate is at 46%, then statistically speaking it's a dead heat.

Let's review what we've learned in this chapter.

1. Inferential Statistics
    - i.) Compare an obtained value against a critical value
    - ii.) Compare a p-value against a significance level
    - iii.) Compare a confidence interval with a population mean
2. The eight-step process in performing a test
    - i.) Null Hypothesis, the status quo, there is no difference in means
        - a.) The null hypothesis is always equality between means, when testing means.
    - ii.) Alternative Hypothesis, there is a difference in means, it can be positive, negative, or not equal.
        - a.) The alternative hypothesis must be stated in declarative form, it is not a question
        - b.) The alternative hypothesis must be testable
        - c.) The alternative hypothesis does not postulate causes
    - iii.) Significance level, $\alpha = .05$ unless otherwise stated
    - iv.) Choose the type of test—1-tailed or 2-tailed—and assumptions
    - v.) Perform the test using SPSS to find obtained values
    - vi.) Find a critical value for comparison with the obtained value
    - vii.) Compare the obtained t-value with the critical t-value and compare the obtained p-value with the significance level. If the test is 2-tailed or if the test is 1-tailed and the obtained value is on the side indicated in step 2 then the following two rules apply.

Rule 1. If the obtained p-value is greater than the significance level, we accept the null hypothesis

Rule 2. If the obtained p-value is less than the significance level, we fail to accept the null hypothesis

vii.) Make a decision with regard to the comparison in part vii. Write the conclusion in complete sentences in context of the problem.

3. Types of tests
   i.) z tests for normal populations
   ii.) z tests for proportions and difference of proportions
   iii.) One-sample T test
   iv.) Independent Samples T test
4. Margin of Error
   i.) For a z test $E = z_c \frac{\sigma}{\sqrt{n}}$, for a t test $E = t_c \frac{s}{\sqrt{n}}$
   ii.) For a proportion $E = z_c \sqrt{\frac{pq}{n}}$, maximum error is when $pq = .25$, hence $E = z_c \frac{.5}{\sqrt{n}}$
5. Find the sample size for a given error of margin.
   i.) If a z test is being used, then $n = \left(\frac{z_c \sigma}{E}\right)^2$
   ii.) If a proportion test is being used, then $n = p(1-p)\left(\frac{z_c}{E}\right)^2 = pq\left(\frac{z_c}{E}\right)^2$
   iii.) The maximum value for n is when $p = .5$, hence $n = .25\left(\frac{z_c}{E}\right)^2$

Class Work 7a

Test if the sample values 12, 15, 9, 10, 11, 12, 14, 9, 11, 15, 8, 10, have a mean different than 13. If so, can we then say the population mean is different than 13?

Class Work 7b

To test whether or not there is a difference in the average yields of two varieties of corn, five acres of each variety are planted and grown under similar conditions. Yields (in bushels) of variety A and variety B are recorded as follows:

Variety A	110	71	64	102	104
Variety B	88	76	79	60	105

Use SPSS to see if there is a difference in the mean yield for variety A and variety B. Be sure to follow the eight-step process. If we input the grouping variable as a string variable using the letters A and B, then automatically SPSS will recognize the variable as a Nominal Variable and list it as such. Then when we define groups, we need to use A and B as our group 1 and group 2.

Homework 7

1. 1. If we are testing a large sample against a fixed mean, our obtained z value is -2.104, and we are doing a two-tailed test at the 5% level, then what should our conclusion be?
2. 2. If we are doing a one-tailed small-sample test and the two-tailed p-value is .086, then what should be our conclusion at the 5% level?
3. 3. If we are testing the difference of two means and the 95% confidence interval is [-0.123, .012], then what conclusion can we reach at the 5% level?
4. 4. If we are testing a small sample of size 15 using a two-tailed test at the 5% level and the obtained t value is -2.14, then should we accept or fail to accept the null hypothesis?
5. 5. If we are testing the difference of means between two small samples and one sample has size 8 and the other sample has size 7 and the obtained t value is -2.17, then should we accept or fail to accept the null hypothesis at the 5% level using a two-tailed test?
6. 6. If we are testing if proportion 1 is higher than proportion 2 and the obtained z value is 2.01, then what should be our conclusion at the 5% level?
7. 7. IQ scores are forced to fit a normal distribution with mean 100 and standard deviation 15. Find the probability that a person selected at random will have an IQ greater than 120.
8. 8. IQ scores are forced to fit a normal distribution with mean 100 and standard deviation 15. Find the probability that a sample of 36 persons selected at random will have a mean IQ greater than 120.
9. 9. Do banks employ more females than males? A random sample of 200 bank employees found 112 female bank employees and 88 male employees. Use a 5% level of significance to test the hypothesis that banks employ more females than males. Assume $\overline{p} = \overline{q} = .5$.
10. 10. In a random sample of 75 people, 42 supported Al Gore for president. If Al can be 95% certain that the percentage of people who would vote for him is greater than 50%, then he'll consider running for president. Find a 95% confidence interval for Al Gore's true voter percentage and check if it is greater than 50%.
11. 11. Open the data set Bodyfat.sav (available via the Journal of Statistical Education, http://www.amstat.org/publicatins/jse) and test at the 5% level if the mean of men's weight is greater than 176 pounds.
12. 12. Open the data set Bodyfat.sav and test at the 5% level if the mean of men's fatperc (fat percentage) is not the same as 23.

13. 13. Open the data set Bodyfat.sav and test at the 5% level if the mean of men's height is less than 5' 9".

14. 14. Open the data set GSS94 (http://www.icpsr.umich.edu/GSS/home.htm) and test if the number of hours watching television is greater than 2.75 hours per day.

15. 15. Open the data set GSS94 (http://www.icpsr.umich.edu/GSS/home.htm) and test if the number of hours watching television is different for men and women.

16. 16. Open the data set GSS94 (http://www.icpsr.umich.edu/GSS/home.htm) and test if the family income is different for men and women.

17. 17. Open the data set GSS94 (http://www.icpsr.umich.edu/GSS/home.htm) and test if the respondents' income is different for men and women.

18. 18. If a population standard deviation is 3 units, then what size sample is needed to find the estimate for the population mean within .5 units at a 95% confidence interval?

19. 19. If a preliminary estimate for a proportion is .60, then what size sample is needed to find an estimate for the population proportion within .03 points at a 95% confidence interval?

20. 20. If a sample of size 1,260 has a standard deviation of 16, then what is the margin of error for a 95% confidence interval?

21. 21. We found that a value of .5 for p, a proportion will maximize the sample size for an estimate of the true population proportion. And if the error is .03, a 95% confidence interval requires a sample of size 1,068. What sample size is needed if we reduce the error to .01 and increase the confidence interval to 99%?

Just for fun: Are people's left foot and right foot the same length? Use the random numbers (Rv. Uniform) generator to generate 60 numbers from 6 to 13 with 2 decimal places. Call these the left foot and then do the same for the right foot. Test the alternative hypothesis that the lengths of the two feet are different.

References

(1) Statlib

Chapter 8

t-Test for Dependent Groups with Significance

Objectives

After completing this chapter, a student should:

- ✓ Understand a Dependent Test
- ✓ Be able to find the obtained values, degrees of freedom and the critical values for a dependent test
- ✓ Make a decision for a dependent test
- ✓ Use confidence intervals to make a decision for a dependent test not centered at 0

Dependent Test

Sometimes we test the same participants twice, a before and after test. In this case we have what is called a dependent test or a paired-samples test.

Example 1.

A group of newly arrived incarcerates in a prison system are found to have trouble controlling their frustration. They are tested through a physical course where part of the course is purposely set to block their progress to gauge a measure of their frustration level. Afterward a psychologist (who has a pony-tail) counsels them through various self-esteem exercises. (The hypothesis is their lack of self-esteem causes them to be quick to anger.) To prove his hypothesis (statistically speaking, to see if the mean of the difference is less than zero) the psychologist tests their anger response time by having the prisoners work through a series of physical problems, which sometimes say the prisoner gave the wrong answer even if it's right, before and after his counseling sessions. (The sessions last for quite some time—you can't build self-esteem in a day.) People with high self-esteem aren't controlled by their environment, they're in control of their environment, hence do not get angry easily. People with low self-esteem snap easily if their environment gets out of their control. He records their time to get angry before and after his counseling in the following table. A score of 15 minutes means the incarcerate made it through the course without getting angry.

Before	After
10	12
15	15
11	15
9	12
12	10
13	12
9	14
7	10
2	5
13	15

Do the data support the psychologist's claim that building self-esteem will help a person control their anger? That is, will a person with higher self-esteem be less likely to get frustrated easily?

Solution: Before we proceed to step 1 in a paired-samples *t*-test, we need to determine what will be the difference value. Normally the difference value is D = Before – After or $D = B - A$. However the difference could also be D = After – Before or $D = A - B$. It doesn't matter which way we choose the difference, but it does matter that we establish the difference before we proceed. We will test the hypothesis that the mean of the differences is negative, that is $D = B - A$. Thus the after scores will be higher than the before scores, which implies the person did not get as frustrated as easily after the sessions as before the sessions.

Open the data set chpt8example1.sav, or enter the data values in two columns of an SPSS data editor. Use Analyze → Compare Means → Paired-Samples T Test; see **Figure 1a**. Highlight both variables before and after and move them to the active window; see **Figure 1b**. Click OK and the output should be as in **Figure 1c**. From the *p*-value we can quickly determine if the psychologist correct in his assessment. However, we always expect to perform eight-step process as follows:

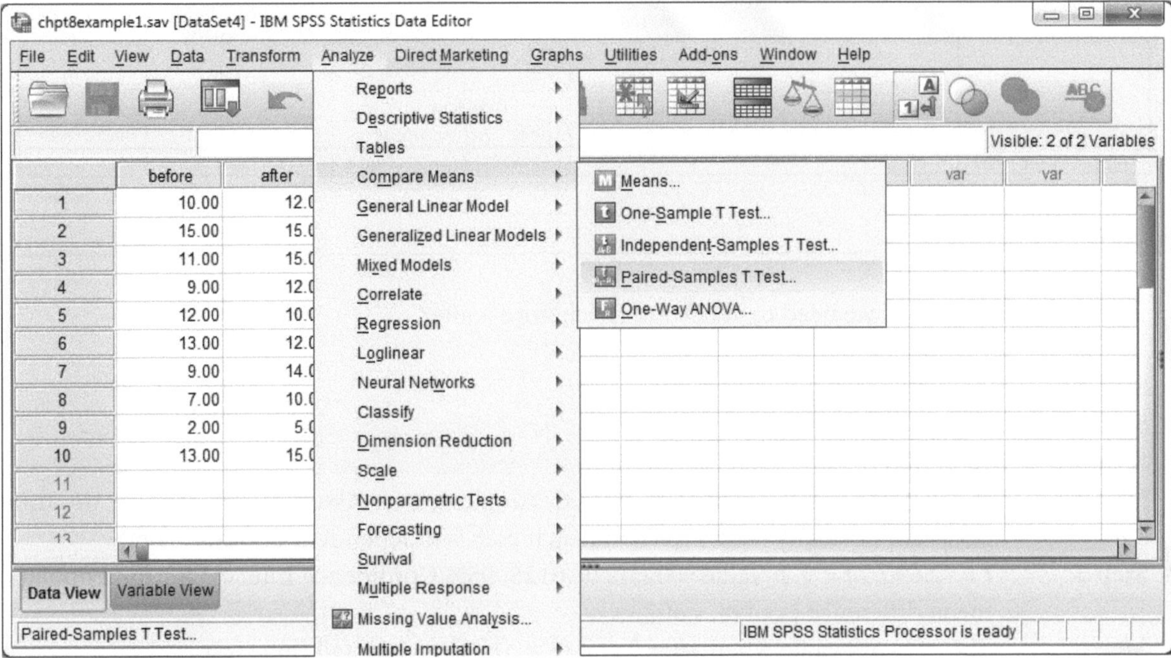

Figure 1a

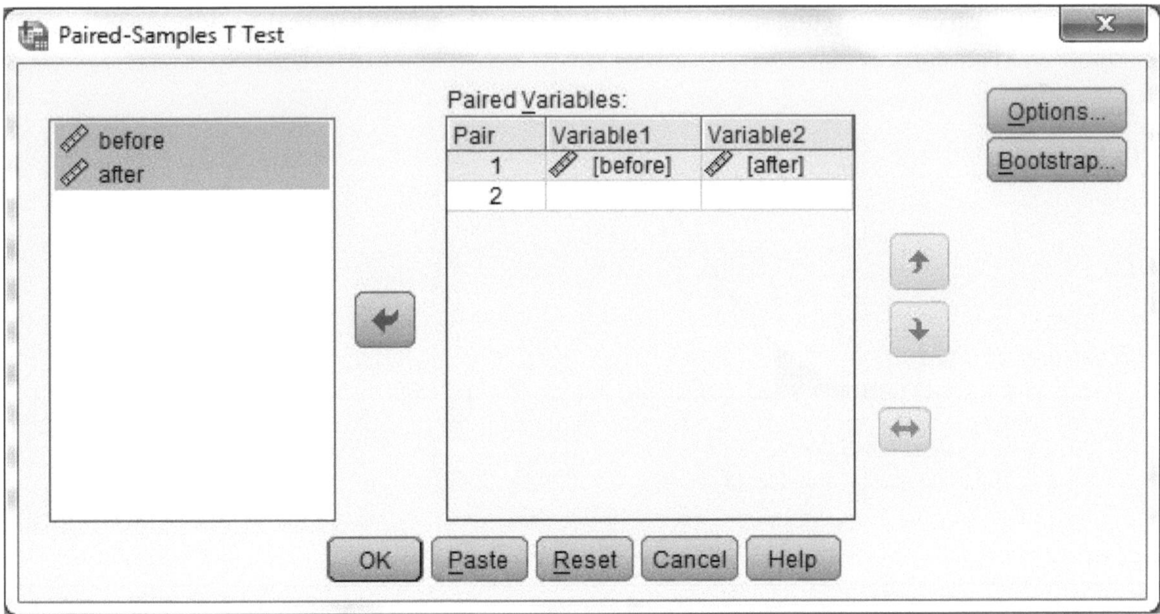

Figure 1b

Paired Samples Test

		Paired Differences					t	df	Sig. (2-tailed)
					95% Confidence Interval of the Difference				
		Mean	Std. Deviation	Std. Error Mean	Lower	Upper			
Pair 1	before - after	-1.90000	2.23358	.70632	-3.49781	-.30219	-2.690	9	.025

Figure 1c

Before we start the test we need to set *D*. Set *D* = before – after

Step 1. $H_0 : \mu_D = 0$

Step 2. $H_1 : \mu_D < 0$

Step 3. $\alpha = .05$

Step 4. We will use a one-tailed (left) paired-samples *t* test. Assumptions are that the differences are normally distributed and each pair is independent of other pairs.

Step 5. *t* = -2.690, *df* = 9, *p*-value =.025/2 = .0125, 95% Confidence Interval for the Population mean difference [-3.498, -.302]

Step 6. The critical *t* value where *df* = 9 and $\alpha = .05$ for a one-tail (left) test is -1.833

Step 7. Our obtained *t*-value = -2.69, which is less than -1.833, which implies our obtained value lies in the critical region; see **Figure 1d**. Alternatively .0125 < .05, therefore we reject the null hypothesis.

Step 8. There appears to be enough statistical evidence to indicate that scores after the self-esteem counseling sessions are higher than the scores before the self-esteem counseling sessions.

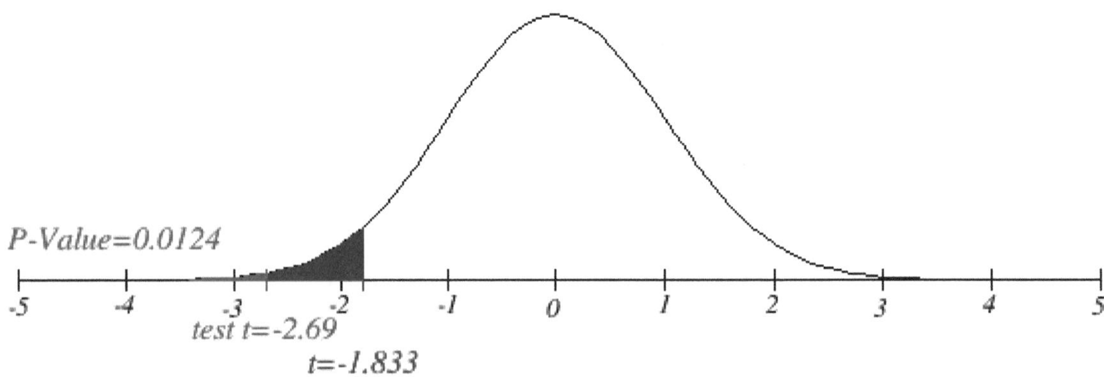

Figure 1d

Remember statisticians do not make conclusions as to what an alterative hypothesis implies, such as the effectiveness of the self-esteem counseling sessions. That is not our job. Our job is to simply report the statistics. It is not our job to infer that the counseling session will have a positive or negative effect on incarcerates. It is simply our job to infer that the scores after the sessions will be higher or lower than scores before the sessions.

The formula SPSS uses to find the t value is $t = \frac{\sum D}{\sqrt{\frac{n\sum D^2 - (\sum D)^2}{n-1}}}$, where $D = B - A$. We can calculate this same value on our calculator by first making a D column in the data editor. Use Transform → Compute Variable, put D in the target variable window, and set the numeric expression to *before – after*. Next find the mean and standard deviation of D and use this alternative formula $t = \frac{\bar{D}}{\frac{s_D}{\sqrt{n}}} = \frac{-1.9}{\left(\frac{2.23358}{\sqrt{10}}\right)} = -2.69$. For dependent samples, SPSS uses the same formula as a calculator so the statistics should be exactly the same.

Before and After Values Using the Same Group

A paired-samples t test tests the same group before and after some kind of learning program or any type of experience. The test finds the difference between each pair of before and after values and then finds the mean of those differences. If there isn't a difference in the before and after values, then the mean of the difference should be 0. However, if the after score is always less than or always greater than the before score, then there should be a difference in the means and we can compare the p-value with the alpha level or compare the t-value with the critical t-value to see if the difference is significant.

Before and After Values Using Different Groups

Sometimes we have a situation where we can't test the same group twice because the test before might alter the results in the test after. Such as, the incarcerates in Example 1 should be better able to maneuver through the course the second time regardless of any counseling sessions. But if we use different groups as in an independent test, then we can't necessarily gauge the effect of any programs if all the data are from diverse sections. This is exactly the scenario our toy company in Chapter 7, Example 8, was forced to cope with. They can't test the same kids twice because small children have a tendency to be influenced by things they first come in contact with, and as we saw, testing independent groups simply shows the diverse character of all the kids. Thus testing identical twins in separate before and after groups will give accurate results in regard to any differences that may exist, since twins have a tendency to behave similarly in similar situations.

Example 2.

Recall the tests of preschool children in different rooms with different educational toys, Chapter 7, Example 8. Since IQ differences and home environment might be uncontrollable factors when testing the preschool children, a group of twins who have the same IQ and the same home environment might give a better representation of the effect of the educational toys. Therefore, randomly select seven pairs of preschool identical twins and perform the same test as in Chapter 7, Example 8 with one twin in group A and the other twin in group B. The results are given in the following table:

Twin Pairs	Pair 1	Pair 2	Pair 3	Pair 4	Pair 5	Pair 6	Pair 7
Group A Scores	8	1	3	2	1	2	7
Group B Scores	10	4	2	7	5	3	9

Do the results indicate the preschoolers in group B, the educational toy group, will consistently score higher than the preschoolers in group A?

Solution: Open a new SPSS data editor and input the scores in two columns labeled A and B, or open chpt8example2.sav. Whenever we do a hypothesis test we use the eight-step process. Our first step is to decide what to use for *D*. In this example we will use $D = B - A$, so our alternative hypothesis will be based on $\overline{D} > 0$. ($\overline{D}$ is the sample mean of the differences and μ_D is the population mean of the differences.)

Step 1. $H_0 : \mu_D = 0$
Step 2. $H_1 : \mu_D > 0$
Step 3. $\alpha = .05$
Step 4. Paired Samples T Test, one-tailed (right), assumes population of paired differences is normal and each pair is independent.
Step 5. *T* value = 3.060, $df = 6$, *p*-value = 0.022/2 = .011, 90% confidence interval [.834, 3.737], see **Figure 2a.**
Step 6. Critical *t* value for $df = 6$ and one-tail test with cumulative probability .95 is 1.943.
Step 7. Comparing the *t*-value against the critical *t*-value we have 3.060 > 1.943, see **Figure 2b.** Alternatively comparing the *p* value with α, .011 < .05, we see that the test value lies in the rejection region.
Step 8. We fail to accept the null hypothesis. There is statistical evidence to indicate that the alterative hypothesis is a better choice than the null hypothesis, which is that the educational toys appear to have a positive influence on preschool children.

Paired Samples Test

	Paired Differences					t	df	Sig. (2-tailed)
				90% Confidence Interval of the Difference				
	Mean	Std. Deviation	Std. Error Mean	Lower	Upper			
Pair 1 B - A	2.286	1.976	.747	.834	3.737	3.060	6	.022

Figure 2a

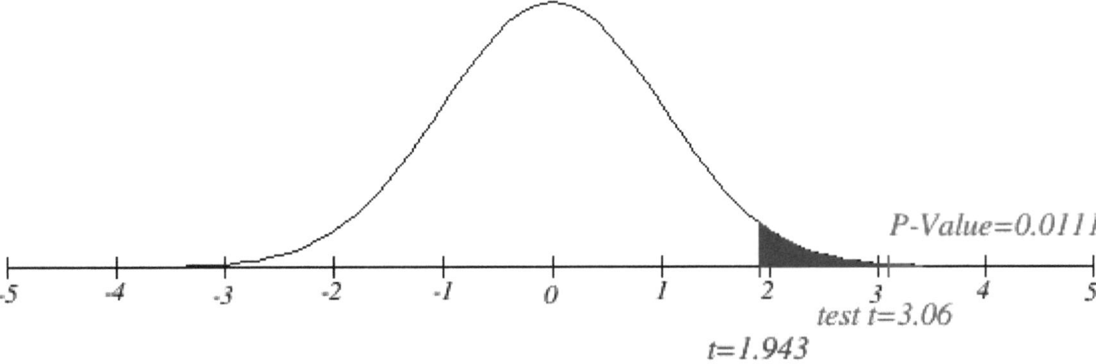

Figure 2b

The educational toy company should be more pleased with these results than with the results obtained in the test from Chapter 7, although the *p*-value is still greater than .01. Playthings help develop a young child's imagination, in fact playthings help develop most young animals' imaginations. But our human goal is to develop a young child's imagination along academic lines because we support the concept that academics is the key to success. Therefore we're always seeking better ways to stimulate young children's imaginations along academic lines.

Degrees of Freedom

The degrees of freedom for dependent samples is the number of pairs minus 1, that is $n - 1$. As we saw in Chapter 7, the degrees of freedom for independent samples is the total number of data values minus 2, that is $n_1 + n_2 - 2$ assuming equal variances, which is usually the case. (Levene's Test for Equality of Variances will always tell us if the variances are equal.) To remember the formula for the degrees of freedom always think of how many means are present. In Chapter 7 we looked at the difference between two means. Hence to find the degrees of freedom, we add up all the data values and subtract 2. In this chapter we look at the mean of the difference. For each pair of data values, we find the difference and the mean of the differences. Hence to find the degrees of freedom, we subtract 1 from the number of pairs. Later, in Chapter 11, the degrees of freedom for the correlation coefficient is the number of pairs of data minus 2, because again there are two means, an x mean and a y mean.

Example 3.

Open the data set chpt8example3.sav. This data set contains individual-event race times for a high school swim meet (1). Test if the swim times in the 50-meter free stroke second heat, fr5002, are faster than the swim times in the 50-meter free stroke first heat, fr5001.

Set $D = fr5001 - fr5002$

Solution: Whenever we do a hypothesis test, we always use the eight-step process.

Step 1. $H_0: \mu_D = 0$

Step 2. $H_1: \mu_D > 0$

Step 3. $\alpha = .05$

Step 4. A paired-samples test is appropriate here because it's the same person tested before and after. In the one-tailed (right) test, assume the population of paired differences is normal and each pair is independent.

Step 5. Obtained t = 4.457, df = 44, p-value = .000/2 = 0, 90% Confidence Interval [1.649, 3.646]. To change the confidence interval from 95% to 90% use Analyze → Compare Means → Paired-Samples T Test, move 50 Freestyle 1 into variable 1 and 50 Freestyle 2 into variable 2, and click on Options; see **Figure 3a**. Change Confidence Interval Percentage to 90%, as in **Figure 3b**. Continue, OK. The output is in **Figure 3c**.

Step 6. Critical t value = 1.68.

Step 7. Compare obtained t-value against critical t-value, 4.457 > 1.68. Alternatively compare the p-value against significance level, 0 < .05.

Step 8. Fail to accept the null hypothesis; it appears the students' times in the second heat are faster than the students' times in the first heat.

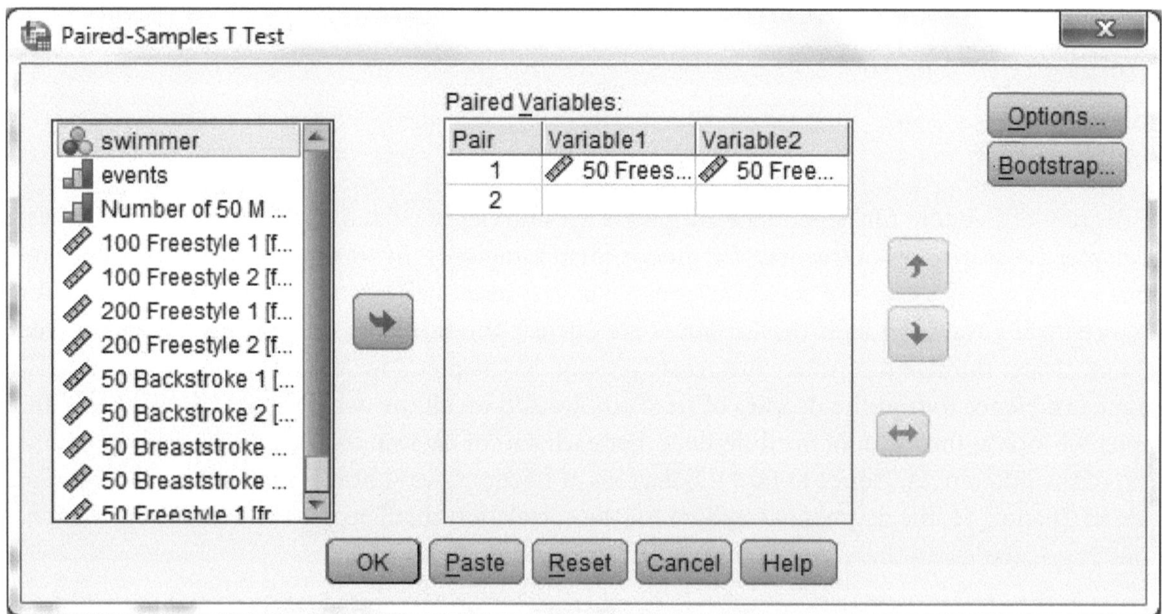

Figure 3a

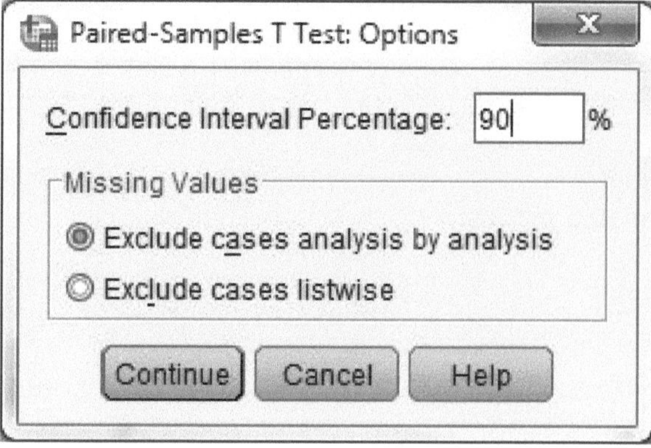

Figure 3b

Paired Samples Test

		Paired Differences					t	df	Sig. (2-tailed)
					90% Confidence Interval of the Difference				
		Mean	Std. Deviation	Std. Error Mean	Lower	Upper			
Pair 1	50 Freestyle 1 - 50 Freestyle 2	2.64756	3.98510	.59406	1.64939	3.64572	4.457	44	.000

Figure 3c

Technique for Comparing Against a Value Other Than Zero.

The *t* test which SPSS does is always against the value zero. That is, the mean of the differences is measured against 0, and the *t* value is obtained with 0 as the mean of the population of differences. Sometimes there is an occasion to test against a value other than 0. The formula for the *t* value is $t = \frac{\bar{D} - \mu_D}{\frac{s_D}{\sqrt{n}}}$. SPSS won't allow us to change the value we will test against, so instead we'll find a confidence interval to see if the test value is in that interval. If we want to do a one-tail test, then we double α and find a confidence interval with the percentage $1 - 2\alpha$. If we want to do a two-tailed test, then the confidence interval will have the percentage α. The error in the confidence interval is $E = t_c \frac{s_D}{\sqrt{n}}$, where t_c is the critical *t* value that is dependent upon α and *df.*

Example 4.

An egg farmer wants to increase the egg output laid by his hens by more than 2 eggs per hen per week, so he tries a new feed mixture and gives his hens more yard time. After a month of this new regimen, he measures the weekly egg production of 12 of his randomly selected hens and compares it to the weekly egg production of these same twelve hens before the new regimen. The results are given in the following table. Do the results indicate mean egg production has increased by more than 2 eggs per hen? Test at the 5% level. chpt8example4.sav

Before	12	15	9	10	11	12	14	9	11	15	8	10
After	12	16	15	12	16	15	15	16	12	14	8	13

Set D = before – after

Solution: We use the eight-step process.

Step 1. $H_0 : \mu_D = -2$

Step 2. $H_1 : \mu_D < -2$

Step 3. $\alpha = .05$

Step 4. Use a paired-samples *t* test. In the one-tailed (right) test, assume the population of paired differences is normal and each pair is independent.

Step 5. We need to find a 90% confidence interval for the mean of the differences since that will give us 5% in both tails and this is a one-tailed test; see **Figure 4a**. Use Analyze → Compare Means → Paired-Samples T Test, move before and after into the paired variables window, and click on the tab Options. Change the confidence interval to 90%; see **Figure 3b**, and click continue and OK. The output is in **Figure 4b**, but all we need is the confidence interval, which is [-3.647, -1.019].

Step 6. Omit since we're not using the *t* value.

Step 7. Since the after values should increase by more than 2, the before minus after should be less than -2. But -2 is included in the CI, and we're 90% confident the population mean is in that interval. Hence we can't reject the null hypothesis; see **Figure 4c**.

Step 8. Accept the null hypothesis: the mean increase in the production of eggs per week per hen is not greater than 2.

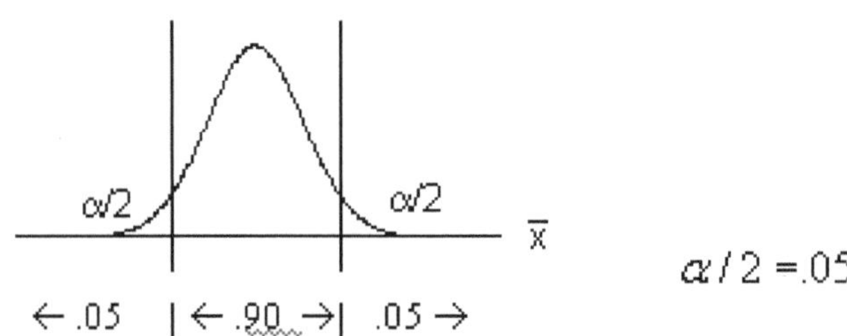

Figure 4a

Paired Samples Test

		Paired Differences					t	df	Sig. (2-tailed)
		Mean	Std. Deviation	Std. Error Mean	90% Confidence Interval of the Difference: Lower	Upper			
Pair 1	before - after	-2.333	2.535	.732	-3.647	-1.019	-3.189	11	.009

Figure 4b

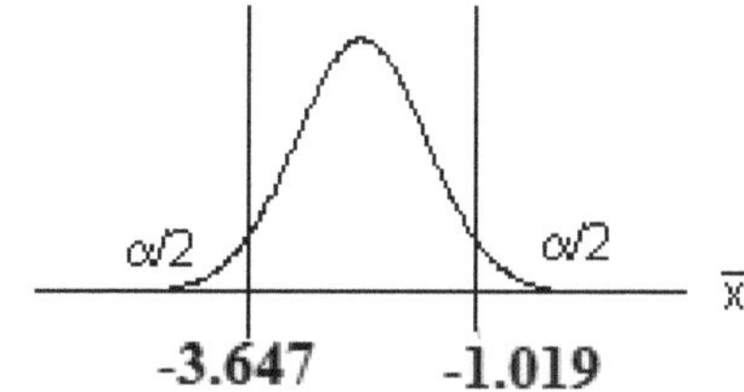

Figure 4c

Let's review what we learned in this chapter.

1. Dependent Tests—There is a dependence between the two tests.
   i.) The same subjects are tested twice, before and after.
   ii.) Identical twins are tested in two different situations.
2. D = Before minus After—The test variable is the difference of before and after; SPSS always subtracts variable 2 from variable 1, so make sure D is stated before the test.
3. Degrees of Freedom—The degree of freedom is the number of pairs minus 1, $n - 1$.
4. A confidence interval can be used to test against any value.
5. We assume that the paired sample differences are from a normal distribution.

Class Work 8

In the psychology lab students are studying what they call Rat Olympics. First, a rat is trained to run a maze with ten major turns and is rewarded only at the end with 20 grams of food pellets. Later, the rat is permitted to run the maze again but this time is immediately rewarded with two grams of food for each correct turn. No reward is given for a wrong turn. Six rats run through the maze with A being rewarded immediately and B being rewarded at the completion of the maze. Using a 5% level of significance, test the claim that immediate rewards tend to shorten the time needed to correctly run the maze.

A	30.2	19.9	46.1	44.7	31.0	42.6
B	35.7	21.2	53.9	44.0	39.5	47.1

Always work with the eight-step process when performing a test.

Homework 8

1. The following are the critical thinking scores of students before and after studying abstract reasoning. Test at the 5% level that their critical thinking skills improved after studying abstract reasoning. chpt8problem1.sav

Before	72	75	69	70	71	82	34	59	61	75	88	90
After	72	76	75	80	76	85	45	66	72	74	88	93

2. The weights of women over 40 were measured before and after an exercise and diet plan. Their weights are listed in the following table. Test if the exercise and diet plan had a mean weight loss greater than 15 pounds for this sample of women over 40. chpt8problem2.sav

Before	172	175	169	170	171	182	134	159	161	175	188	190
After	155	161	148	139	145	169	126	130	136	160	162	163

3. A company decides to try a new advertising program for their twelve stores. The average number of daily customers is given in the following table. Test if the average increase is greater than 5 customers per store per day. chpt8problem3.sav

Before	364	275	169	365	171	182	234	359	461	275	188	190
After	390	291	248	393	195	219	226	430	436	260	262	213

4. Identical preschool twins are separated and randomly selected to go to one of two rooms. One room has alphabet coloring books, preschool readers, and alphabet puzzles. The other room has non-educational toys. Both groups spend 2 hours a day in their respective rooms for 6 months, after which a record is kept as to when they first start to read a primary text. Their ages in months are given in the following table. Is there evidence to suggest that alphabet toys will help a child read sooner? Test at the 5% level. chpt8problem4

Basic toys	70	71	69	63	64	62	63	75	65	52	64	60
Alphabet toys	69	70	68	61	65	58	62	71	60	53	61	58

5. In Alberta, Canada, the highway that runs from Jasper Park to Banff Park has several wildlife overpasses, locations where wildlife can safely overpass the highway; however, some wildlife still prefer to cross the highway where it's dangerous to cross. One way to measure the success of the overpasses is to count the number of elk in square-mile units along sections of the highway before and after the highway was built. The following table gives the number of elk for a period of a year before the highway was built and for a period of a year after the highway was built. The data are from random square-mile units. Is there evidence to suggest the elk population has declined? chpt8problem5

Before Highway	36	71	45	94	126	90	75	118	122	110
After Highway	15	89	36	65	87	106	64	212	121	120

6. Twelve married couples were selected to participate in a random-sample study. Their ages are as follows:

Husband	37	20	25	31	26	19	44	35	36	22	27	52
Wife	27	19	23	28	32	18	43	30	38	23	24	48

Test, at the 5% level, if there is a difference between the ages of the husbands and wives.

7. Using the data from class work 5, test if there is a difference in means between the Poisson distribution and the actual percentage of deaths caused by horse kicks. From Chapter 5 class work we have

this problem: A classic example of the Poisson distribution is the number of deaths caused by horse kicks of men in the Prussian Army between 1875 and 1894. In 1898 it was reported that in the 20 years between 1875 and 1894, there were 196 deaths among the members of 14 corps in the Prussian Army due to horse kicks. Find the mean value per year and find the probability of 0, 1, 2, 3, and 4 deaths per year. Also find how close the Poisson distribution is to the actual distribution of 0 deaths in 144 corps-years; 1 death in 91 corps-years; 2 deaths in 32 corps-years; 3 deaths in 11 corps-years; and 4 deaths in 2 corps-years. A corps-year is the number of corps times the number of years, hence 14 * 20 = 280 total corps-years. chpt8problem7.sav (See Chapter 5 regarding the Poisson distribution.) (Use a paired-sample T-test.)

8. From Chapter 5 problem 42, we are given that 116 homicide deaths occurred in Richmond, VA in 1991. We would like to test if the Poisson distribution is a good indicator for the actual number of deaths. The data set is given below and also as chpt5problem42. Use a dependent t-test to test if the mean difference is 0, x is the number of homicides, p is the Poisson distribution value, and a is the actual value.

x	p	a
.00	265.63	268.00
1.00	84.42	79.00
2.00	13.41	17.00
3.00	1.42	1.00

Just for fun: Does adding plus or minus to grades make a difference in grade point average? We will assume plus (+.3 grade points) and minus (-.3 grade points) are randomly assigned to grades and then will test if the mean difference is 0. Open a new Data Editor and enter the number 1 to 30 in the first column, and call it x. Set the Random Number Generator to Random. Use Transform → Compute → Arithmetic → Trunc → Random Numbers → Rv.Uniform[0,5] to generate 30 random grades from 0 to 4; call it g. Change the decimal value to 0. In the next column create a new variable and add random values of {-.3, 0, +.3} to the old variable. (Use Transform → Compute label a new variable and set it equal to g + (Trunc(Rv.Uniform(0,3))-1)*.3. Do a paired-samples test on those two variables.

References

(1) Brian Carver 2006

Chapter 9

ANOVA

Objectives

After completing this chapter, a student should know:

- ✓ When to use ANOVA and why
- ✓ The ratio of the F-value and how to compute it
- ✓ How the F-value relates to the Barplots
- ✓ How to find the degrees of freedom for a critical F-value
- ✓ Which decision to choose based on the F-value
- ✓ When to use a Post Hoc test and which Post Hoc test to use

If we have more than two groups, then several independent t-tests would not feasible because it would lead to a greater chance of a type I error. In this case we use ANOVA or Analysis of Variance. The theory of Analysis of Variance is to compare the variance between several groups with the variances within several groups to see if there is a marked difference. This example should help clear it up.

Example 1.

Open the data set chpt9example1.sav. Use Graphs → Legacy Dialog → Boxplot, click define, put redcomsc in the Variable dialog box and station in the Category Axis dialog box, and hit OK. In the output, scroll down until you see the 5 box plots as shown in **Figure 1a**. Redcomsc is the reading comprehension scores of eight students in five rooms each listening to one of five different radio stations for a total of 40 data values. The radio stations are classical music, rock and roll, news, talk, and silence (silence really isn't a radio station, it's just the radio turned off). The students are given a magazine article and asked to answer 10 reading comprehension questions related to the article. They do this while sitting in one of the five rooms playing one of the five types of radio stations. The students were randomly selected and the rooms were randomly assigned. You might recall your mother always told you the best study environment was a quiet atmosphere, well here's an opportunity to prove she was correct. As we can see in the box plots, the median values (the heavy black line in the middle of each of the boxes) vary widely among the different rooms, while the variances within the rooms (the length of each of the boxes can represent the variance within the rooms) are mostly the same. The question is will the ratio of the variances between the rooms over the variance within the rooms be much larger than 1? If it is then we reject the null hypothesis, if it isn't then we accept the null hypothesis.

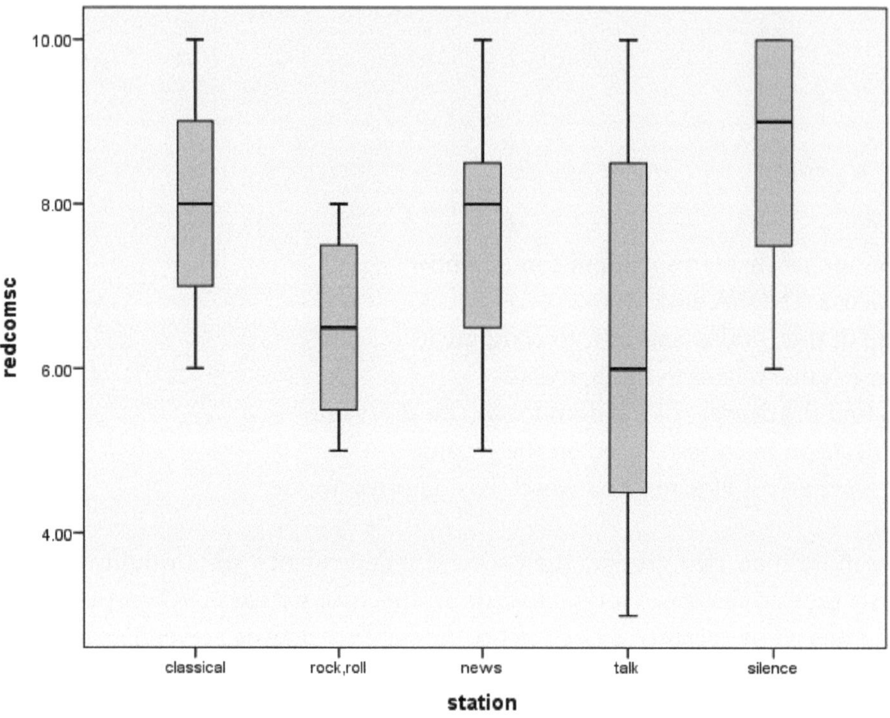

Figure 1a

Test Value for One-Way ANOVA

Solution: The statistic we wish to study is called the *F* value and is defined as $F = \frac{\text{var}_{between}}{\text{var}_{within}}$. The actual mathematical formula is quite involved and will be omitted. SPSS will compute the *F* value for us quite nicely. Use Analyze → Compare Means → One-way ANOVA, and move redcomsc into the Dependent List and station into the Factor window. Click on the Options tab and check Homogeneity of variance test. (This is needed to be certain any differences suggested by the *F* test are valid.) Also check Descriptive so we'll have an output of the statistics for each room; see **Figure 1b**. Click continue and OK; the output should read as shown in **Figure 1c**. The *F* value is 2.702, which is larger than 1 and the

p-value is .046, which is less than .05, so it appears we reject the null hypothesis. However, we will work the eight-step process.

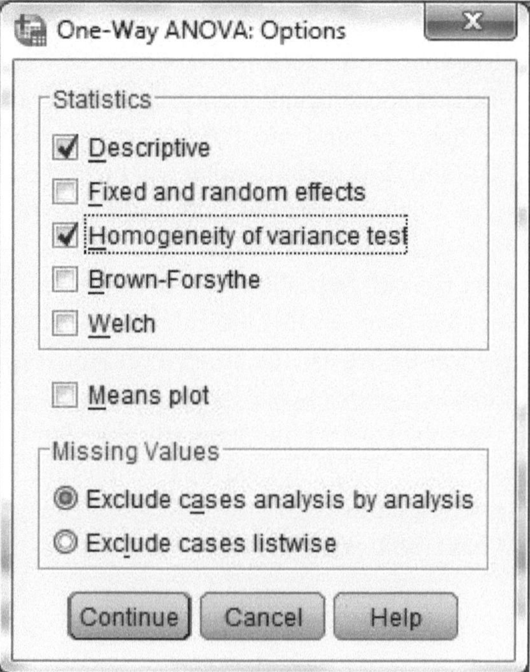

Figure 1b

ANOVA

redcomsc

	Sum of Squares	df	Mean Square	F	Sig.
Between Groups	30.150	4	7.538	2.702	.046
Within Groups	97.625	35	2.789		
Total	127.775	39			

Figure 1c

Step 1. $H_0 : \mu_{classical} = \mu_{rock,roll} = \mu_{news} = \mu_{talk} = \mu_{silence}$. All means are the same. (In ANOVA this will always be the null hypothesis.)

Step 2. H_1 : At least one mean is different. (In ANOVA this will always be the alternative hypothesis.)

Step 3. $\alpha = .05$

Step 4. Use ANOVA, one-way because we only have one factor. The different rooms are the groups and the different radio stations are the factor. Assumptions are the different groups have a normal population, each group is independent of the other groups, and the variances within groups are about the same.

Step 5. Perform the test. The obtained F value is 2.702 and the p-value is .046; see **Figure 1c**.

Step 6. Find a critical *F* value. Since *F* values are always positive, our critical *F* value will have the entire rejection region to the right. So if our significance level is 5%, then we want 95% of the area under the *F* curve to lie to the left of the critical value; see **Figure 1d**. The degrees of freedom are 4 and 35, df1 = 4 and df2 = 35. Open a new SPSS data editor and put .95 in the first column, call it *a* for area. Use Transform → Compute Variable, call the target variable *cF*, and put IDF.F(*a*, 4, 35) into the numeric expression; see **Figure 1e**. Hit OK and the critical *F* value will be in the data editor. It is 2.64.

Step 7. Compare, 2.702 > 2.64 alternatively .046 < .05. The obtained value is in the critical region and the *p*-value is less than α.

Step 8. Conclusion: reject the null hypothesis. Since the obtained value is in the critical region and the *p*-value is less than α, this indicates that at least one mean is different. To find where the differences lie, we use the Tukey Post Hoc test.

(Side note: John Tukey was the subject of a Final Jeopardy clue, January 12, 2011. The category was Computer Science. The clue was "John Tukey coined this compound word in 1958 saying it was as important as "'tubes, transistors, wires, tapes....'"

The correct response was "What is software?") (1)

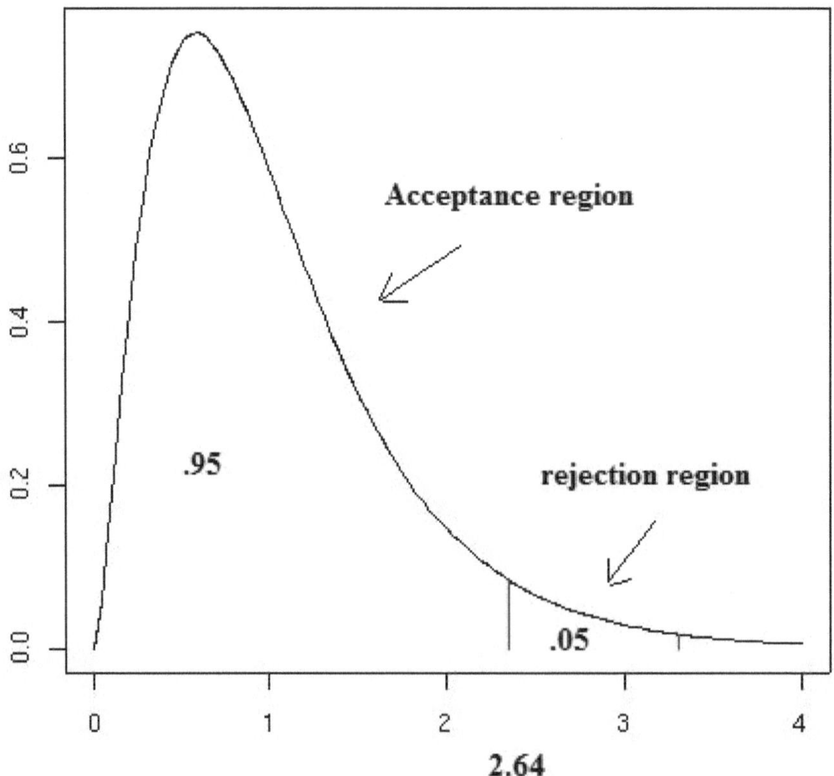

Figure 1d

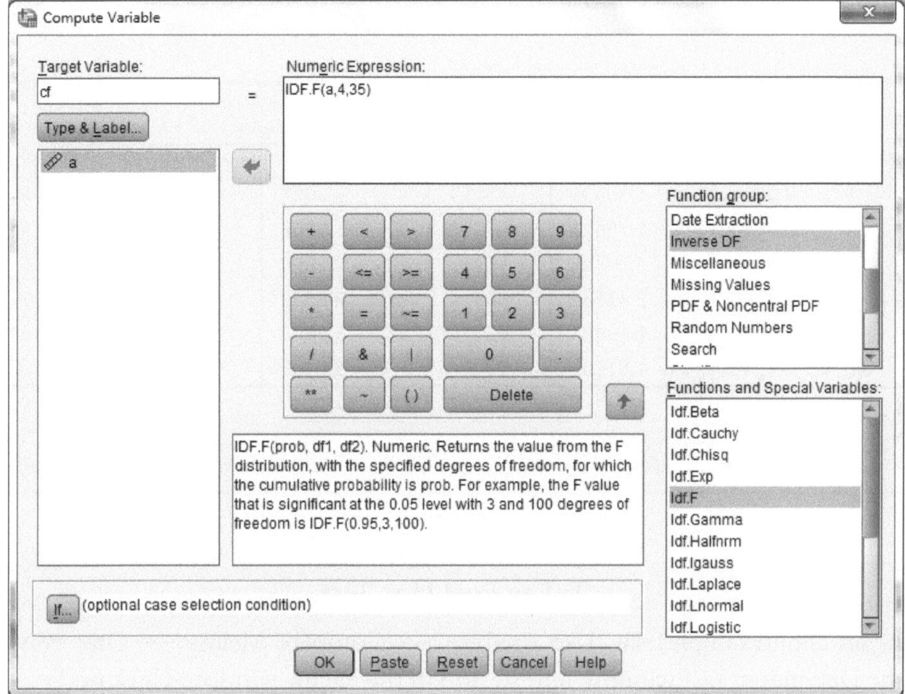

Figure 1e

Step 9. If we reject the null hypothesis in step 8 of an ANOVA test, then we are required to do step 9. This step is to find where the differences lie. Since our test for homogeneity of variance is .180, see **Figure 1f**, we can be certain any difference suggested by the *F* test is valid. We can see by the Descriptives table the means for silence and talk are furthest apart; see **Figure 1g**. Thus we expect the Tukey Post Hoc test to verify the difference lies between these two means.

Test of Homogeneity of Variances

redcomsc

Levene Statistic	df1	df2	Sig.
1.665	4	35	.180

Figure 1f

Descriptives

redcomsc

	N	Mean	Std. Deviation	Std. Error	95% Confidence Interval for Mean		Minimum	Maximum
					Lower Bound	Upper Bound		
classical	8	8.0000	1.30931	.46291	6.9054	9.0946	6.00	10.00
rock,roll	8	6.5000	1.19523	.42258	5.5008	7.4992	5.00	8.00
news	8	7.6250	1.59799	.56497	6.2890	8.9610	5.00	10.00
talk	8	6.3750	2.44584	.86474	4.3302	8.4198	3.00	10.00
silence	8	8.6250	1.50594	.53243	7.3660	9.8840	6.00	10.00
Total	40	7.4250	1.81005	.28619	6.8461	8.0039	3.00	10.00

Figure 1g

Tukey Post Hoc Test

Open the data set chpt9example1.sav. Use Analyze → Compare Means → One Way ANOVA, put redcomsc in the Dependent List window, and station in the Factor window. Click on Post Hoc and click Tukey, continue, OK. The output should be as in **Figure 1h**. The column headed by Sig. is our between significance level and the smallest value is .075, which is the value that compares silence and talk. So we will say the difference in means lies between silence and talk radio (our moms were correct).

It sometimes happens that our post hoc test will give a value higher than .05 when it really is significant, as in this case. We'll use this **Rule:** whenever ANOVA states the *F* value is significant and no post hoc significant value is less than .05, then we will use the smallest significance value in the post hoc test to establish where the differences lie. Otherwise the differences lie between all groups that have a between significant value less than .05.

Degrees of Freedom

If we wish to find a critical *F* value from SPSS, we need to know the degrees of freedom for the numerator and the denominator. Remember the numerator is the variance between the groups, so the value for the degrees of freedom for the numerator is the number of groups minus 1. If we let k equal the number of groups, then $df_{num} = k - 1$. The degrees-of-freedom value for the denominator is the total number of data values minus the number of groups. If we let n be the total number of data values, then $df_{den} = n - k$. Together the degrees of freedom for the denominator plus the degrees of freedom for the numerator give the total degrees of freedom. Thus $df_{total} = df_{den} + df_{num} = n - k + k - 1 = n - 1$. In the above example there were 40 data values and 5 groups, so *df* between groups is 5 – 1 = 4, *df* within groups is 40 – 5 = 35 and *df* total is 40 – 1 = 39. SPSS uses df1 for degrees of freedom between groups and df2 for degrees of freedom within groups; see **Figure 1f**. Note that the sizes of each group need not be the same; some groups can be larger or smaller than other groups.

Post Hoc Tests

Multiple Comparisons

Dependent Variable: redcomsc
Tukey HSD

(I) station	(J) station	Mean Difference (I-J)	Std. Error	Sig.	95% Confidence Interval	
					Lower Bound	Upper Bound
classical	rock,roll	1.50000	.83506	.392	-.9008	3.9008
	news	.37500	.83506	.991	-2.0258	2.7758
	talk	1.62500	.83506	.313	-.7758	4.0258
	silence	-.62500	.83506	.943	-3.0258	1.7758
rock,roll	classical	-1.50000	.83506	.392	-3.9008	.9008
	news	-1.12500	.83506	.664	-3.5258	1.2758
	talk	.12500	.83506	1.000	-2.2758	2.5258
	silence	-2.12500	.83506	.104	-4.5258	.2758
news	classical	-.37500	.83506	.991	-2.7758	2.0258
	rock,roll	1.12500	.83506	.664	-1.2758	3.5258
	talk	1.25000	.83506	.571	-1.1508	3.6508
	silence	-1.00000	.83506	.753	-3.4008	1.4008
talk	classical	-1.62500	.83506	.313	-4.0258	.7758
	rock,roll	-.12500	.83506	1.000	-2.5258	2.2758
	news	-1.25000	.83506	.571	-3.6508	1.1508
	silence	-2.25000	.83506	.075	-4.6508	.1508
silence	classical	.62500	.83506	.943	-1.7758	3.0258
	rock,roll	2.12500	.83506	.104	-.2758	4.5258
	news	1.00000	.83506	.753	-1.4008	3.4008
	talk	2.25000	.83506	.075	-.1508	4.6508

Figure 1h

Assumptions Regarding One-Way ANOVA

1. The data values come from populations that have a normal distribution.
2. The test is similar to a *t* test of independent groups; we assume each group is independent of the other groups.
3. The different groups have approximately the same variances and standard deviations. This assumption will be tested each time we perform ANOVA.

ANOVA is Always a One-Tailed Test

Since the *F* value is the ratio of variances and variances are always positive, the ratio will always be greater than 0. If the ratio is between 0 and 1, then the variances between the groups are smaller than the variances within the groups. If the ratio is larger than 1, then the variances between will be larger than the variance within and if they're large enough, we can say the difference between at least two means is significant. The next example will illustrate these three cases.

Critical F Value

To find the critical *F*-value we put the cumulative area, which is usually .95 into an area variable labeled *a*, see **Figure 2c**, and use Transform → Compute Variable. Enter cf for critical *F*-value, click on Inverse DfF under Function Group and double-click Idf.F under Functions and special Variables. Put *a*, df1, df2 for the three parameters, see **Figure 1e**, click OK. The critical *F*-value will be in a new column labeled cf, see **Figure 2c**.

Example 2a.

Open the data set chpt9example2.sav. Draw the boxplot for test 1 and find the *F* value. Was the *F* value what you expected from the boxplots?

Solution: Use Graphs → Legacy Dialog → Boxplot, click define. Move test 1 under Variable and move groups under Category Axis, click OK. The boxplots should be as in **Figure 2a**. Notice the medians are all close to the same value and hence the means will be close to the same values as well, which means the null hypothesis is satisfied. We will use the eight-step process to test if our *F* value is significant.

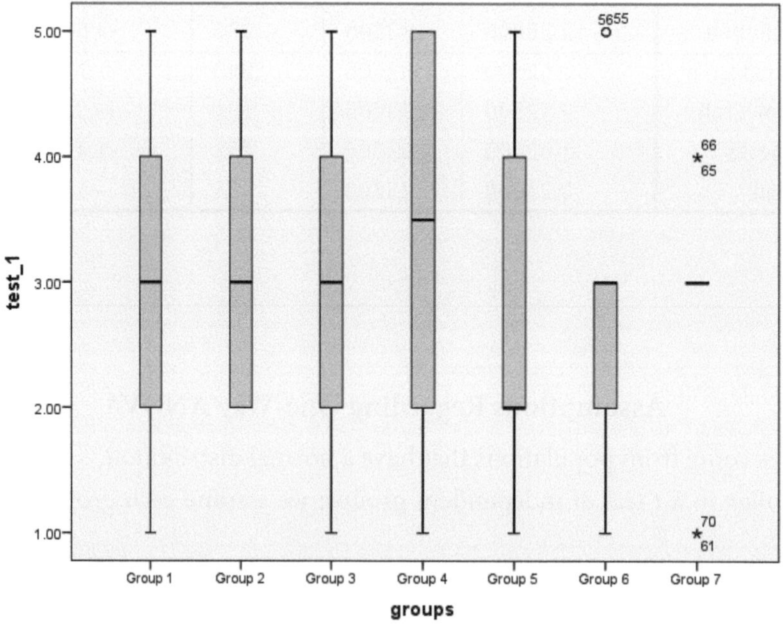

Figure 2a

Step 1. H_0 : All the means are equal.

Step 2. H_1 : At least one mean is different.

Step 3. $\alpha = .05$

Step 4. The test will be One-Way ANOVA because we have several groups and one factor—the radio stations. Assumptions are the different groups have a normal population, each group is independent of the other groups, and the variances within groups are about the same.

Step 5. The obtained *F* value is .191, df1 = 6, df2 = 63 and the *p*-value is .978; see **Figure 2b**. Use Analyze → Compare Means → One-way ANOVA, move test 1 into Dependent List and groups into Factor.

Step 6. To find the critical *F* value, we use Inverse DF function (IDF.F). Use Transform → Compute Variable, call the target variable *cF* and put IDF.F(*a*, 6, 63) into the numeric expression. The critical *F* value = 2.25; see **Figure 2c**.

Step 7. Compare the obtained *F* value .191 against the critical *F* value 2.25 to get .191 < 2.25, or alternatively compare the obtained the *p*-value .978 against α to get .978 > .05—we accept the null hypothesis; see **Figure 2d**. (The *x*-axis is the *F* value and the *y*-axis is the probability density value.)

Step 8. Our decision is to accept the null hypothesis; that is, we have no reason to believe any of the means in the groups for the test 1 data set are different.

ANOVA

test_1

	Sum of Squares	df	Mean Square	F	Sig.
Between Groups	2.143	6	.357	.191	.978
Within Groups	117.800	63	1.870		
Total	119.943	69			

Figure 2b

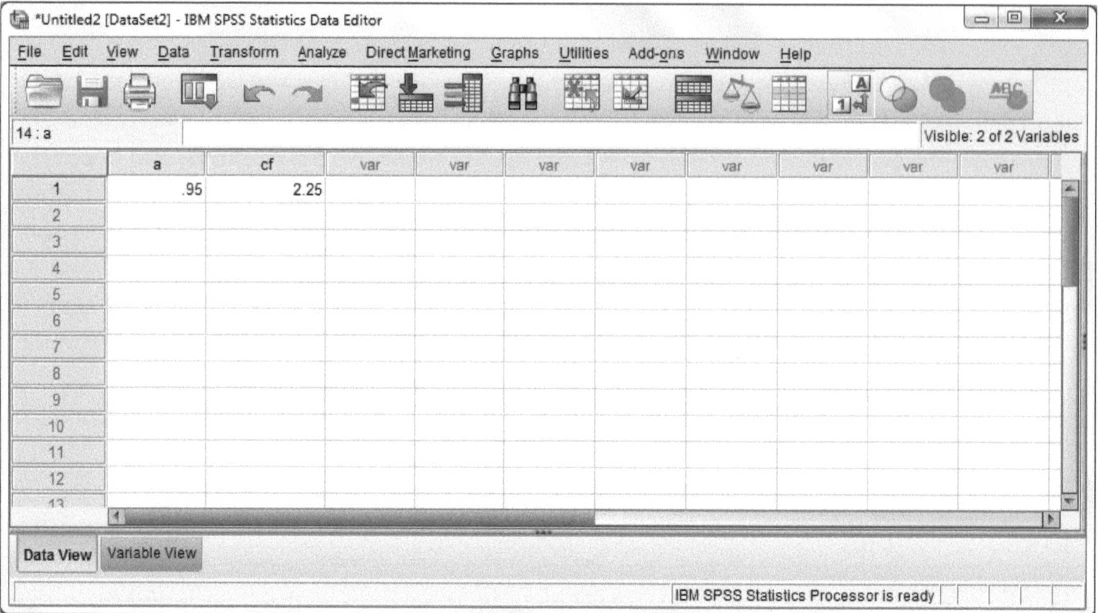

Figure 2c

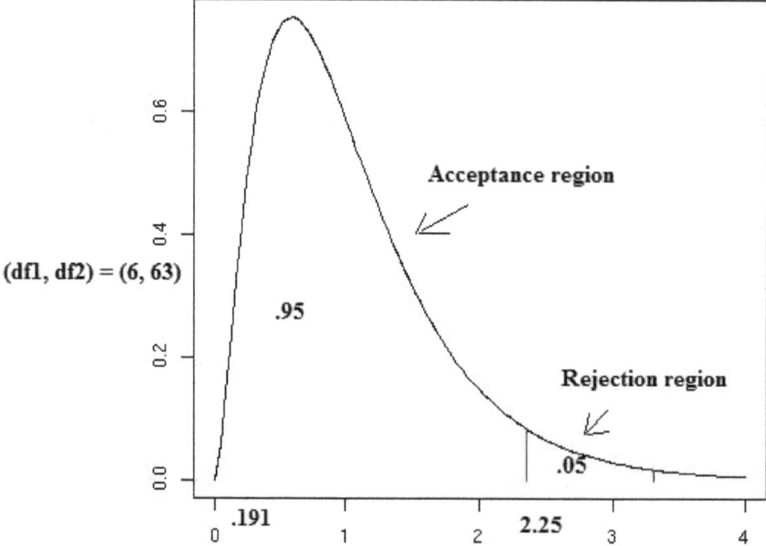

Figure 2d

If the variances between is large and the variance within is also large, then the *F* ratio will be close to 1, so we can't really say the means are statistically different since the variances are too big to nail down any differences in the mean.

Example 2b.

Open the data set chpt9example2.sav. Draw the boxplot for test 2 and find the *F* value. Was the *F* value what you expected from the boxplots?

Solution: Use Graphs → Legacy Dialog → Boxplot, click define. Move test 2 under Variable and move groups under Category Axis, click OK. The boxplots should be as in **Figure 2e**.

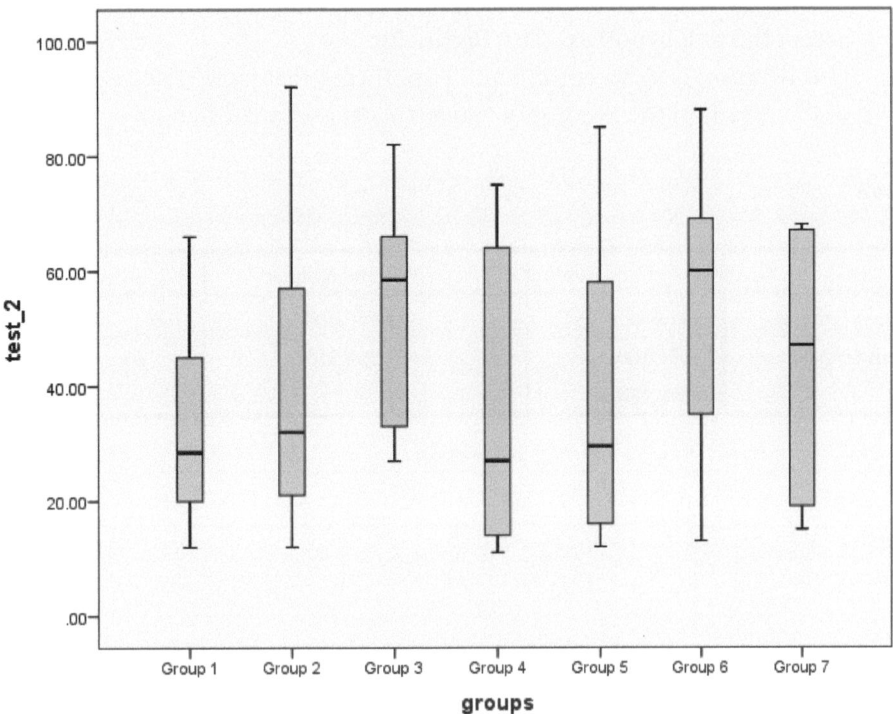

Figure 2e

In this case the medians (the thick black lines in the middle of the boxes) between groups vary about as much as the average length of the boxes, so we might expect the *F* value to be about 1. We will use the eight-step process to test the *F* value.

Step 1. H_0: All the means are equal.

Step 2. H_1 : At least one mean is different.

Step 3. $\alpha = .05$

Step 4. The test will be One-Way ANOVA because we have several groups and one factor—the radio stations. Assumptions are the different groups have a normal population, each group is independent of the other groups, and the variances within groups are about the same.

Step 5. Use Analyze → Compare Means → One-way ANOVA, move test 2 into Dependent List and groups into Factor. The obtained *F* value is 1.185, df1 = 6, df2 = 63 and the obtained *p*-value = .326; see **Figure 2f.**

Step 6. The critical *F* value is the same as in Example 2a since all the parameters are the same. The critical *F* value = 2.25.

Step 7. Compare the obtained *F* value 1.185 against the critical *F* value 2.25 to get 1.185 < 2.25, or alternatively compare the obtained the *p*-value .326 against α to get .326 > .05—we accept the null hypothesis; see **Figure 2g**.

Step 8. Our decision is to accept the null hypothesis; that is, we have no reason to believe any of the means in the groups for the test 2 data set are different.

ANOVA

test_2

	Sum of Squares	df	Mean Square	F	Sig.
Between Groups	3941.686	6	656.948	1.185	.326
Within Groups	34934.100	63	554.510		
Total	38875.786	69			

Figure 2f

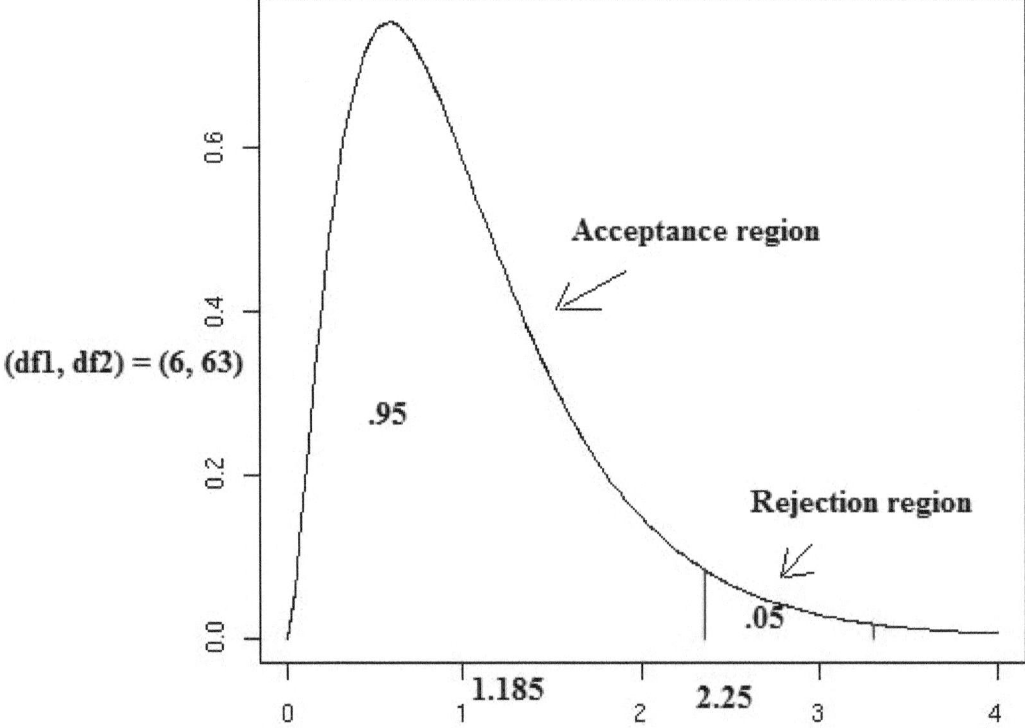

Figure 2g

Only if the variance between is large in comparison to the variance within will we get a statistically significant result, as in **Example 2c**, and the equality of variance test must be satisfied.

Example 2c.

Open the data set chpt9example2.sav. Draw the boxplot for test 3 and find the F value. Was the F value what you expected from the boxplots?

Solution: As above we'll first draw the boxplots. Use Graphs → Legacy Dialog → Boxplot, click define. Move test 2 under Variable and move groups under Category Axis, click OK. The boxplots should be as in **Figure 2h**.

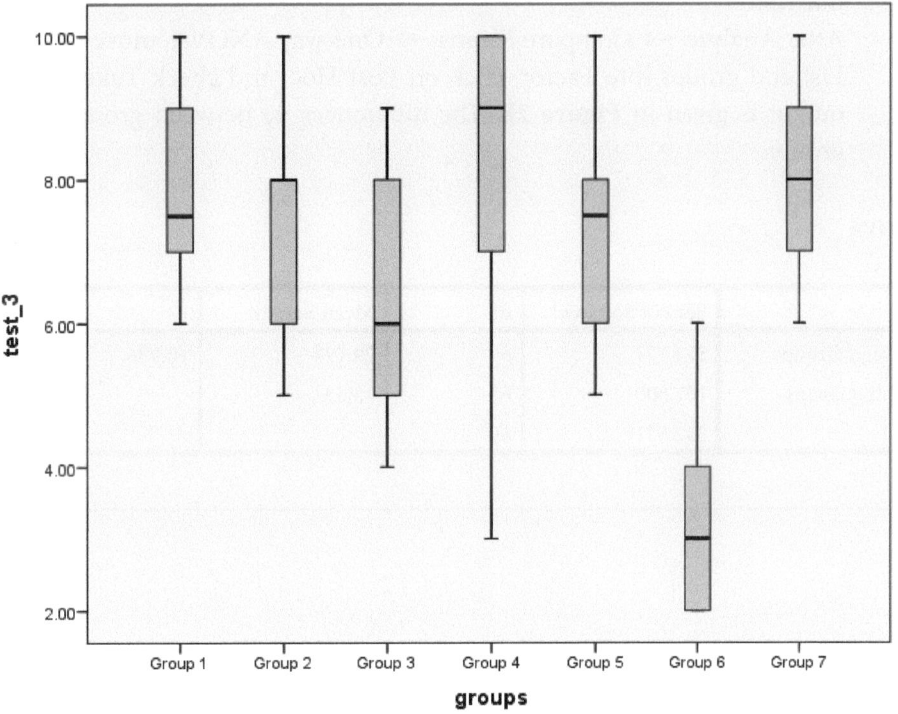

Figure 2h

Since the variance between seems to be much larger than the average variance within, we should expect the F value to be significant. We will do a hypothesis test using the eight-step process.

Step 1. H_0 : All the means are equal.

Step 2. H_1 : At least one mean is different.

Step 3. $\alpha = .05$

Step 4. The test will be One-Way ANOVA because we have several groups and one factor—the radio stations. Assumptions are the different groups have a normal population, each group is independent of the other groups, and the variances within groups are about the same.

Step 5. Use Analyze → Compare Means → One-way ANOVA, move test 3 into Dependent List and groups into Factor. The obtained *F* value is 10.936, df1 = 6, df2 = 63 and the obtained *p*-value = .0; see **Figure 2i**.

Step 6. The critical *F* value is the same as in Example 2a since all the parameters are the same. The critical *F* value = 2.25.

Step 7. Compare the obtained *F* value 10.936 against the critical *F* value 2.25 to get 10.936 > 2.25, or alternatively compare the obtained the *p*-value .0 against α to get .0 < .05—we reject the null hypothesis; see **Figure 2j**.

Step 8. Our decision is to reject the null hypothesis; it appears there is a difference of the means in the groups for the test 3 data set. To find where the difference(s) lie we will do a post hoc test.

Step 9. After Analyze → Compare Means → One-way ANOVA, move test 3 into Dependent List and groups into Factor, click on Post Hoc, and check Tukey, continue, OK. The output is given in **Figure 2k**. The differences lie between group 6 and all the other groups.

ANOVA

test_3

	Sum of Squares	df	Mean Square	F	Sig.
Between Groups	174.771	6	29.129	10.936	.000
Within Groups	167.800	63	2.663		
Total	342.571	69			

Figure 2i

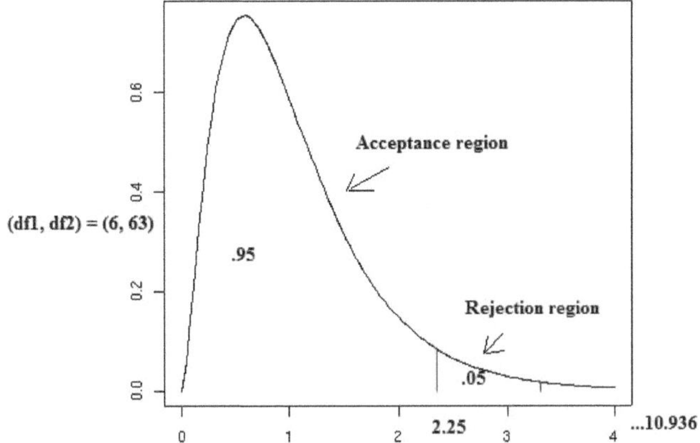

Figure 2j

Post Hoc Tests

Multiple Comparisons

Dependent Variable: test_3
Tukey HSD

(I) groups	(J) groups	Mean Difference (I-J)	Std. Error	Sig.	95% Confidence Interval	
					Lower Bound	Upper Bound
Group 1	Group 2	.30000	.72986	1.000	-1.9229	2.5229
	Group 3	1.30000	.72986	.565	-.9229	3.5229
	Group 4	-.50000	.72986	.993	-2.7229	1.7229
	Group 5	.40000	.72986	.998	-1.8229	2.6229
	Group 6	4.50000*	.72986	.000	2.2771	6.7229
	Group 7	-.10000	.72986	1.000	-2.3229	2.1229
Group 2	Group 1	-.30000	.72986	1.000	-2.5229	1.9229
	Group 3	1.00000	.72986	.815	-1.2229	3.2229
	Group 4	-.80000	.72986	.927	-3.0229	1.4229
	Group 5	.10000	.72986	1.000	-2.1229	2.3229
	Group 6	4.20000*	.72986	.000	1.9771	6.4229
	Group 7	-.40000	.72986	.998	-2.6229	1.8229
Group 3	Group 1	-1.30000	.72986	.565	-3.5229	.9229
	Group 2	-1.00000	.72986	.815	-3.2229	1.2229
	Group 4	-1.80000	.72986	.189	-4.0229	.4229
	Group 5	-.90000	.72986	.878	-3.1229	1.3229
	Group 6	3.20000*	.72986	.001	.9771	5.4229
	Group 7	-1.40000	.72986	.476	-3.6229	.8229
Group 4	Group 1	.50000	.72986	.993	-1.7229	2.7229
	Group 2	.80000	.72986	.927	-1.4229	3.0229
	Group 3	1.80000	.72986	.189	-.4229	4.0229
	Group 5	.90000	.72986	.878	-1.3229	3.1229
	Group 6	5.00000*	.72986	.000	2.7771	7.2229

	Group 7	.40000	.72986	.998	-1.8229	2.6229
Group 5	Group 1	-.40000	.72986	.998	-2.6229	1.8229
	Group 2	-.10000	.72986	1.000	-2.3229	2.1229
	Group 3	.90000	.72986	.878	-1.3229	3.1229
	Group 4	-.90000	.72986	.878	-3.1229	1.3229
	Group 6	4.10000*	.72986	.000	1.8771	6.3229
	Group 7	-.50000	.72986	.993	-2.7229	1.7229
Group 6	Group 1	-4.50000*	.72986	.000	-6.7229	-2.2771
	Group 2	-4.20000*	.72986	.000	-6.4229	-1.9771
	Group 3	-3.20000*	.72986	.001	-5.4229	-.9771
	Group 4	-5.00000*	.72986	.000	-7.2229	-2.7771
	Group 5	-4.10000*	.72986	.000	-6.3229	-1.8771
	Group 7	-4.60000*	.72986	.000	-6.8229	-2.3771
Group 7	Group 1	.10000	.72986	1.000	-2.1229	2.3229
	Group 2	.40000	.72986	.998	-1.8229	2.6229
	Group 3	1.40000	.72986	.476	-.8229	3.6229
	Group 4	-.40000	.72986	.998	-2.6229	1.8229
	Group 5	.50000	.72986	.993	-1.7229	2.7229
	Group 6	4.60000*	.72986	.000	2.3771	6.8229

*. The mean difference is significant at the 0.05 level.

Figure 2k

Another Reason for Using ANOVA

If we were to compare 7 different groups using 2 at a time and doing a *t* test for each pair, we would have to do 21 different tests. (Drawing diagonals on a circle is like making a pair between two groups, such as pairing group 1 with group 2 is like drawing a line (diagonal) between point 1 and point 2. So to find the number of pairings between groups, we count the number of diagonals between 7 points drawn on a circle. From the first point we can draw 6 diagonals to each of the other six points. From the second point we can draw 5 diagonals to each of the remaining points other than the first point, because a diagonal was already drawn from the first point to the second point. From the third point we don't need to draw diagonals to the first and second points, we only need to draw 4 diagonals to the other four points, etc. The sum total of all the diagonals is 6 + 5 + 4 + 3 + 2 + 1 = 21.) If we perform 21 tests and each test has a 5% chance of making a type I error, then using the mean of a binomial distribution we should have 21(.05) = 1.05 tests with a type I error. Thus one test should have a type I error, which implies the mean was rejected when it should not have been rejected. ANOVA can avoid this scenario by running all the tests at once.

Example 2d.

Open data set chpt9example2.sav. Use Analyze → Compare Means → One-Way ANOVA, move test 2 into the Dependent List and groups in the Factor window. Click on the options tab and check Descriptives on the One-Way ANOVA: Options window; see **Figure 3a**.

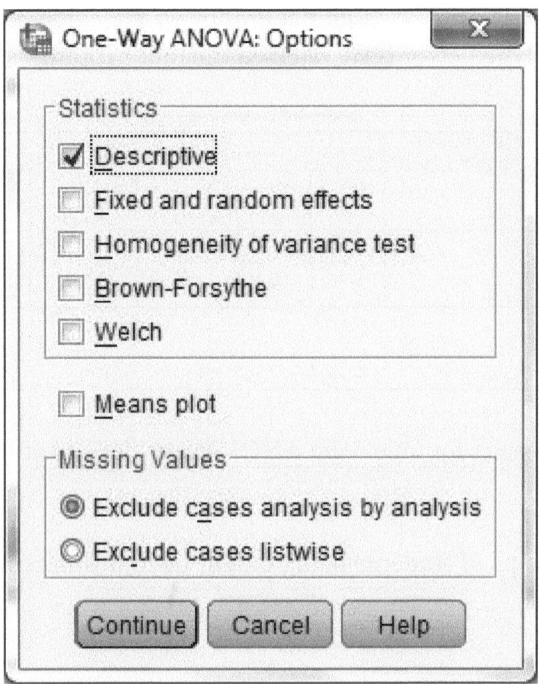

Figure 3a

Click continue and OK. A Descriptives box will appear in the output, see **Figure 3b**. Between which two groups would a type I error exist?

Descriptives

test_2

	N	Mean	Std. Deviation	Std. Error	95% Confidence Interval for Mean		Minimum	Maximum
					Lower Bound	Upper Bound		
Group 1	10	32.9000	18.38145	5.81273	19.7507	46.0493	12.00	66.00
Group 2	10	39.7000	25.29405	7.99868	21.6057	57.7943	12.00	92.00
Group 3	10	52.1000	19.70025	6.22977	38.0073	66.1927	27.00	82.00
Group 4	10	38.6000	26.15424	8.27070	19.8904	57.3096	11.00	75.00
Group 5	10	37.7000	25.02465	7.91349	19.7984	55.6016	12.00	85.00
Group 6	10	55.3000	25.30283	8.00146	37.1994	73.4006	13.00	88.00
Group 7	10	43.2000	23.77580	7.51857	26.1918	60.2082	15.00	68.00
Total	70	42.7857	23.73641	2.83704	37.1260	48.4455	11.00	92.00

Solution: The mean of group 1 is 32.9, which is the lowest mean and the mean of group 6 is 55.3, which is the highest mean, so we might expect a t test to indicate a difference of means between these two groups. Indeed an Independent-Samples T Test yields a t-value of -2.265 and a p-value of .036, which implies we should reject the null hypothesis; see **Figure 3c**. Since the ANOVA test has an F-value of 1.185 and a p-value of .326, in context of all the groups we conclude there is no difference in means. Thus we would have made a type I error if we rejected the null hypothesis that all means are equal.

Independent Samples Test

		Levene's Test for Equality of Variances		t-test for Equality of Means						
									95% Confidence Interval of the Difference	
		F	Sig.	t	df	Sig. (2-tailed)	Mean Difference	Std. Error Difference	Lower	Upper
test_2	Equal variances assumed	.789	.386	-2.265	18	.036	-22.40000	9.88995	-43.17801	-1.62199
	Equal variances not assumed			-2.265	16.430	.037	-22.40000	9.88995	-43.32125	-1.47875

Figure 3c

Now that we have all the basics for One-Way ANOVA tests, let's do another example to reinforce the concepts.

Example 3.

In this example we'll test groups of students being taught by four different methods. The results on the final are as follows:

Table 1

Method 1	Method 2	Method 3	Method 4
92	87	94	78
92	85	91	76
90	82	90	75
90	79	88	74
89	79	88	72
86	77	87	72
75	77	77	72

Do the results on the final indicate the methods are different?

Solution: Input the data into an SPSS data editor, chpt9example3.sav. Since we have 7 students in each method, we want to enter 1 seven times in the first column, followed by entering 2 seven times in the first column, followed by entering 3 seven times, and 4 seven times; see **Figure 4a**. In the next column put the values corresponding to the different methods next to the group numbers, as in **Figure 4a**. Change the variable names in variable view to method and finalscores, also change the decimals to 0, and change the measure of methods to Nominal and measure of finalscores to scale. Now we will run the F-test.

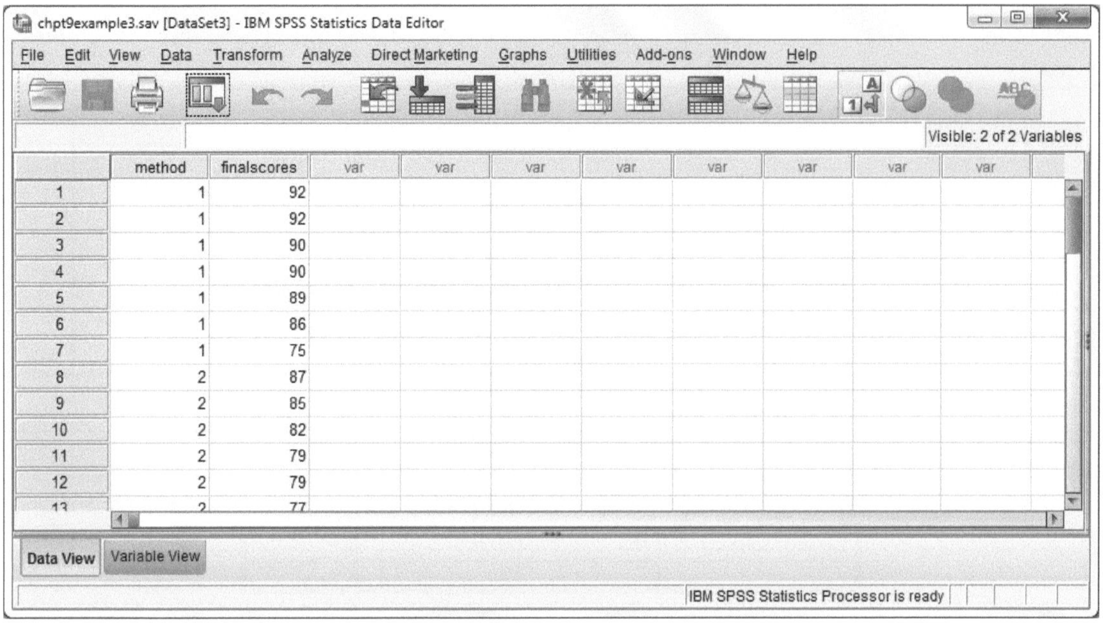

chpt9example3.sav [DataSet3] - IBM SPSS Statistics Data Editor

File Edit View Data Transform Analyze Direct Marketing Graphs Utilities Add-ons Window Help

Visible: 2 of 2 Variables

	method	finalscores	var	var	var	var	var	var	var	var
1	1	92								
2	1	92								
3	1	90								
4	1	90								
5	1	89								
6	1	86								
7	1	75								
8	2	87								
9	2	85								
10	2	82								
11	2	79								
12	2	79								
13	2	77								

Data View | Variable View

IBM SPSS Statistics Processor is ready

Figure 4a

Step 1. H_0: All means are the same.

Step 2. H_1: At least one mean is different.

Step 3. $\alpha = .05$

Step 4. We will use One-Way ANOVA, which is a right-tailed test. We assume the data in each group are independent and the data come from populations that are normal. We also assume variances in each group are equal.

Step 5. Find the *F*-value, df1, df2 and the *p*-value. Use Analyze → Compare Means → One-Way ANOVA, move finalscores into the Dependent List and Method into the Factor window. Click on the tab Options and check Descriptive and Homogeneity of variance test, continue, OK. The *p*-value for test of Homogeneity of variance is .569, which is greater than .05 so we can apply ANOVA. (Even if the test of Homogeneity of variance *p*-level is less than .05, we still can oftentimes use the ANOVA result because the ANOVA test is very robust and can still detect differences for the *F*-value if Homogeneity of variance is less than .05.) The obtained *F*-value is 14.093, df1 = 3, df2 = 24, and the obtained *p*-value is .000; see **figures 4b, 4c, 4d.**

Step 6. Find a critical *F*-value. The *df* between groups is 3, (4 – 1), and the *df* within groups is 24, (28 – 4). Using Transform → Compute Variable to find a critical *F*-value we get cf = 3.01.

Step 7. Compare the obtained *F*-value, 14.093 against the critical *F*-value, 3.01 to get 14.093 > 3.01. Alternatively comparing the *p*-value .000 with .05 we get .000 < .05. Either way we reject the null hypothesis.

Step 8. Our decision is to reject the null hypothesis; this implies at least one mean is different. We will do a post hoc test to see where the differences lie.

Descriptives

finalscores

	N	Mean	Std. Deviation	Std. Error	95% Confidence Interval for Mean		Minimum	Maximum
					Lower Bound	Upper Bound		
1	7	87.71	5.964	2.254	82.20	93.23	75	92
2	7	80.86	3.934	1.487	77.22	84.50	77	87
3	7	87.86	5.336	2.017	82.92	92.79	77	94
4	7	74.14	2.340	.884	71.98	76.31	72	78
Total	28	82.64	7.222	1.365	79.84	85.44	72	94

Figure 4b

Test of Homogeneity of Variances

finalscores

Levene Statistic	df1	df2	Sig.
.687	3	24	.569

Figure 4c

ANOVA

finalscores

	Sum of Squares	df	Mean Square	F	Sig.
Between Groups	898.429	3	299.476	14.093	.000
Within Groups	510.000	24	21.250		
Total	1408.429	27			

Figure 4d

Step 9. Post Hoc test. Use Analyze → Compare Means → One-Way ANOVA, the variable finalscores should be in the dependent list and the variable method in the factor window. Click on the tab Post Hoc and check Tukey, then Continue and OK. The output for Tukey post hoc test tells us that only methods 1 and 3 are the same, otherwise all the methods are different. See **Figure 4e**.

Multiple Comparisons

Dependent Variable: finalscores

Tukey HSD

(I) method	(J) method	Mean Difference (I-J)	Std. Error	Sig.	95% Confidence Interval	
					Lower Bound	Upper Bound
1	2	6.857*	2.464	.047	.06	13.65
	3	-.143	2.464	1.000	-6.94	6.65
	4	13.571*	2.464	.000	6.77	20.37

2	1	-6.857*	2.464	.047	-13.65	-.06
	3	-7.000*	2.464	.042	-13.80	-.20
	4	6.714	2.464	.054	-.08	13.51
3	1	.143	2.464	1.000	-6.65	6.94
	2	7.000*	2.464	.042	.20	13.80
	4	13.714*	2.464	.000	6.92	20.51
4	1	-13.571*	2.464	.000	-20.37	-6.77
	2	-6.714	2.464	.054	-13.51	.08
	3	-13.714*	2.464	.000	-20.51	-6.92

*. The mean difference is significant at the 0.05 level.

Figure 4e

These differences could've been deduced by observing the descriptive output; see **Figure 4b**. The means of methods 1 and 3 are very close, while the means of all the other two methods are below these two means. In this case the Tukey post hoc test merely confirms these observations.

The Homogeneous Subsets test compares those groups that have close means to see if they are significantly close or significantly different. In this case we get the same significance we had in the post hoc test; see **Figure 4f**. We would accept the null hypothesis that methods 1 and 3 are the same and accept the null hypothesis that methods 2 and 4 are the same.

Homogeneous Subsets

finalscores

Tukey HSD

method	N	Subset for alpha = 0.05	
		1	2
4	7	74.14	
2	7	80.86	
1	7		87.71
3	7		87.86
Sig.		.054	1.000

Means for groups in homogeneous subsets are displayed.

a. Uses Harmonic Mean Sample Size = 7.000.

Figure 4f

Let's review what we've learned in this chapter.

1. One-Way ANOVA is used when we have more than 2 groups and one factor.
2. One-Way ANOVA is more dependable than several *t*-tests.

3. The *F*-value is the test value for One-Way ANOVA.
4. The *F*-value is tested against the critical value or the *p*-value against α.
5. If the test value is significant we perform a post hoc test to see where the differences lie.
6. The *F*-value has two degrees of freedom, one for the numerator and one for the denominator.
7. A critical *F*-value can be found using the inverse function group in SPSS.
8. The sizes of each group need not be the same; some groups can be larger or smaller than other groups.

Class Work 9

Another test of preschool children was performed on three groups. Group 1 used arbitrary play toys, group 2 used the educational toys, and group three used any play toys and interaction with a trained adult counselor. The adult counselor played with the child while stimulating the child's curiosity and creativity with questions and interaction. Recall playtime stimulates a child's imagination. The results of the test are given in the following table. chpt9classwork

Table 2

Group 1	Group 2	Group 3
7	6	8
5	9	10
9	5	7
1	5	6
6	8	8
4	1	5
7	3	6
8	8	8
9	8	7
2	6	7
4	7	6
5	2	4
3	3	4
8	5	7
9	4	6
1	2	3

Test at the 5% level if one mean is different from the others.

Homework 9.

1. 1. If a data set has 36 data values placed into 3 different groups, then what is *df* between and *df* within?

2. 2. If a data set has 45 data values placed into 4 different groups, then what is *df* between and *df* within?

3. 3. If a data set has 64 data values placed into 3 different groups, then what is *df* between and *df* within?

4. 4. If a data set has 72 data values placed into 6 different groups, then what is *df* between and *df* within?

5. 5. Test at the 5% level if these three groups have the same mean. chpt9problem5.sav

Group A	Group B	Group C
52	78	24
31	93	38
61	88	23
59	77	50
62		39
		90

6. 6. Test at the 1% level if these four treatments have the same mean. chpt9problem6.sav

Treatment 1	3	2	3	4	
Treatment 2	5	6	7	5	6
Treatment 3	1	7	5	4	6
Treatment 4	2	2	1		

7. 7. Studies in how we remember are ongoing and changing with new methods being developed continually to help us learn to remember. Studies also reveal that different methods are more appropriate for different age levels. The following method was applied to different age groups with the results given in the table. The data values are new words remembered using the new method. Using Analysis of Variance for which age group would the method be most appropriate, or is the method the same for all age groups? Use $\alpha = .05$ chpt9problem7.sav

age 7 to 14	10	11	16	15	12	11	10	12	13	17
age 15 to 22	21	19	17	15	16	22	22	18	21	22
age 41 to 48	16	16	15	11	12	14	15	18	19	14
age 49 to 56	10	19	9	10	11	12	10	11	11	14

8. 8. The following problem is taken from Statistical Methods for Psychology by David C. Howell. He says, to investigate the maternal behavior of laboratory rats, we move the rat pup a fixed distance from the mother and record the time (in seconds) required for the mother to retrieve the pup to the nest. We run the study with 5-, 20-, and 35-day-old pups. The data are given below for the six pups per group.

5 days:	15	10	25	15	20	18
20 days:	30	15	20	25	23	20
35 days:	40	35	50	43	45	40

Do the data suggest the means are different? If they are different, what implications might it have for human behavior? chpt9problem8.sav

9. 9. In problem 5 the p-value was significant. Do a post hoc test to see where the differences lie. Use a 5% level of significance.

10. 10. In problem 6 the p-value was significant. Do a post hoc test to see where the differences lie. Use a 1% level of significance.

11. 11. In problem 7 the p-value was significant. Do a post hoc test to see where the differences lie. Use a 5% level of significance.

12. 12. In problem 8 the p-value was significant. Do a post hoc test to see where the differences lie. Use a 5% level of significance.

13. 13. Find the critical F-value for the data in problem #5.

14. 14. Find the critical F-value for the data in problem #6.

15. 15. Find the critical F-value for the data in problem #7.

16. 16. Find the critical F-value for the data in problem #8.

17. 17. Explain why an ANOVA test is more beneficial than several different independent t tests.

18. 18. Using the data file labeled Airline.sav, test if the means for millions of flight miles (flights) are the same for each General Geographic region (region). Since one of the greatest contributors to global warming is jet exhaust in the upper atmosphere, what does this say about which region has more responsibility to offset the onslaught of global warming?

Just for fun: Paul, Henry, Sue, and Ellen all participated in four events: pie-eating contest, hotdog-eating contest, sack race, and egg toss. Each person took the positions 1 through 4, each contest gave the positions 1 through 4, and no person or contest repeated a position. Henry came in first in the pie-eating contest, Paul came in second in the egg toss, Sue came in third in the pie-eating contest, and Ellen came in last in the egg toss. If Paul was either first or fourth in the sack race, and Sue was first or second in the sack race, then who came in last in the sack race?

A.) Paul B.) Henry C.) Sue D.) Ellen E.) Cannot be determined.

Reference

(1) Jeopardy

Chapter 10

MANOVA

Objectives

After completing this chapter, a student should know:

- ✓ The difference between ANOVA and MANOVA
- ✓ How Boxplots can indicate a difference in means
- ✓ The steps to perform a MANOVA test.

Multivariate Analysis of Variance (MANOVA) is similar to ANOVA except we have one factor (the groups) and two or more dependent variables (scale variables of interest).

Research Questions

1. What are the main effects of the independent variables?
2. What are the interactions among the independent variables?
3. What is the importance of the dependent variables?
4. What is the strength of association between dependent variables?
5. What are the effects of covariates? How may they be used?

Assumptions

- ✓ One independent variable (with nominal measure) consisting of two or more groups (categories).

✓ Two or more dependent variables that have scale measure.
✓ We assume the variables are random and from populations that are normal.
✓ We assume equality of variance between the independent groups (homogeneity of variance).
✓ We assume that cases (rows in a data editor) are independent.

MANOVA can provide a clearer view of the differences between groups over several variables than ANOVA can, which provides the difference between groups over one variable. MANOVA enables the researcher to examine the significance between different dependent variables and the groups in the factor, a process which is not possible doing several ANOVA tests for each dependent variable. The following example will demonstrate multivariate analysis of variance (MANOVA) as compared to univariate analysis of variance (ANOVA). (Up to this point we are concerned only with one-way ANOVA and one-way MANOVA. That is, we use only one factor or one independent variable. In chapter 15 we will include discussions with two factors or two independent variables. At that point we'll evaluate the interactions among the independent variables as well as their effect on the dependent variable.)

Example 1.

Preschool children are randomly selected and placed in one of three groups. The groups are children who have free access to play with any type of toys, children who are placed in rooms that only have educational developing toys, and children who are placed in an environment that has adult instructors who help guide their activities. They are then given a ten-question test on critical thinking. (A type of question is "If the red clown gets a blue balloon and the green clown gets a red balloon, then what color balloon should the blue clown get?" The "correct" answer is "green balloon.") After each question they are asked how sure they are that their answer is correct, either 0, not sure; or 1, sure. The data are given in chpt10example1.sav. Test if there is a difference in means.

Solution: Open the data set chpt10example1.sav. The three groups are the same as in classwork 9; that is, Group 1 used arbitrary play toys, group 2 used the educational toys, and group 3 used any play toys and interaction with a trained adult counselor. The adult counselor played with the child while stimulating the child's curiosity and creativity with questions and interaction. First we'll explore the data by creating boxplots; see **Figure 1a**.

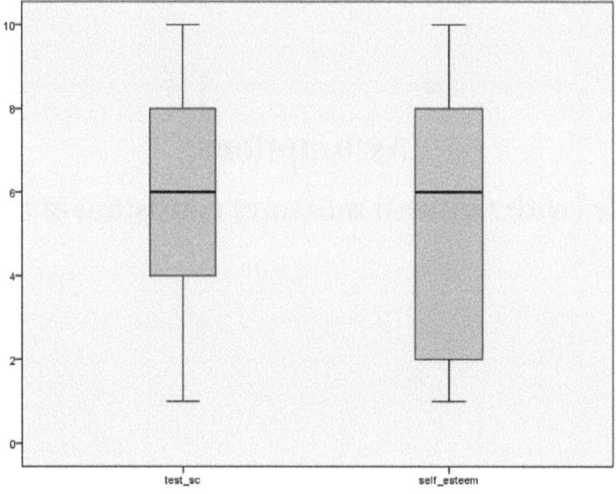

Figure 1a

To get these boxplots use (IBM SPSS) Graphs → Legacy Dialogs → Boxplot, check Summaries of separate variables, click Define, move test_sc and self_esteem into the Boxes Represent dialog box and the nominal variable group into the Label Case by: dialog box; see **Figure 1b**. Click OK. The boxplots will be displayed as above. From the boxplots it appears there is no difference between the medians of the test score variable and the self-esteem variable; hence we might conclude playthings or adult interaction do not have an effect on the critical-thinking development of children. But when we test the hypothesis that there is a difference between the mean vectors using MANOVA, we will find that a difference exists. As always we use the eight-step process in hypothesis testing. (The mean vector for each dependent variable is a single-column vector where each entry is the mean for that subgroup in the factor.)

Step 1. H_0 : the population mean vectors are all equal

Step 2. H_1 : at least one mean vector is different

Step 3. $\alpha = .05$

Step 4. We will test using MANOVA because there are two dependent variables, test_sc and self_esteem and one factor: group. We assume: the variables are random and from populations that are normal; equality of variance between the independent groups; and that cases are independent.

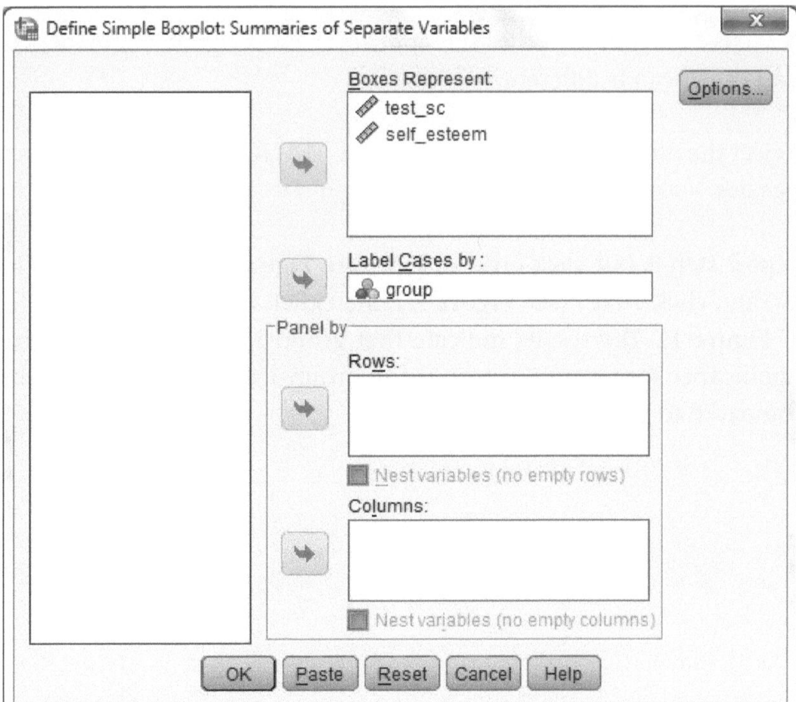

Figure 1b

Step 5. Use Analyze → General Linear Model → Multivariate. Move test_sc and self_esteem into the Dependent Variables dialog box and group into the Fixed Factor box; see **Figure 1c**. Click on Options, move group into Display Means for: and check Descriptive Statistics and Homogeneity Tests; see **Figure 1d**. Click OK and the results will be printed out as in **Figure 1e**. As can be seen in the output the *F*-value for all the tests, Pillai's, Wilks' Lambda, and Hotelling's Trace are all about the same. Roy's Largest Root is the largest eigenvalue and gives an upper bound for the *F*-value and also a lower bound for the significance level if all the other tests are significant. Since Box's test of equality is not significant (.379 > .05), then the *F*-value we will use is Wilks' Lambda. Wilks' Lambda is also the most commonly reported MANOVA statistic. (1) The *F*-value for group is 3.150 and the significance level is .018.

Step 6. Find the critical *F*-value. To find the critical *F*-value we need to find df1 and df2. The calculations for df1 and df2 for the different test statistics are rather complex so we'll forgo their derivations, and instead we'll use the results given in the Multivariate Tests output, **Figure 1e**. df1 is the Hypothesis df and df2 is the Error df. In another column of the data set put .95 as the first entry and call the variable *a* for area. Use Transform → Compute Variable, call the target variable *cf* for critical *F*-value, and enter IDF.F(*a*, 4, 88) in the numeric expression; see **Figure 1g**. Click OK and the critical *F*-value is 2.48; see **Figure 1h**.

Step 7. Compare. The obtained *F*-value 3.150 is greater than the critical *F*-value 2.48. Alternatively the *p*-value is .018, which is less than the significance level of .05.

Step 8. We reject the null hypothesis; it appears there is significant evidence to indicate that at least one mean is different.

Whenever we reject the null hypothesis in ANOVA or MANOVA, we always do a post hoc test to see where the difference lies.

Step 9. Repeat step 5, but click on Post Hoc (see Figure 1c), move group into Post Hoc Tests for: and click Tukey (see **Figure 1i**), then click Continue and OK. The results are given in **Figure 1j**. The results indicate that group 3 is different from groups 1 and 2. The significance for group 3 compared to group 1 is .010 and the significance for group 3 compared to group 2 is .008.

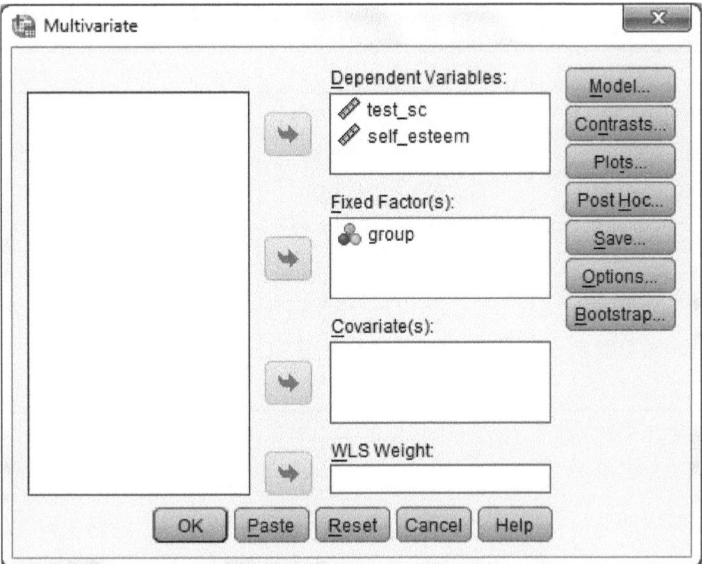

Figure 1c

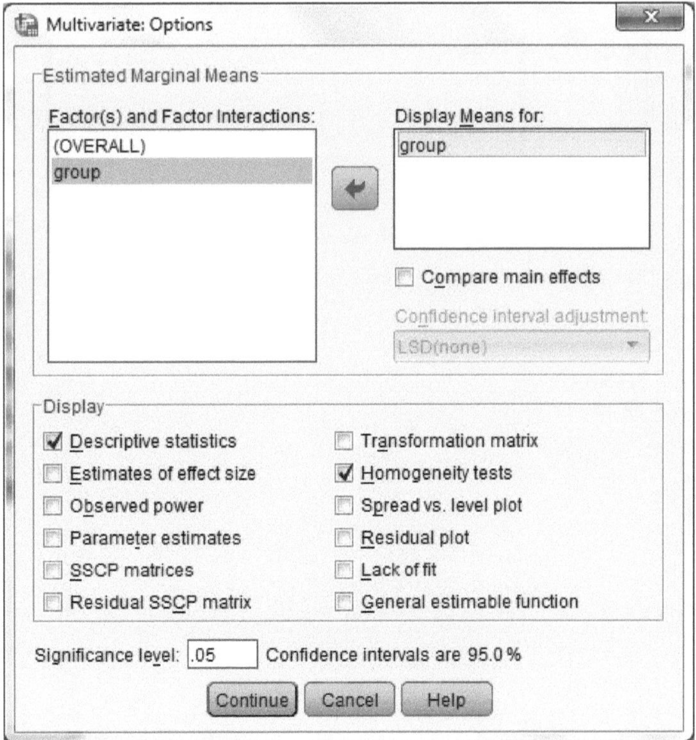

Figure 1d

Multivariate Tests[a]

Effect		Value	F	Hypothesis df	Error df	Sig.
Intercept	Pillai's Trace	.902	203.461[b]	2.000	44.000	.000
	Wilks' Lambda	.098	203.461[b]	2.000	44.000	.000
	Hotelling's Trace	9.248	203.461[b]	2.000	44.000	.000
	Roy's Largest Root	9.248	203.461[b]	2.000	44.000	.000
group	Pillai's Trace	.236	3.006	4.000	90.000	.022
	Wilks' Lambda	.765	3.150[b]	4.000	88.000	.018
	Hotelling's Trace	.306	3.287	4.000	86.000	.015
	Roy's Largest Root	.302	6.796[c]	2.000	45.000	.003

a. Design: Intercept + group

b. Exact statistic

c. The statistic is an upper bound on F that yields a lower bound on the significance level.

Figure 1e

Box's Test of Equality of Covariance Matrices[a]

Box's M	6.854
F	1.069
df1	6
df2	50469.231
Sig.	.379

Tests the null hypothesis that the observed covariance matrices of the dependent variables are equal across groups.

a. Design: Intercept + group

Figure 1f

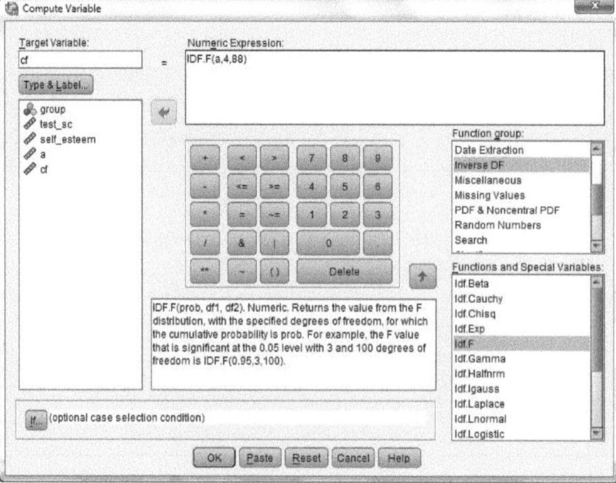

Figure 1g

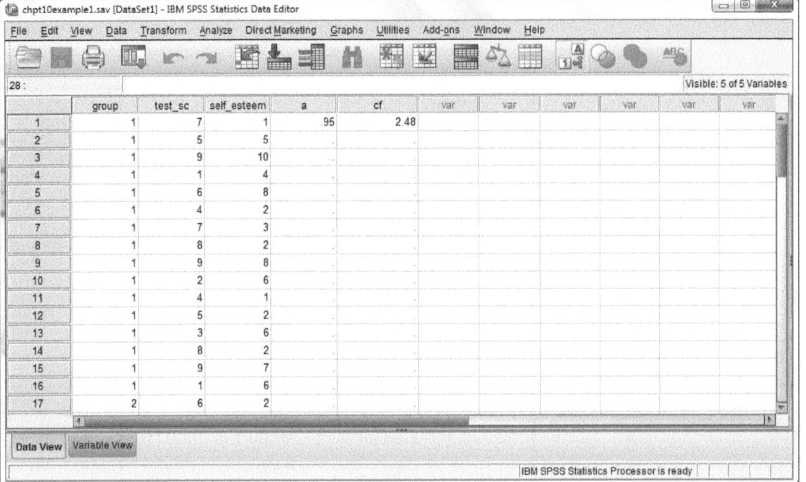

	group	test_sc	self_esteem	a	cf
1	1	7	1	.95	2.48
2	1	5	5	.	.
3	1	9	10	.	.
4	1	1	4	.	.
5	1	6	8	.	.
6	1	4	2	.	.
7	1	7	3	.	.
8	1	8	2	.	.
9	1	9	8	.	.
10	1	2	6	.	.
11	1	4	1	.	.
12	1	5	2	.	.
13	1	3	6	.	.
14	1	8	2	.	.
15	1	9	7	.	.
16	1	1	6	.	.
17	2	6	2	.	.

Figure 1h

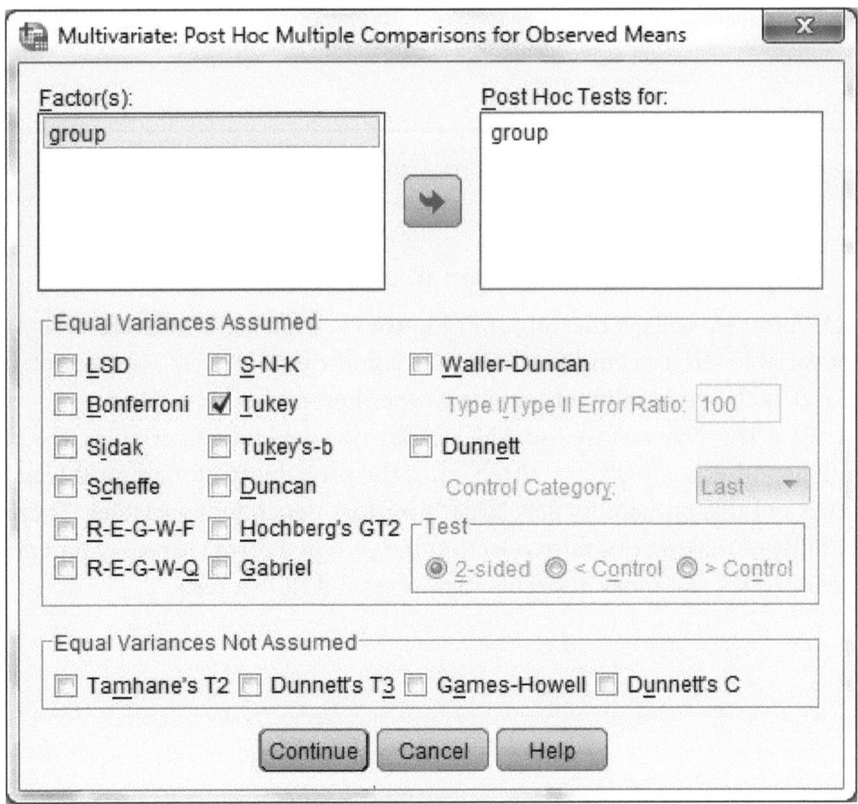

Figure 1i

Multiple Comparisons

Tukey HSD

Dependent Variable	(I) group	(J) group	Mean Difference (I-J)	Std. Error	Sig.	95% Confidence Interval	
						Lower Bound	Upper Bound
test_sc	1	2	.37	.846	.898	-1.67	2.42
		3	-.88	.846	.559	-2.92	1.17
	2	1	-.37	.846	.898	-2.42	1.67
		3	-1.25	.846	.311	-3.30	.80
	3	1	.88	.846	.559	-1.17	2.92
		2	1.25	.846	.311	-.80	3.30
self_esteem	1	2	.06	.918	.997	-2.16	2.29
		3	-2.81*	.918	.010	-5.04	-.59
	2	1	-.06	.918	.997	-2.29	2.16
		3	-2.88*	.918	.008	-5.10	-.65
	3	1	2.81*	.918	.010	.59	5.04
		2	2.88*	.918	.008	.65	5.10

Based on observed means.
The error term is Mean Square(Error) = 6.749.
*. The mean difference is significant at the .05 level.

Figure 1j

Another reason to use MANOVA

If we use an ANOVA test we will get the output in **Figure 1k**, which also indicates that a difference exists in the dependent variable self_esteem. Recall that the significance level, α, is the probability of making a type I error and β is the probability of making a type II error.

If $\alpha = .05$, then $1 - \alpha = .95$, so if we use ANOVA for two dependent variables, the probability of accepting a true null hypothesis is $(.95)^2 = .9025$, thus the probability of making at least one type I error is 1 – .9025 = .0975 and this probability gets larger for more dependent variables. Therefore MANOVA lessens the probability of making one false rejection (i.e., a type I error) for several dependent variables. This makes one MANOVA test more desirable than several ANOVA tests.

ANOVA

		Sum of Squares	df	Mean Square	F	Sig.
test_sc	Between Groups	13.167	2	6.583	1.150	.326
	Within Groups	257.500	45	5.722		
	Total	270.667	47			
self_esteem	Between Groups	86.292	2	43.146	6.393	.004
	Within Groups	303.688	45	6.749		
	Total	389.979	47			

Figure 1k

The Correlation Coefficient

When we use MANOVA, the correlation coefficient between the dependent variables must be moderate. There is no exact cut-off point of the moderate range; however, the range 0.4 to 0.7 is often used. In particular, correlation coefficients close to 1 have high correlation, and correlation coefficients close to 0 have low correlation. To find the correlation coefficient between the two dependent variables in Example 1, use Analyze → Correlate → Bivariate and move the two variables test_sc and self_esteem into the Variables: dialog box; see **Figure 1l**. Click OK. The output is given in **Figure 1m**. The correlation coefficient is .239, which is moderately weak.

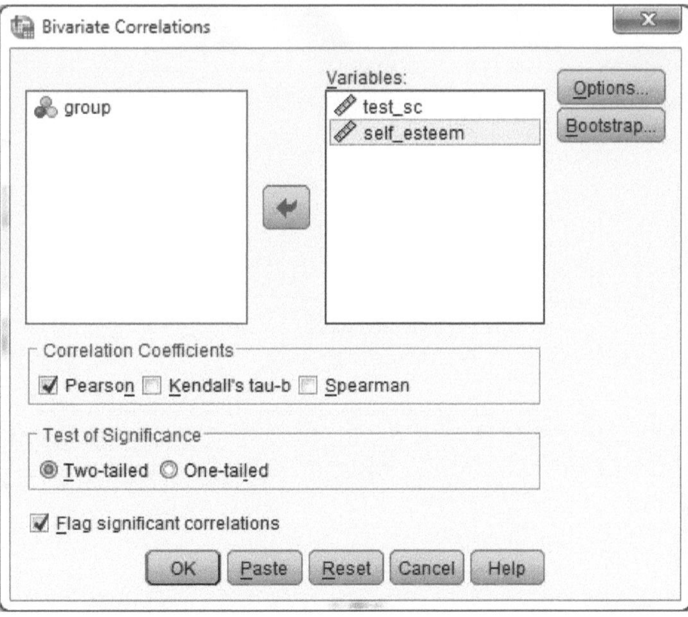

Figure 1l

Correlations

		test_sc	self_esteem
test_sc	Pearson Correlation	1	.239
	Sig. (2-tailed)		.102
	N	48	48
self_esteem	Pearson Correlation	.239	1
	Sig. (2-tailed)	.102	
	N	48	48

Figure 1m

Example 2.

Autism spectrum disorders tend to be highly comorbid with other disorders. That is, the traits of ASD often overlap with symptoms of other disorders, making traditional diagnosis procedures difficult. (2) The data set Autism.sav lists 40 autism patients in four different treatment groups and clinicians measuring on a scale of 1 to 100 the appearance of the other disorders, Tourette syndrome (tics), learning disabilities, obsessive-compulsive disorder (OCD), and attention deficit hyperactivity disorder (ADHD). We will test is to see if the other disorders appear with the same frequency over the different groups and the different variables.

Solution: Open the data set Autism.sav. Use Graphs Legacy Dialogs Boxplots, leave Simple Boxplot checked, and check Summaries for groups of cases; click Define. Move the four scale variables; Tourette syndrome (tics), learning disabilities, obsessive-compulsive disorder and attention deficit hyperactivity disorder into Boxes Represent: dialog box and group into Label Cases by; see **Figure 2a**, and click OK. The boxplots are in **Figure 2b**.

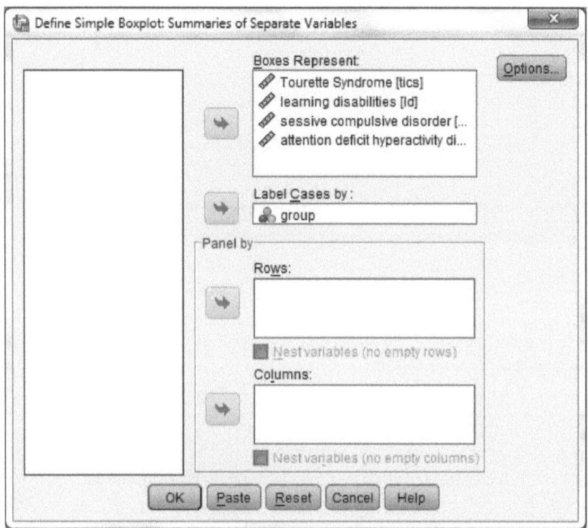

Figure 2a

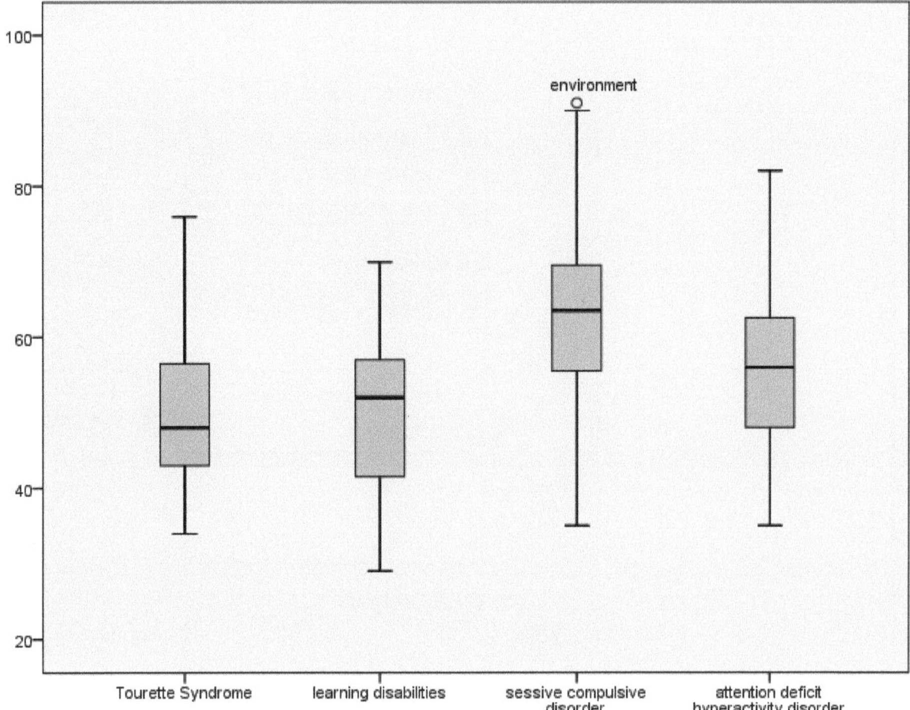

Figure 2b

Note there appears to be a difference in medians (hence means) and the OCD environment mean is the max value for that boxplot, which would indicate that OCD is aggravated more so by the environment. Nonetheless our job as statisticians is not to give reasons for the existence of a difference in means but only to report that a difference in means exists.

Step 1. H_0 : the population mean vectors are all equal

Step 2. H_1 : at least one mean vector is different

Step 3. $\alpha = .05$

Step 4. We will test using MANOVA because there are four dependent variables and one independent variable (factor). If we perform a correlations test for all the dependent variables, we see that the correlation coefficient is moderate; see **Figure 2h**. We assume the variables are random and from populations that are normal, we assume equality of variance between the independent groups and we assume that cases are independent.

Step 5. Find the obtained F-value and the p-value. Use Analyze → General Linear Model → Multivariate. Move the four scale variables, Tourette syndrome (tics), learning disabilities, obsessive-compulsive disorder, and attention deficit hyperactivity disorder, into the Dependent Variables dialog box and into the Fixed Factor dialog box; see **Figure 2c**. Click on Post Hoc, check Tukey, click on Options, and check Descriptive statistics and Homogeneity tests; see **figures 1c** and **1i**. The output is given in **Figure 2d**.

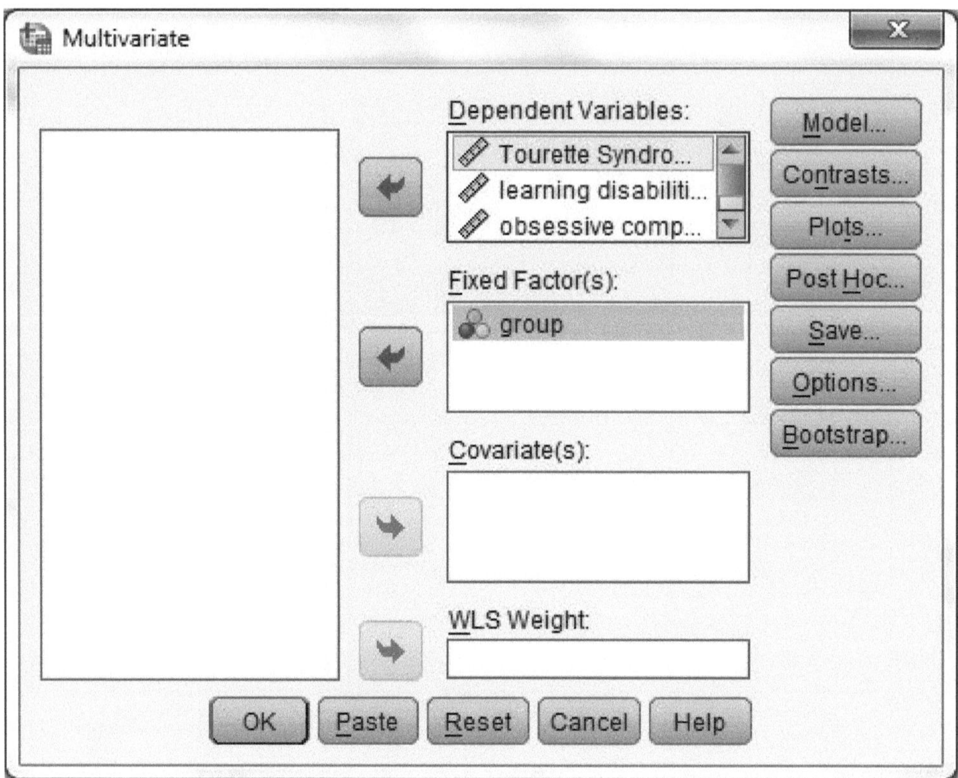

Figure 2c

Multivariate Tests[a]

Effect		Value	F	Hypothesis df	Error df	Sig.
Intercept	Pillai's Trace	.979	378.958[b]	4.000	33.000	.000
	Wilks' Lambda	.021	378.958[b]	4.000	33.000	.000
	Hotelling's Trace	45.934	378.958[b]	4.000	33.000	.000
	Roy's Largest Root	45.934	378.958[b]	4.000	33.000	.000
group	Pillai's Trace	.508	1.783	12.000	105.000	.060
	Wilks' Lambda	.531	1.974	12.000	87.601	.036
	Hotelling's Trace	.810	2.139	12.000	95.000	.021
	Roy's Largest Root	.707	6.188[c]	4.000	35.000	.001

a. Design: Intercept + group

b. Exact statistic

c. The statistic is an upper bound on F that yields a lower bound on the significance level.

Figure 2d

Since Box's test of equality of Covariance Matrices is not significant, see **Figure 2e**, we can use Wilks' Lambda. The *F*-value is 1.974 and the *p*-value is .036. The Hypothesis df (df1) is 12 and the Error df (df2) is 87.601.

Box's Test of Equality of Covariance Matrices[a]

Box's M	31.158
F	.823
df1	30
df2	3563.227
Sig.	.739

Tests the null hypothesis that the observed covariance matrices of the dependent variables are equal across groups.

a. Design: Intercept + group

Figure 2e

Step 6. Find the critical *F*-value. Put .95 into a new variable and call it *a* for area. Use Transform → Compute Variable, call the target variable *cf* for critical *F*-value, and enter IDF.F(*a*, 12, 87.601) in the numeric expression; see **Figure 2f**. Click OK and the critical *F*-value is 1.86.

Step 7. Compare 1.974 > 1.86, alternatively .036 < .05.

Step 8. We reject the null hypothesis; it appears at least one mean vector is different from the others.

Step 9. Post Hoc test. Whenever we reject the null hypothesis in ANOVA or MANOVA, we perform the Tukey Post Hoc test. The output is given in **Figure 2g**. The differences lie in the obsessive compulsive disorder variable. It appears the mean for the Control group is different from the mean for intravenous drugs and the environment.

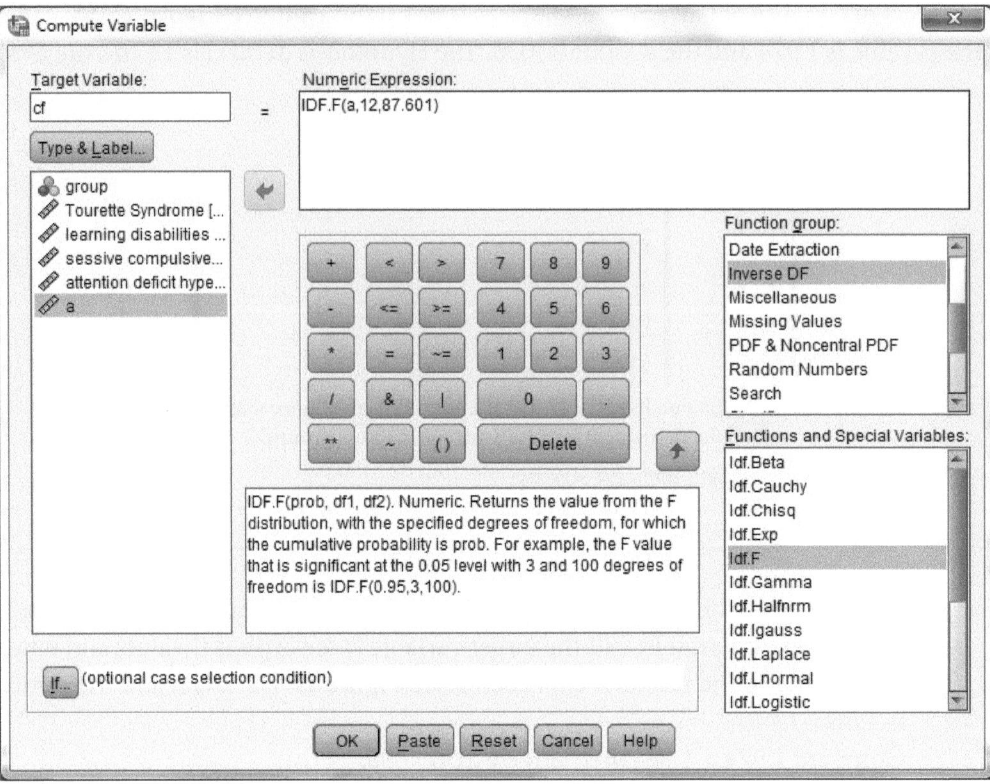

Figure 2f

Multiple Comparisons

Tukey HSD

Dependent Variable	(I) group	(J) group	Mean Difference (I-J)	Std. Error	Sig.	95% Confidence Interval	
						Lower Bound	Upper Bound
Tourette Syndrome	Control	oral drugs	.40	4.609	1.000	-12.01	12.81
		intravenous drugs	.90	4.609	.997	-11.51	13.31
		environment	2.20	4.609	.964	-10.21	14.61
	oral drugs	Control	-.40	4.609	1.000	-12.81	12.01
		intravenous drugs	.50	4.609	1.000	-11.91	12.91
		environment	1.80	4.609	.979	-10.61	14.21
	intravenous drugs	Control	-.90	4.609	.997	-13.31	11.51
		oral drugs	-.50	4.609	1.000	-12.91	11.91
		environment	1.30	4.609	.992	-11.11	13.71
	environment	Control	-2.20	4.609	.964	-14.61	10.21
		oral drugs	-1.80	4.609	.979	-14.21	10.61
		intravenous drugs	-1.30	4.609	.992	-13.71	11.11
learning disabilities	Control	oral drugs	-2.40	4.480	.950	-14.47	9.67
		intravenous drugs	-1.90	4.480	.974	-13.97	10.17
		environment	-.20	4.480	1.000	-12.27	11.87
	oral drugs	Control	2.40	4.480	.950	-9.67	14.47
		intravenous drugs	.50	4.480	.999	-11.57	12.57
		environment	2.20	4.480	.961	-9.87	14.27
	intravenous drugs	Control	1.90	4.480	.974	-10.17	13.97
		oral drugs	-.50	4.480	.999	-12.57	11.57
		environment	1.70	4.480	.981	-10.37	13.77
	environment	Control	.20	4.480	1.000	-11.87	12.27
		oral drugs	-2.20	4.480	.961	-14.27	9.87
		intravenous drugs	-1.70	4.480	.981	-13.77	10.37

obsessive compulsive disorder	Control	oral drugs	-10.90	4.745	.118	-23.68	1.88
		intravenous drugs	-14.10*	4.745	.026	-26.88	-1.32
		environment	-15.90*	4.745	.010	-28.68	-3.12
	oral drugs	Control	10.90	4.745	.118	-1.88	23.68
		intravenous drugs	-3.20	4.745	.906	-15.98	9.58
		environment	-5.00	4.745	.719	-17.78	7.78
	intravenous drugs	Control	14.10*	4.745	.026	1.32	26.88
		oral drugs	3.20	4.745	.906	-9.58	15.98
		environment	-1.80	4.745	.981	-14.58	10.98
	environment	Control	15.90*	4.745	.010	3.12	28.68
		oral drugs	5.00	4.745	.719	-7.78	17.78
		intravenous drugs	1.80	4.745	.981	-10.98	14.58
attention deficit hyperactivity disorder	Control	oral drugs	-1.00	4.814	.997	-13.96	11.96
		intravenous drugs	-3.90	4.814	.849	-16.86	9.06
		environment	-6.00	4.814	.602	-18.96	6.96
	oral drugs	Control	1.00	4.814	.997	-11.96	13.96
		intravenous drugs	-2.90	4.814	.931	-15.86	10.06
		environment	-5.00	4.814	.728	-17.96	7.96
	intravenous drugs	Control	3.90	4.814	.849	-9.06	16.86
		oral drugs	2.90	4.814	.931	-10.06	15.86
		environment	-2.10	4.814	.972	-15.06	10.86
	environment	Control	6.00	4.814	.602	-6.96	18.96
		oral drugs	5.00	4.814	.728	-7.96	17.96
		intravenous drugs	2.10	4.814	.972	-10.86	15.06

Based on observed means.
The error term is Mean Square(Error) = 115.853.
*. The mean difference is significant at the .05 level.

Figure 2g

Correlations

		Tourette Syndrome	learning disabilities	sessive compulsive disorder	attention deficit hyperactivity disorder
Tourette Syndrome	Pearson Correlation	1	.650**	.493**	.458**
	Sig. (2-tailed)		.000	.001	.003
	N	40	40	40	40
learning disabilities	Pearson Correlation	.650**	1	.448**	.729**
	Sig. (2-tailed)	.000		.004	.000
	N	40	40	40	40
sessive compulsive disorder	Pearson Correlation	.493**	.448**	1	.567**
	Sig. (2-tailed)	.001	.004		.000
	N	40	40	40	40
attention deficit hyperactivity disorder	Pearson Correlation	.458**	.729**	.567**	1
	Sig. (2-tailed)	.003	.000	.000	
	N	40	40	40	40

**. Correlation is significant at the 0.01 level (2-tailed).

Figure 2h

Let's review what we've learned in this chapter.

1. One-way MANOVA is used when we have more than two groups and one factor and more than one dependent variable. The correlation coefficient for the dependent variables must be moderate.
2. One-way MANOVA is more dependable than several univariate (ANOVA) tests.
3. The *F*-value is the test value for one-way MANOVA.
4. The *F*-value is tested against the critical value or the *p*-value against α.

5. If the test value is significant, then we perform a post hoc test to see where the differences lie.
6. The *F*-value has two degrees of freedom, one for the numerator and one for the denominator. The numerator degrees of freedom is the Hypothesis df and the denominator degrees of freedom is the Error df.
7. A critical *F*-value can be found using the inverse function group in SPSS.
8. The sizes of each group need not be the same; some groups can be larger or smaller than other groups.

Class Work 10

Use the data set Chpt10classwork.sav to test if the mean vectors for the dependent variables test_sc, self_esteem, and happiness are all the same. Use alpha = 10%. If there is a difference, then use the Tukey Post Hoc test to see where the differences lie.

Homework 10

1. Use the data set chpt10problem1.sav to test if the means of the dependent variables, score, rate, and number are all the same. Test with group as the factor.

2. Use the data set Bowling.sav to test if there is a significant difference among the variables game 1, game 2, and game 3. Test with the factor lane (the factor need not always be a nominal variable).

3. Use the data set Pheresis.sav to test if there is a significant difference among the variables ph and platelet. Test with the factor Machine Code (maccode).

4. Use the data set chpt10problem4.sav to test if there is a significant difference in means of the variables recall, keyword, and time over the factor depth. The depth is different utilization tools for Alzheimer's patients to maintain their cognitive functions as their disease progresses.

5. Draw the boxplots for the data set chpt10problem1.sav in problem 1. Decide if there appears to be a difference in means.

6. Draw the boxplots for the data set Bowling.sav in problem 2. Decide if there appears to be a difference in means.

7. Draw the boxplots for the data set Pheresis.sav in problem 3. Decide if there appears to be a difference in means.

8. Draw the boxplots for the data set chpt10problem4.sav in problem 4. Decide if there appears to be a difference in means.

9. Find the correlation coefficient for the variables dependent variables, score, rate, and number from problem 1 to see if MANOVA is appropriate.

10. Find the correlation coefficient for the variables dependent variables game 1, game 2, and game 3 from problem 2 to see if MANOVA is appropriate.

11. Find the correlation coefficient for the variables dependent variables ph and platelet from problem 3 to see if MANOVA is appropriate.

12. Find the correlation coefficient for the dependent variables recall, keyword, and time from problem 4 to see if MANOVA is appropriate.

13. MANOVA is extremely sensitive to outliers. Open chpt10problem13.sav and run MANOVA and Tukey's Post Hoc test. Notice MANOVA indicates significance but the Post Hoc tests do not indicate significance. Draw a boxplot and notice there are several outliers. These outliers affect the results.

Just for fun. Negate the following statement attributed to Abraham Lincoln: "You can fool some of the people all of the time, and all of the people some of the time, but you cannot fool all of the people all of the time."

References

(1) Advanced and Multivariate Statistical Methods, Mertler and Vannatta, Pyrczak
(2) Wikipedia

Chapter 11

Linear Regression with Correlation Coefficient

Objectives

After completing this chapter, a student should be able to:

- ✓ Draw a scatter/dot plot
- ✓ Find the correlation coefficient
- ✓ Determine if a correlation coefficient is strong enough to imply a correlation
- ✓ Draw the best-fit line
- ✓ Know the difference between a direct and indirect relationship
- ✓ Find the equation of the best-fit line
- ✓ Understand Coefficient of Determination and Coefficient of Alienation
- ✓ Understand Residuals
- ✓ Find a Predicted value and a Confidence Interval for the predicted value
- ✓ Know how Confounding and Lurking variables affect a relationship

One of the goals of a research analyst is to find a regression line that best fits the data of two variables, such as height of son as a function of height of father, or number of sports injuries as a function of hours in training. The statistic that tells us how strongly two variables are related is called the Pearson Product-Moment Correlation Coefficient. The correlation coefficient tells us how strongly the two variables correlate or how close the points are to the best-fit line. Our first step is to draw a scatter plot that relates two variables.

Scatter plots

A scatter plot is simply a location of points in two-dimensional space. Each point has two coordinates, an *x* coordinate and a *y* coordinate. The *x* coordinate is the independent coordinate and the *y* coordinate is the dependent coordinate. The *x*-axis is the horizontal axis and the *y*-axis is the vertical axis. It is expected the reader should know a little about Cartesian coordinates. A scatter plot is a set of points in two-dimensional coordinate space.

Example 1.

Use the data file labeled Breast Cancer Survival and draw a scatter plot for the variables age and time (in months). Use age as the independent variable and time as the dependent variable. Also find the correlation coefficient. Do the points seem to form a line? Draw the best-fit line.

Solution: Open the data file labeled Breast Cancer Survival. To make a Scatter/Dot plot we use Graphs → Legacy Dialogs → Scatter/Dot, click Define the scatter plot as simple scatter; see **Figure 1a**. Put time (months) in the Y axis and Age (years) in the X axis, click OK and the scatter plot will appear; see **Figure 1b**.

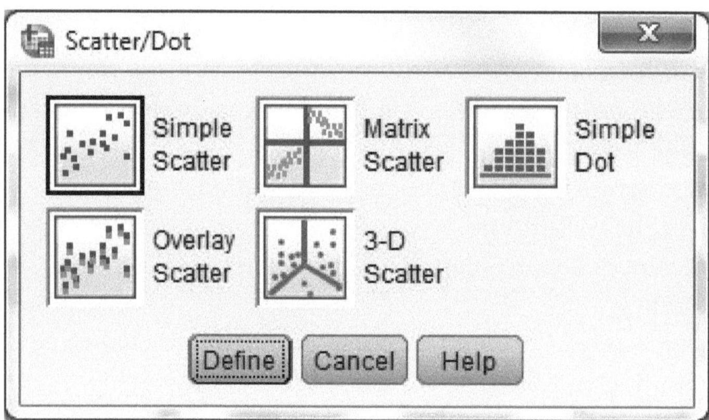

Figure 1a

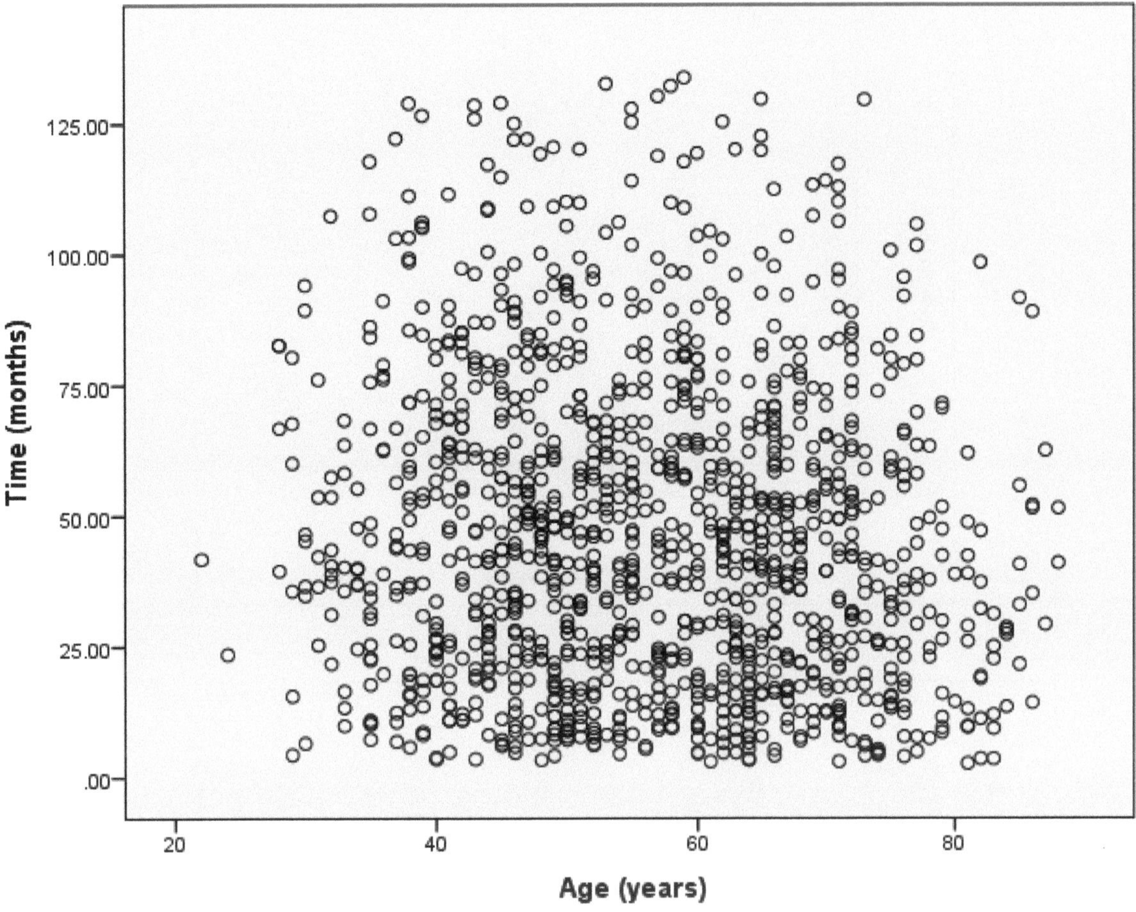

Figure 1b

Since there are a lot of points it's hard to tell which way the dots tend to align, but if we look closely at the scatter plot, we can see there are fewer dots higher up on the extreme right, and more dots on the lower right. It appears there might be a tendency for the line to go down toward the right. This indicates the best-fit line probably has a negative regression. The coefficient of the independent variable will be negative, and the correlation coefficient will also be negative.

To draw the best-fit line, move the cursor over the graph anywhere and right-click on the mouse (or double-click, which will open the chart editor). A pop-up menu will appear; see **Figure 1c**. Select Edit Content and In Separate Window, and a chart editor will appear. In the Chart Editor click on the icon that says "Add Fit Line at Total"; see **Figure 1d**. Or use Elements → Fit Line at Total. Now close the chart editor and the line will be on the graph in the output. As we can see, the line does indeed slope down to the right; see **Figure 1e**.

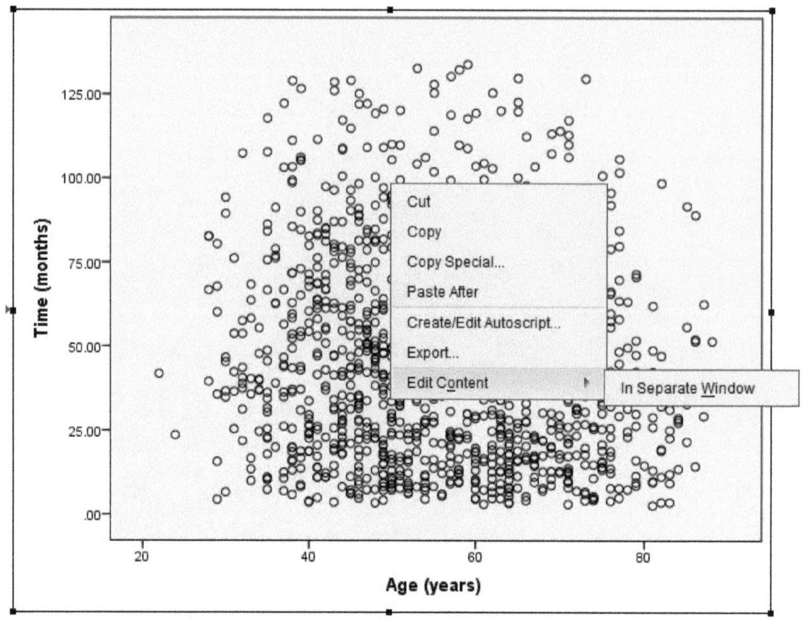

Figure 1c

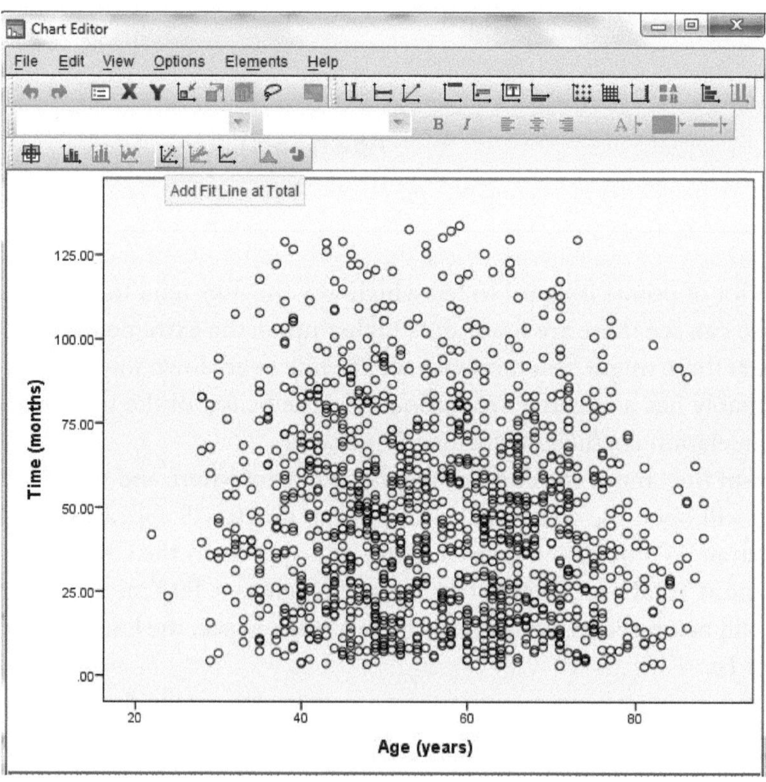

Figure 1d

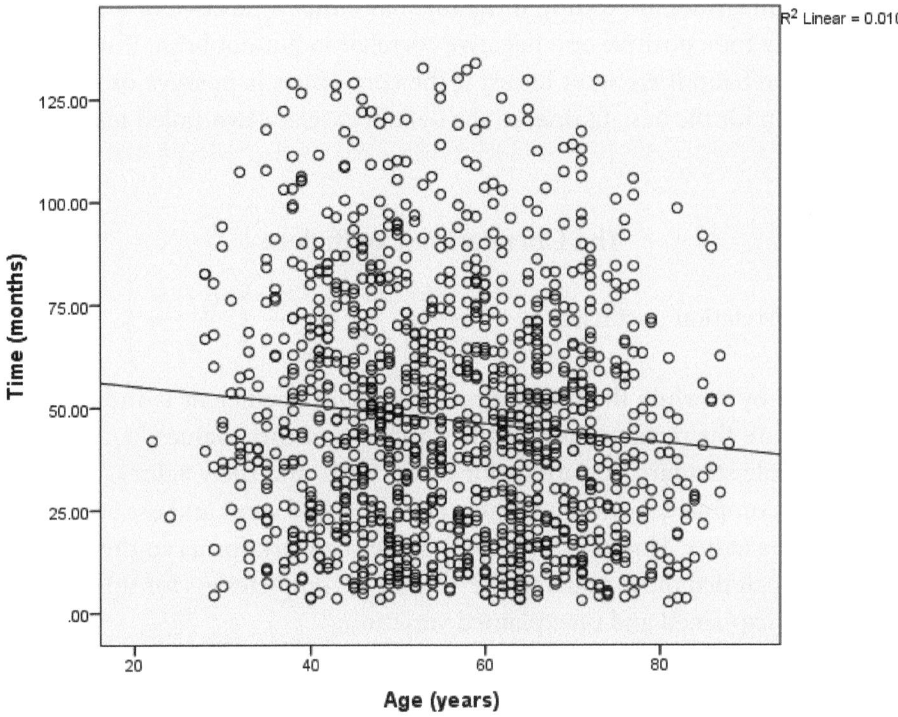

Figure 1e

To find the correlation coefficient we use Analyze → Correlate → Bivariate, move the variables age and time into the variables window, make sure Pearson is checked for the Correlation Coefficient (Pearson is the default mode for the correlation coefficient so it should be checked), and hit OK. The output will be as in **Figure 1f**.

Correlations

		Age (years)	Time (months)
Age (years)	Pearson Correlation	1	-.102**
	Sig. (2-tailed)		.000
	N	1207	1207
Time (months)	Pearson Correlation	-.102**	1
	Sig. (2-tailed)	.000	
	N	1207	1207

**Correlation is significant at the 0.01 level (2-tailed).

Figure 1f

The Pearson correlation coefficient is -.102 and the *p*-value is .0. As we predicted from looking at the scatter plot, the correlation coefficient is negative. SPSS allows us to use a two-tailed or a one-tailed test. A two-tailed test means we are testing if the line has either a positive or a negative correlation. A one-tailed test tests only for a positive or a negative correlation but not both. If we use a one-tailed test, we have to decide beforehand if we want to test if the correlation is positive or negative. Normally we don't assume a direction for the best-fit line, so the default case is a two-tailed test.

The Correlation Coefficient

The formula for the correlation coefficient is $r=\frac{1}{n-1}\sum_{i=1}^{n}\left(\frac{x_i-\overline{x}}{s_x}\right)\left(\frac{y_i-\overline{y}}{s_y}\right)$. The sample correlation coefficient is indicated by *r*, while the population correlation coefficient is indicated by ρ, rho. The sample size is *n*; $\overline{x},\overline{y}$ are the mean values for the *x* values and the *y* values; x_i, y_i are the data values; and s_x, s_y are the sample standard deviations for the *x* values and the *y* values. To find the correlation coefficient without the computer, we need to make tables and find sums and use some short cuts to make the computations a little easier. However, SPSS will do all the work for us so there's no need to use the formula; however, we included the formula in the interest of completeness for this text, and also because it will help understand explained and unexplained variation.

Note the correlation coefficient will remain the same if we add the same constant to either variable, since the mean will also change by the same constant and in taking the difference, that constant will subtract out. If we multiply all values of one variable by the same nonzero constant, then the standard deviation will be multiplied by that same constant, as we saw in Chapter 4, so again the correlation coefficient will remain the same, since that constant will cancel from both the numerator and the denominator. And if we interchange the variables, the formula for the correlation coefficient remains unchanged because of the commutative property of multiplication. Thus adding a constant, multiplying by a constant, or interchanging variables will not affect the correlation coefficient.

The Decision Process for the Correlation Coefficient

In any test we always use the eight-step decision-making process; it is the same for deciding if the correlation coefficient is statistically significant. However, SPSS does not have an inverse correlation coefficient function. Therefore we need to change the correlation coefficient into a *t*-value.

The formula to find *t* is $t=r\sqrt{\frac{n-2}{1-r^2}}$ with *n* – 2 degrees of freedom. Keep in mind that a large sample size will yield a large *t* value, which makes that statistic more significant. As in the Breast Cancer Survival, the correlation coefficient was small, close to 0, which usually indicates there is no correlation. But since the sample size was large, 1,207 data values, then the *t*-value will equal $-.102\sqrt{\frac{1207-2}{1-(-.102)^2}}=-.102\sqrt{\frac{1205}{1-.010404}}\approx-3.6$, which is large in absolute value. The absolute value of this *t*-value is larger than the critical *t*-value of 1.96 (which can be obtained from Table 2 in Chapter 6 since *n* is large). But SPSS also gives us the *p*-value and from the *p*-value we can make a decision. If the

p-value is less than 5%, we reject the null hypothesis and if the p-value is greater than 5%, we accept the null hypothesis.

Degrees of Freedom

The degrees of freedom in a correlation coefficient test is $n - 2$. That is, because we have two means, one for the x variable and one for the y variable. Therefore to find the critical t-value, we use $n - 2$ for the degrees of freedom.

The Null Hypothesis

The status quo assumption for testing if two variables are related is that they are not related: the two variables do not affect each other. For example, we might expect fathers' heights to have somewhat of an effect on sons' heights. That is, the taller the father, the taller the son or the shorter the father, the shorter the son. The null hypothesis will say it doesn't matter if the father is tall or short, the height of the son will be independent of the height of the father. That is, that a father is tall does not imply the son will be tall or short. Also that a father is short does not imply the son will be tall or short. There is no relationship between these two variables. The height of the father will have no effect on the height of the son. The null hypothesis says there is no relationship between two variables. In symbols we write $H_0 : \rho = 0$. ρ, rho, is the population parameter for the correlation coefficient. In words, the null hypothesis says the correlation coefficient for the population will be zero, that there is no relationship between these two variables.

The Alternative Hypothesis

The null hypothesis can be either one-tailed or two-tailed, but SPSS, by default, will always give the p-value of a two-tailed test unless we change it to a one-tailed test. The direction of the alternative hypothesis states which direction the line will flow. If we state the line will go up, then we're saying the correlation coefficient will be positive, the line will have positive slope. Increasing independent values will give increasing dependent values. If we state the line will go down, then we're saying the correlation coefficient will be negative, the line will have negative slope and decreasing independent values will give decreasing dependent values. These will be one-directional one-tailed tests. If we have no idea which way the line will flow, then we use a non-directional two-tailed alternative hypothesis. Symbolically the three choices can be written as follows:

$$H_1 : \rho > 0$$
$$H_1 : \rho < 0$$
$$H_1 : \rho \neq 0$$

The Alpha (Significance) Level

If alpha is not specified, then always choose $\alpha = .05$. If alpha is specified then use the value given for alpha.

Assumptions

The assumptions for the Pearson Correlation Coefficient are that the data for both variables are drawn from populations with a normal distribution and the data do not have outliers.

The Test

We use the correlation coefficient whenever we are trying to establish a correlation between two variables such as a father's height and his son's height. But we also need to decide if it's a one-tailed or two-tailed test, and if it's one-tailed test, we need to decide if it is a direct (positive) correlation or an indirect (negative) correlation.

Perform the Test and Find the Obtained Values

The obtained values we're interested in are the r-value, the t-value, the degrees of freedom, and the p-value. SPSS does not give the t-value in the Correlations output so it needs to be calculated by the above formula or we can use Linear Regression in SPSS and find the t-value in that output. When we use Linear Regression we need to decide which variable will be used for the independent variable and which variable will be used for the dependent variable.

Find the Critical t-Value, Degrees of Freedom

To find the critical t-value we open a new data file and put .025 and .975 into a variable called a for area. (The values .025 and .975 represent the cumulative area for the critical t-values, as in **Figure 2a**. The rejection regions are shaded.) Then we use Transform ⟶ Compute Variable, call the target variable *ct*, and put IDF.T(a, n – 2) into the Numeric Expression window. Click OK. Using 1205 for n – 2; the output is given in **Figure 2b**.

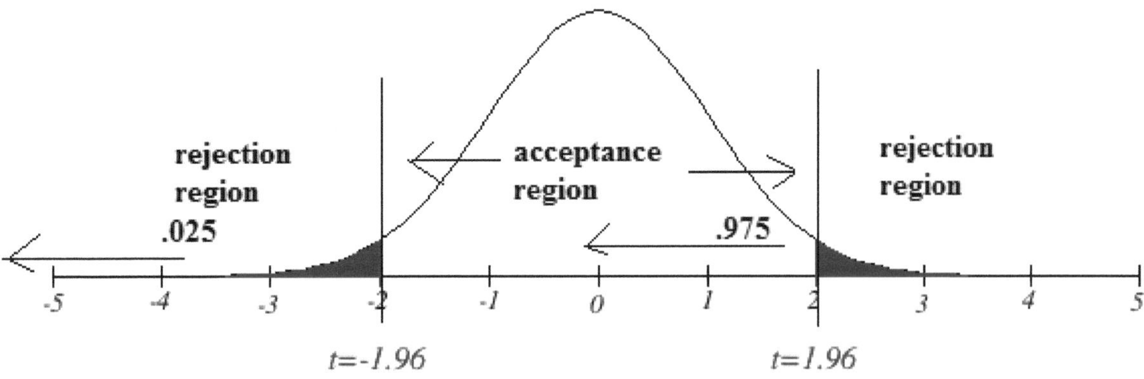

Figure 2a

*Untitled2 [DataSet2] - IBM SPSS Statistics Data Editor

File Edit View Data Transform Analyze Direct Marketing Graphs Utilities Add-ons Window Help

11 : | Visible: 2 of 2 Variables

	a	ct	var	var	var	var	var	var	var	var
1	.025	-1.96								
2	.975	1.96								

Data View | Variable View

IBM SPSS Statistics Processor is ready

Figure 2b

Compare

We always begin by comparing the obtained *t*-value against the critical *t*-value. Alternatively we compare the obtained *p*-value with the significance level, alpha level. Remember SPSS always gives a two-tailed *p*-value so when doing a one-tailed test, be sure to divide the SPSS *p*-value by two or in the Bivariate Correlations window check the box labeled One-tailed; see **Figure 2c**. Then the obtained *p*-value will be for a one-tailed test in the correlations output; see **Figure 2d**.

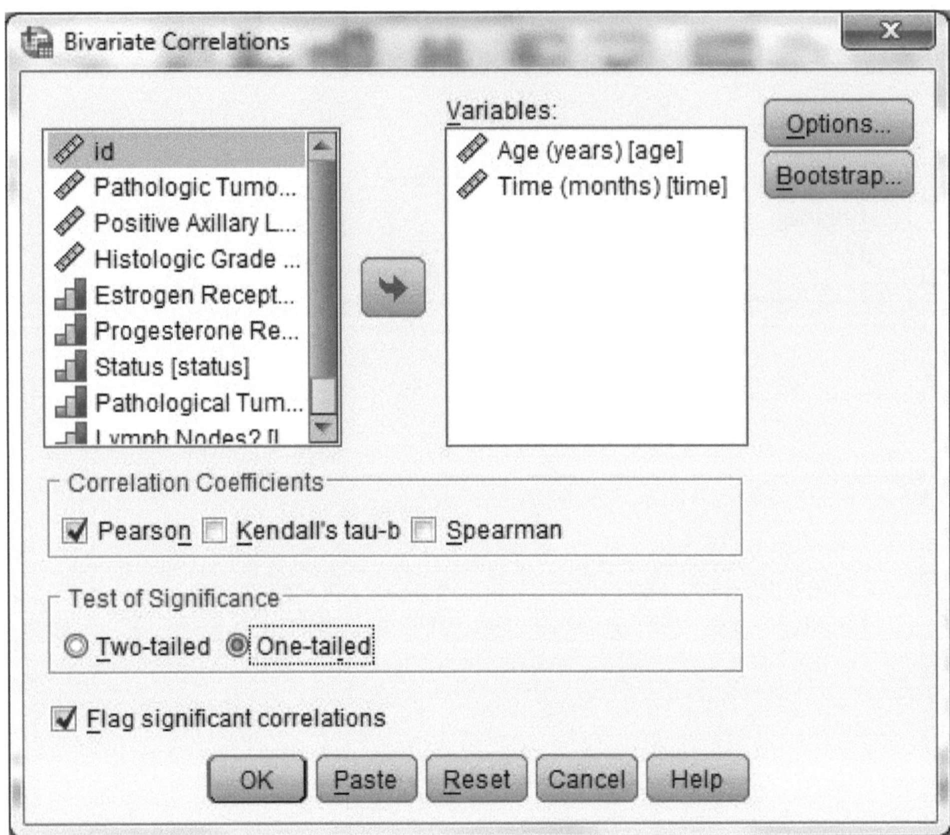

Figure 2c

Correlations

		Age (years)	Time (months)
Age (years)	Pearson Correlation	1	-.102**
	Sig. (1-tailed)		.000
	N	1207	1207
Time (months)	Pearson Correlation	-.102**	1
	Sig. (1-tailed)	.000	
	N	1207	1207

**Correlation is significant at the 0.01 level (1-tailed).

Figure 2d

Decision

Based on the comparison above, make a decision with regard to the problem. Be sure to always write your decision in a complete sentence in context of the problem.

The following three examples compare anxiety levels with twelve patients for different trials. Use Anxiety 2.sav.

Example 2.

Compare Trial 1 with Trial 2. Find the correlation coefficient and test if it is significant. Use either trial as the independent or dependent variable. Alpha = .05.

Solution: Using the eight-step process, we have

Step 1. $H_0: \rho = 0$

Step 2. $H_1: \rho \neq 0$

Step 3. $\alpha = .05$

Step 4. Test the Pearson product-moment correlation coefficient at the 5% level. This is a two-tailed test because we don't know if the coefficient will be positive or negative. Assumptions are the data values for both variables come from normal populations.

Step 5. Use Analyze → Regression → Linear, move Trial 1 into the dependent window and Trial 2 into the independent window (see **Figure 3a**), and hit OK. The output is given in **Figure 3b**. The obtained Pearson correlation coefficient is .488 (under Standardized Coefficients, Beta), the obtained *t*-value is 1.770 (the *t*-value in the same row as Trial 2), degrees of freedom = 10 (Residual df in **Figure 3c**) and the *p*-value is .107.

Step 6. Since $df = 10$, alpha = 5% = .05 and this is a two-tailed test, then the critical *t* value is ± 2.23; see **Figure 3d**.

Step 7. Compare the *t* value, -2.23 < 1.77 < 2.23, we see that the obtained value is in the acceptance region; see **Figure 3e**. Compare the *p*-value, .107 > .05, we see the *p*-value is greater than alpha.

Step 8. We accept the null hypothesis; the anxiety level in Trial 1 is not related to the anxiety level in Trial 2.

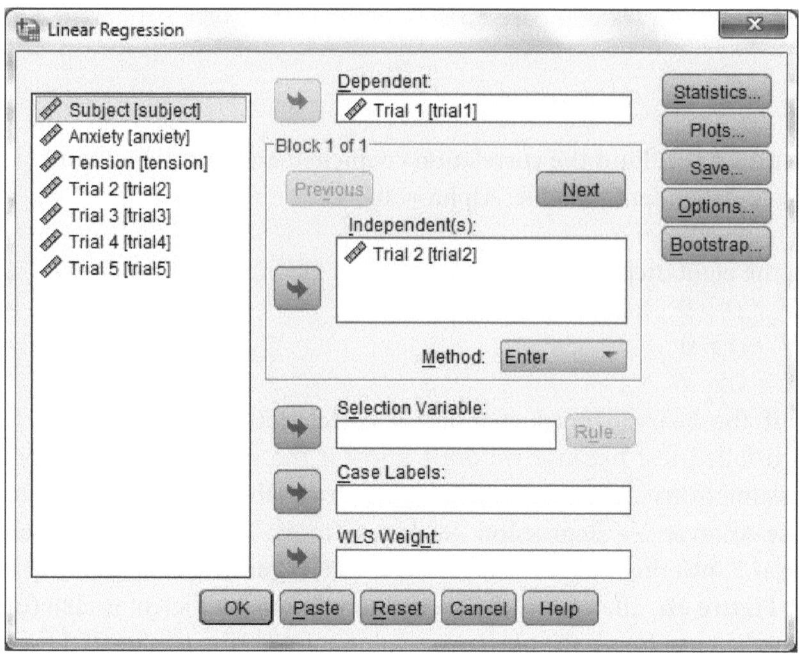

Figure 3a

Coefficients[a]

Model		Unstandardized Coefficients		Standardized Coefficients	t	Sig.
		B	Std. Error	Beta		
1	(Constant)	11.723	2.753		4.258	.002
	Trial 2	.415	.235	.488	1.770	.107

a. Dependent Variable: Trial 1

Figure 3b

ANOVA[a]

Model		Sum of Squares	df	Mean Square	F	Sig.
1	Regression	11.215	1	11.215	3.134	.107[b]
	Residual	35.785	10	3.578		
	Total	47.000	11			

a. Dependent Variable: Trial 1

b. Predictors: (Constant), Trial 2

Figure 3c

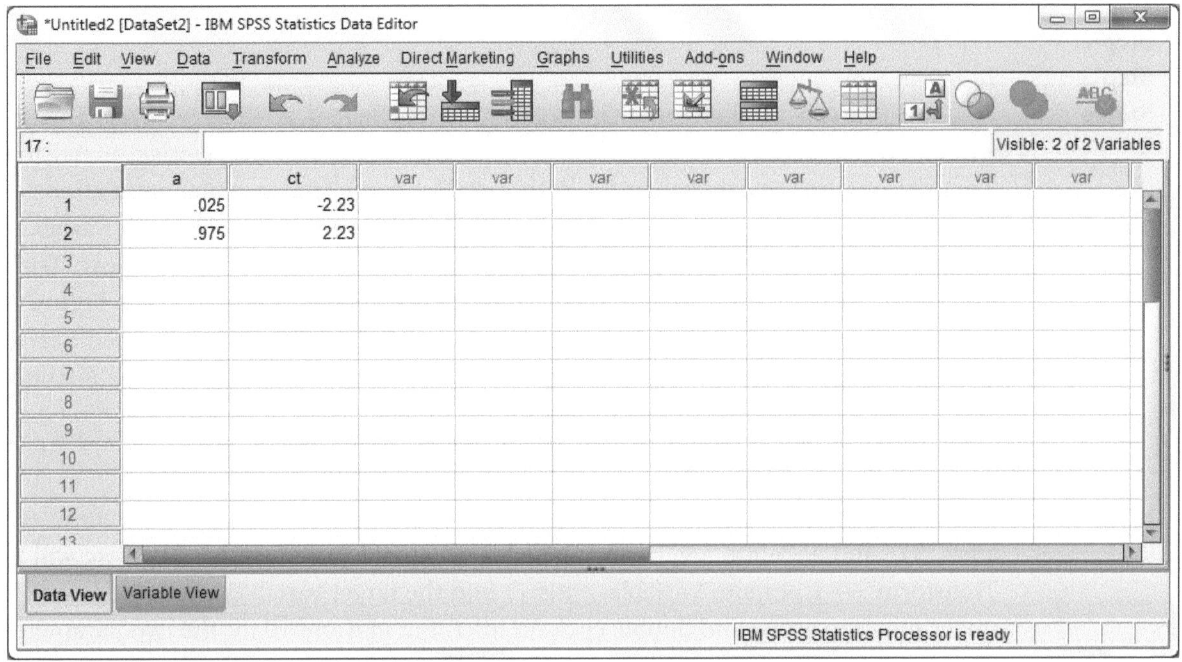

Figure 3d

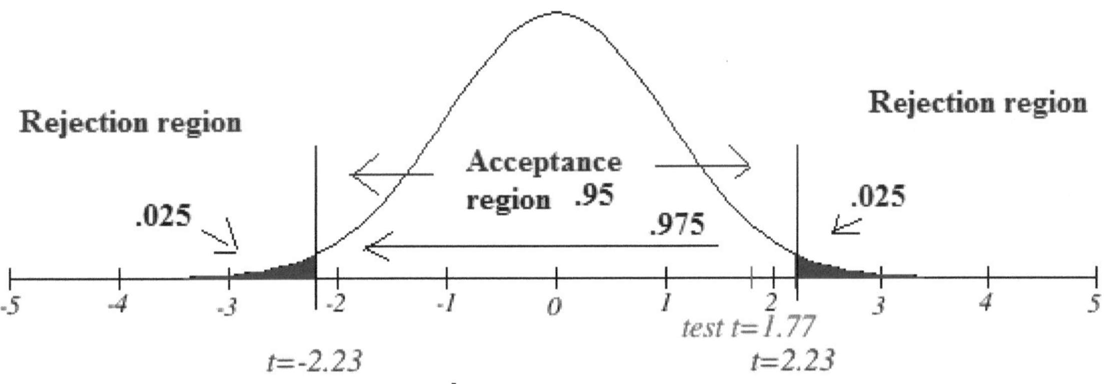

Figure 3e

Example 3.

Compare Trial 3 with Trial 2. Test the correlation coefficient and determine if it is significant. Use either trial as the independent or dependent variable. Alpha = .05.

Solution: Using the eight-step process, we have

Step 1. $H_0 : \rho = 0$

Step 2. $H_1: \rho \neq 0$

Step 3. $\alpha = .05$

Step 4. Test the Pearson product-moment correlation coefficient at the 5% level. This is a two-tailed test because we don't know if the coefficient will be positive or negative. Assumptions are the data values for both variables come from normal populations.

Step 5. Use Analyze → Regression → Linear. Since we have to choose which variable is dependent and which variable is independent, then we will choose to move Trial 3 into the dependent window and Trial 2 into the independent window; see **Figure 4a**, but which is which doesn't really matter in this case. Click OK. The obtained Pearson correlation coefficient is .812, the obtained *t*-value is 4.406, df = 10 and the *p*-value is .001. The obtained *t* value can also be calculated as $t = .812\sqrt{\frac{12-2}{1-.812^2}} = 4.4$. The output is given in **Figure 4b**.

Step 6. To find the critical *t*-value we will open a new data editor and enter the cumulative areas .025 and .975 into a variable labeled *a*, as we did in **Figure 3d**. Then we will use Transform → Compute Variable, enter *ct* into the target variable, click on Inverse DF under function group and double click on Idf.T. Put in *a* and 10 for the two parameters, click OK. The critical *t*-values are ± 2.23; see **Figure 3d**.

Step 7. We compare the *t*-value 4.40 against the critical *t*-value 2.23 to get 4.40 > 2.23. Alternatively we compare the *p*-value against alpha to get .001 < .05. Either way we conclude that the obtained value is in the rejection region; see **Figure 4c**.

Step 8. We reject the null hypothesis; the anxiety level in Trial 3 is directly related to the anxiety level in Trial 2.

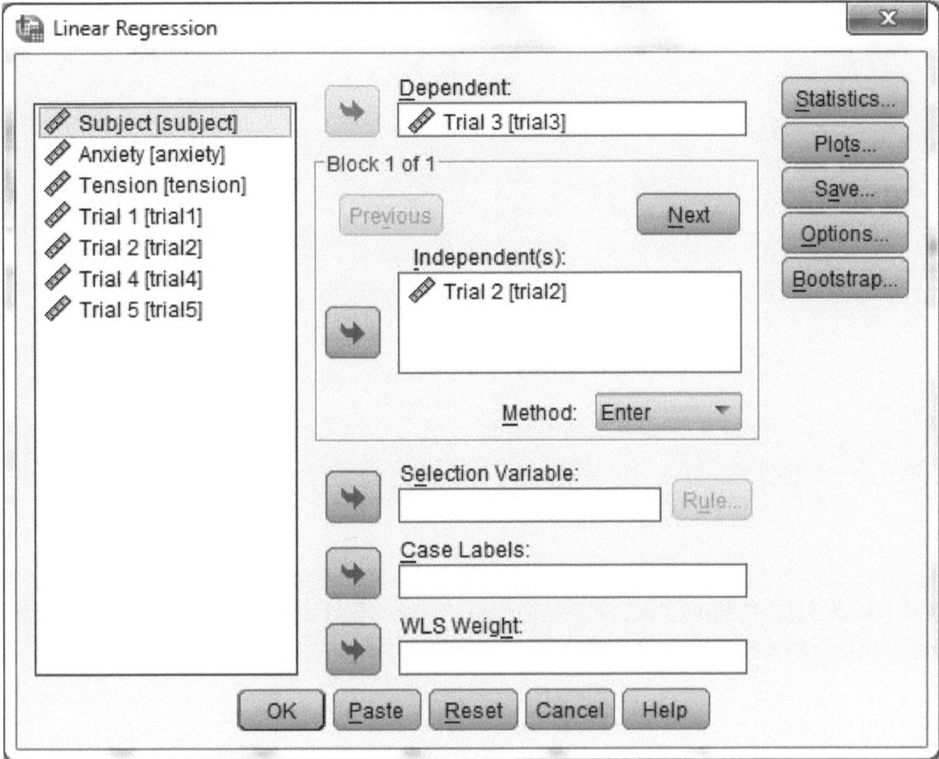

Figure 4a

ANOVA[a]

Model		Sum of Squares	df	Mean Square	F	Sig.
1	Regression	42.404	1	42.404	19.410	.001[b]
	Residual	21.846	10	2.185		
	Total	64.250	11			

a. Dependent Variable: Trial 3

b. Predictors: (Constant), Trial 2

Coefficients[a]

Model		Unstandardized Coefficients		Standardized Coefficients	t	Sig.
		B	Std. Error	Beta		
1	(Constant)	-1.538	2.151		-.715	.491
	Trial 2	.808	.183	.812	4.406	.001

a. Dependent Variable: Trial 3

Figure 4b

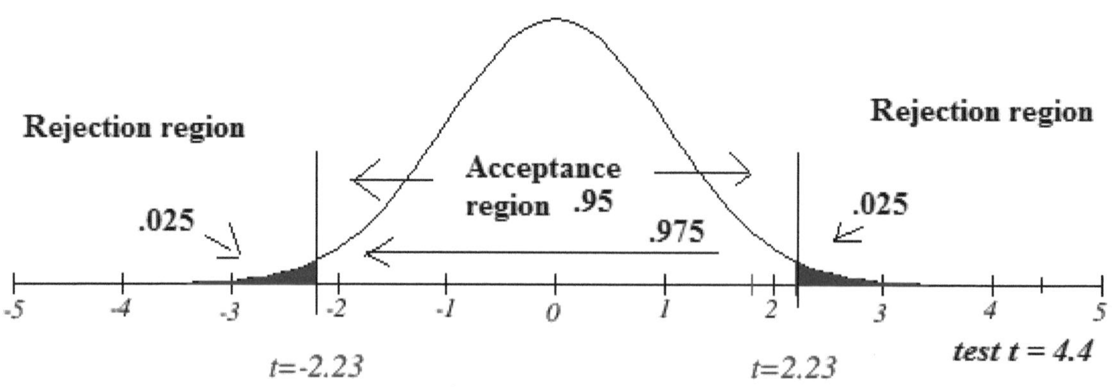

Figure 4c

Example 4.

Compare Trial 4 with Trial 5. Find the correlation coefficient and determine if it is significant. Use either trial as the independent or dependent variable. Alpha = .05.

Solution: Using the eight-step process, we have

Step 1. $H_0 : \rho = 0$

Step 2. $H_1 : \rho \neq 0$

Step 3. $\alpha = .05$

Step 4. Test the Pearson product-moment correlation coefficient at the 5% level. This is a two-tailed test because we don't know if the coefficient will be positive or negative. Assumptions are the data values for both variables come from normal populations.

Step 5. Use Analyze → Regression → Linear. Since we have to choose which variable is dependent and which variable is independent, then we will choose to move Trial 5 into the dependent window and Trial 4 into the independent window; see **Figure 5a**, but which is which doesn't really matter in this case. Click OK. The obtained Pearson correlation coefficient is -.884, the obtained *t*-value is -5.986, df = 10, and the *p*-value is .0. Read the obtained *t* value in the output in the column headed by Trial 4; see **Figure 5b**. The obtained *t* value can also be calculated with the formula $t = -.884\sqrt{\frac{12-2}{1-(-.884)^2}} = -5.97975$. The obtained *t*- value of -5.986 is slightly different from the calculated *t*-value because of rounding.

Step 6. To find the critical *t*-value, we will open a new data editor and enter the cumulative areas .025 and .975 into a variable labeled *a* as we did in **Figure 3d**. Then we will use Transform → Compute Variable, enter *ct* into the target variable, click on Inverse DF under function group, and double click on Idf.T. Put in *a* and 10 for the two parameters, click OK. The critical *t*-values are ± 2.23; see **Figure 3d**.

Step 7. We compare the *t*-value -5.986 against the critical *t*-value -2.23 to get -5.986 < -2.23. Alternatively we compare the *p*-value against alpha to get .0 < .05. Either way we conclude that the obtained value is in the rejection region; see **Figure 5c**.

Step 8. We reject the null hypothesis; the anxiety level in Trial 5 is inversely related to the anxiety level in Trial 4.

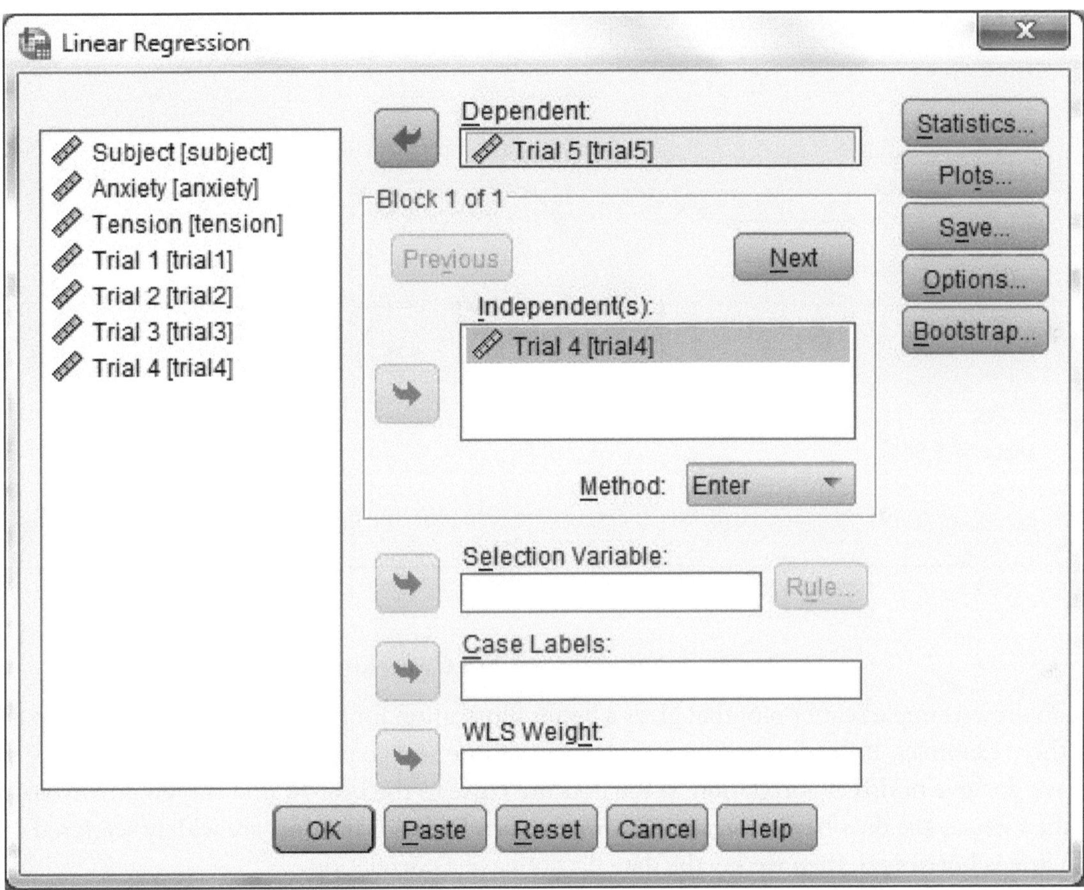

Figure 5a

ANOVA[a]

Model		Sum of Squares	df	Mean Square	F	Sig.
1	Regression	38.244	1	38.244	35.836	.000[b]
	Residual	10.672	10	1.067		
	Total	48.917	11			

a. Dependent Variable: Trial 5

b. Predictors: (Constant), Trial 4

Coefficients[a]

Model		Unstandardized Coefficients		Standardized Coefficients	t	Sig.
		B	Std. Error	Beta		
1	(Constant)	6.350	.550		11.545	.000
	Trial 4	-.651	.109	-.884	-5.986	.000

a. Dependent Variable: Trial 5

Figure 5b

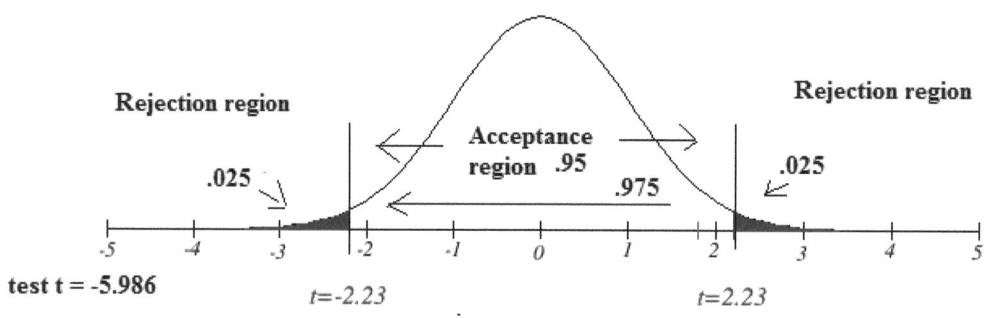

Figure 5c

Positive, Negative, or No Correlation

SPSS can draw a simple scatter plot that gives a better illustration for the decision process in step 8 of the above three examples. If the dots are close to the best-fit line and go upward to the right, then we say the data have positive or direct correlation. If the dots are close to the best-fit line and go downward to the right, then we say the data have a negative or inverse correlation. If the dots are widely scattered and the best-fit line is horizontal, then we say the data do not have a correlation.

The Best-Fit Line

Algebraists have developed a technique for finding the best-fit line in a scatter plot. The line has the equation $\hat{y} = b_0 + b_1 x$, where x is the independent variable and $\hat{y}$ (y hat), is the dependent variable.

The slope of the line, which statisticians call b_1, is given by the formula $b_1 = \dfrac{\sum_{i=1}^{n}(x_i - \overline{x})(y_i - \overline{y})}{\sum_{i=1}^{n}(x_i - \overline{x})^2}$, where $\overline{x}, \overline{y}$ are the means for the two variables x and y, and x_i, y_i are the data values. The constant term, which statisticians call b_0, can be found by substituting the mean values into the equation of the line. This is because $\overline{x}, \overline{y}$ will always be on the best-fit line. Substituting we have $b_0 = \overline{y} - b_1\overline{x}$. Another formula for the slope is $b_1 = r\frac{s_y}{s_x}$ where *r* is the sample correlation coefficient and s_x, s_y are the sample standard deviations for the *x* values and the *y* values. Since the standard deviations are always positive, the variable coefficient b_1 will have the same sign as the correlation coefficient, *r*.

Example 5.

Draw a scatter plot for the above three examples and show that our decisions in the examples is consistent with the scatter plot. Also draw the best-fit line in the scatter plot.

Solution: In Example 2 we compared Trial 1 with Trial 2. Open the data set Anxiety 2.sav. Use Graphs → Legacy Dialogs → Scatter/Dot, Define Simple Scatter, pick Trial 2 for the *X*-Axis, Trial 1 for the *Y*-Axis. (In a linear equation either variable can usually be chosen for the *x*-axis or the *y*-axis; however, if we are given a *dependent* relationship, that is, one variable is dependent upon the other variable, such as Son's height is dependent upon Father's height, then we must be careful to choose the dependent variable for the *y*-axis.) Hit OK, and the scatter plot will appear in the output, see **Figure 6a**. Double-click on the graph to open the SPSS Chart Editor. Click on the Add Fit Line at Total icon and close the chart editor; see **Figure 6b**. The best-fit line will appear on the graph in the output, see **Figure 6c**. The value in the upper-right-hand corner of the graph is R^2 Linear, which is equal to 0.239. R^2 Linear is the square of the correlation coefficient we found in Example 2. (.488 * .488 = .238144, differences are due to rounding.) R^2 Linear gives us the percentage of the variance in the *y*-values that can be explained by the variance in the *x*-values.

The decision in Example 2 was to accept the null hypothesis. This is consistent with the graph, even though the dots have a tendency to flow upward to the right but are too spread out to say there is a definite relationship between these two trials. Dots that have a uniform scattering throughout the graph are said to have no relationship between the two variables. Similarly we can draw the scatter plot for Example 3 and Example 4 and add the best-fit line. To make the scatter plot from Example 3, we use Trial 2 as the *X*-Axis and Trial 3 as the *Y*-Axis and after adding the best-fit line, the output should be the same as in **Figure 6d**. The decision was to reject the null hypothesis, and as can be seen in the figure, the dots have more of tendency to form a straight line upward to the right. R^2 Linear = .66. Follow the same procedure to make the scatter plot from Example 4 using Trial 4 as the *X*-Axis and Trial 5 as the *Y*-Axis; see **Figure 6e**. The decision was to reject the null hypothesis, and as can be seen in the figure, the dots have more of tendency to form a straight line downward to the right. R^2 Linear = .782.

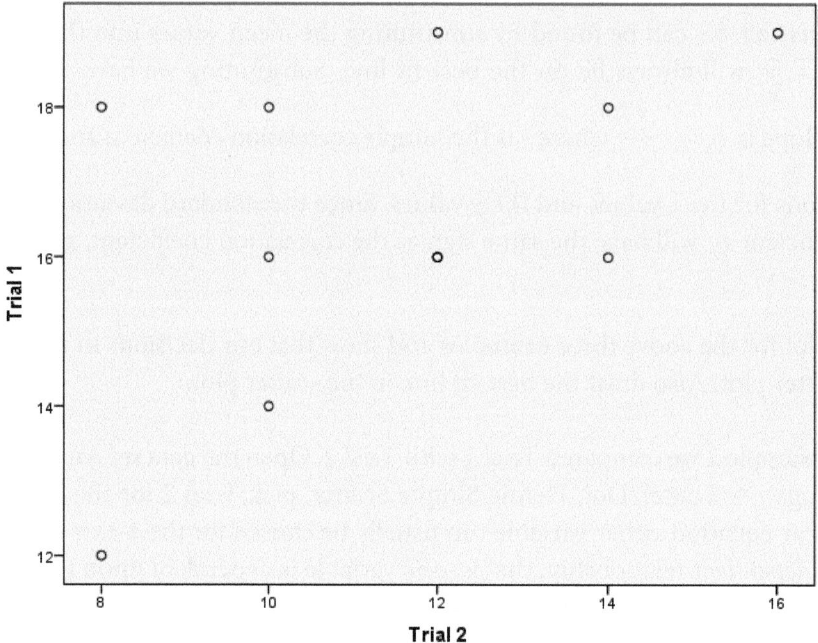

Figure 6a

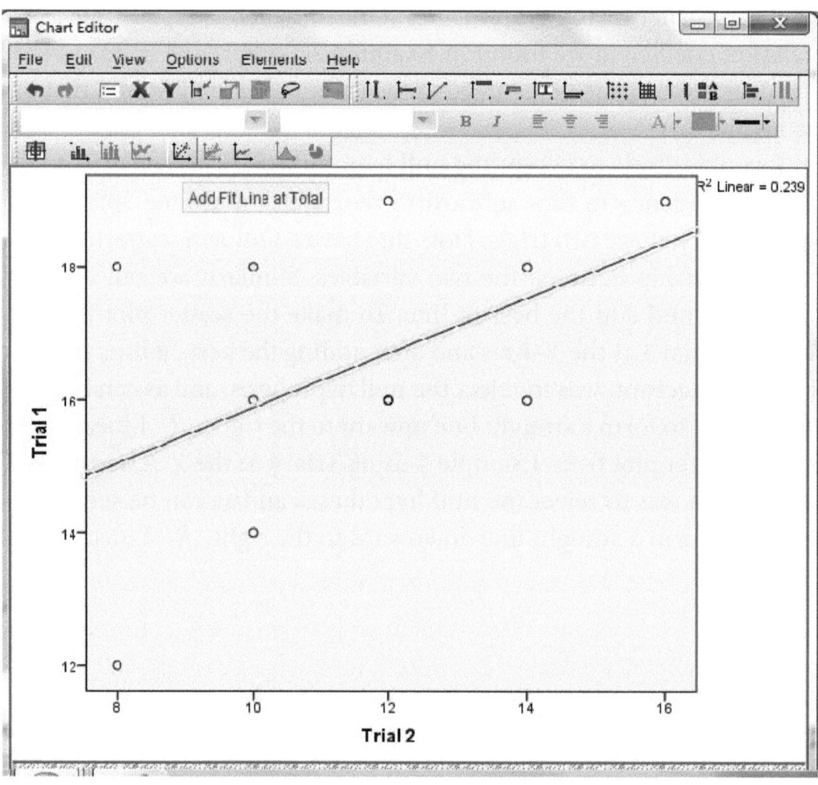

Figure 6b

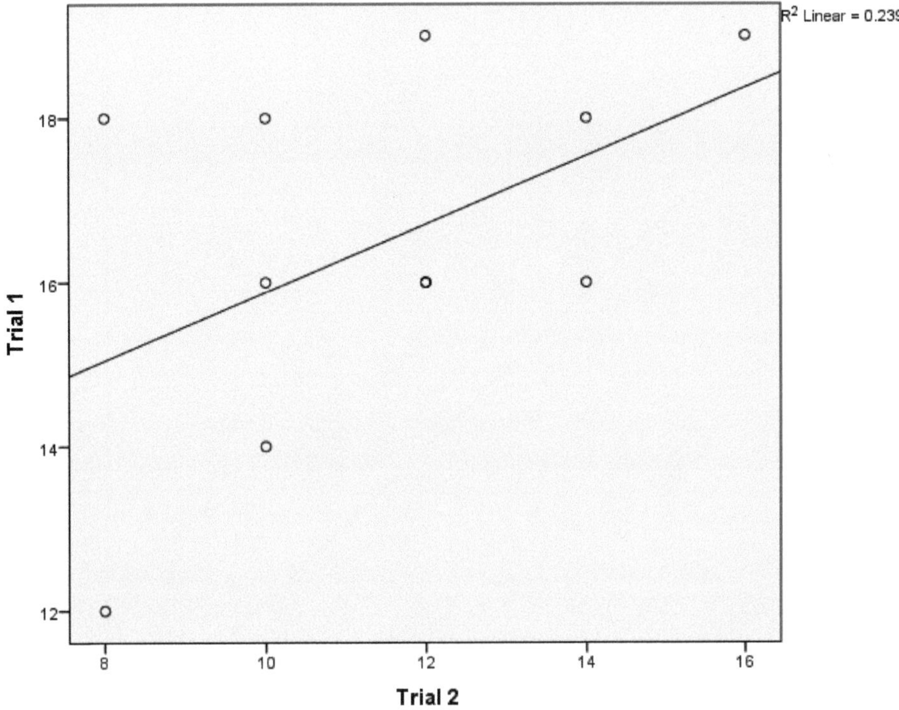

Figure 6c

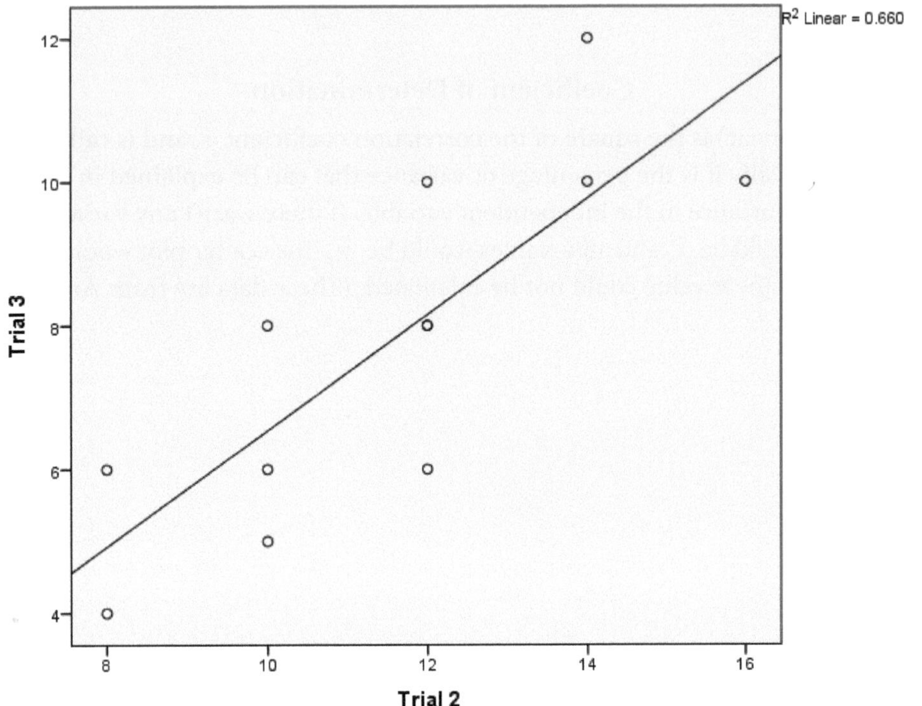

Figure 6d

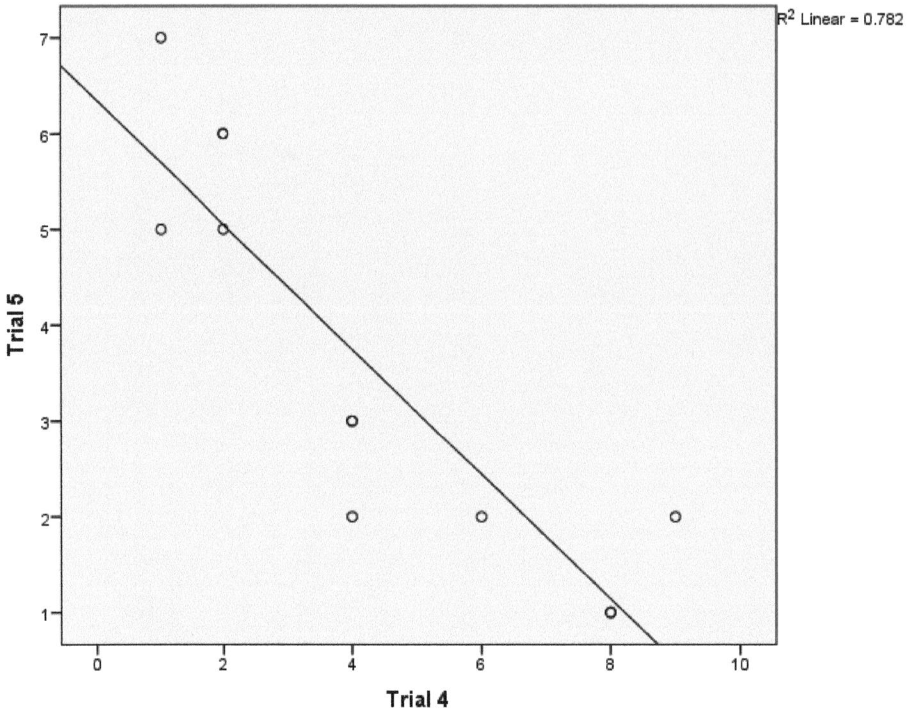

Figure 6e

Coefficient of Determination

R^2 Linear (*R* square linear) is the square of the correlation coefficient, *r*, and is called the coefficient of determination. Technically it is the percentage of variance that can be explained in the dependent variable by the amount of variance in the independent variable. If there wasn't any variance in the variables, then all the *x*-values would be $\overline{x}$ and all *y*-values would be $\overline{y}$. The scatter plot would be a single dot, see **Figure 7a**, and an R^2 Linear value could not be calculated. (These data are from Anxiety 3.sav.)

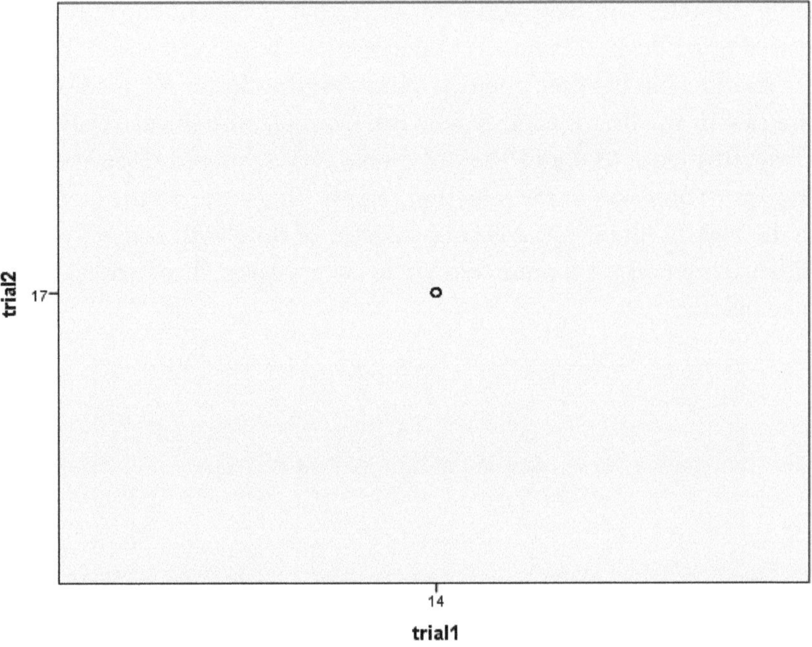

Figure 7a

If the y values were completely dependent upon the x values, then the variance in y would be completely dependent upon the variance in x and the dots would all be on one line; see **Figure 7b**. The R^2 Linear value is 1.

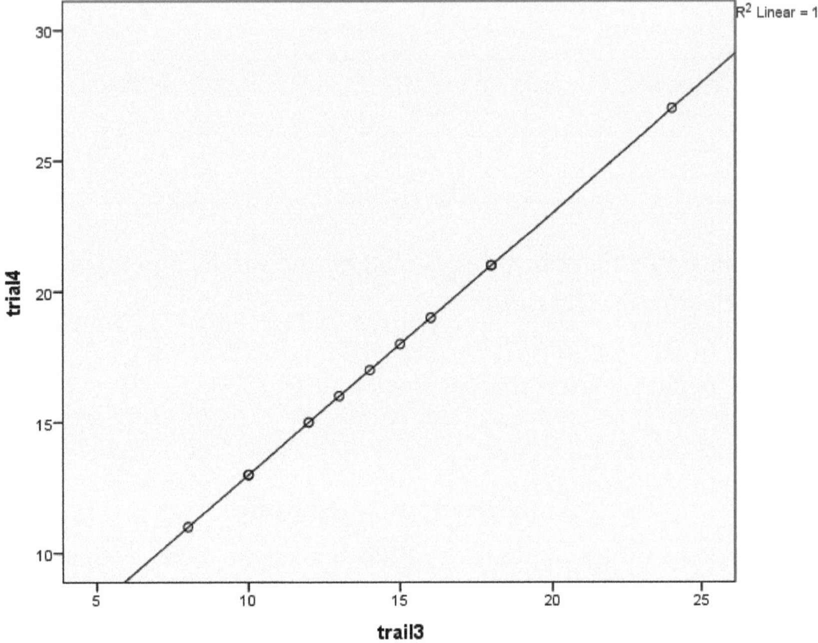

Figure 7b

But usually the y variables will have some variance that is dependent upon x and some variance independent of x; see **Figure 7c**. In this case R^2 Linear will be between 0 and 1. The closer R^2 is to 1 the stronger the linear relationship between the two variables; the closer R^2 is to 0, the weaker the linear relationship. As we saw in the Breast Cancer Survival example, the linear relationship was very weak. R^2 Linear = .01, which is close to 0 and the dots were very scattered. However, we rejected the null hypothesis because our t-value was in the rejection region. This is due to the fact the sample size is very large. In this case the best-fit line is not a good predictor of time with respect to age for an individual person. The confidence interval for a predicted value is very large. The process to find the confidence interval is explained below.

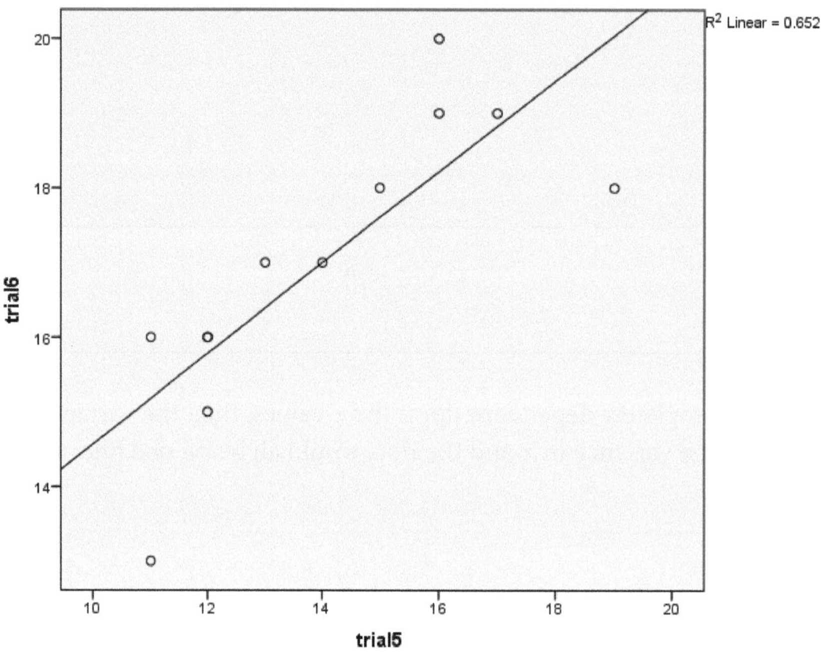

Figure 7c

The amount of variance in y that can be explained by the variance in x is called the coefficient of determination. $r^2 = \frac{explained \quad variation}{total \quad variation}$. With $r^2 \leq 1$, $-1 \leq r \leq 1$, which implies the absolute value of the correlation coefficient is less than or equal to 1, $|r| \leq 1$.

Coefficient of Alienation

The coefficient of alienation is the complement of the coefficient of determination. It is the percentage of variance in the dependent variable that cannot be explained by the amount of variance in the independent variable. The formula for the coefficient of alienation is $1 - r^2$. Thus the coefficient of determination plus the coefficient of alienation equals 1.

Example 6a.

Open the data set Anxiety 3 and draw scatter plots for Trial 1 in the *X*-Axis, Trial 2 in the *Y*-Axis, and find the coefficient of determination. Then do the same for Trial 3 compared to Trial 4 and then Trial 5 compared to Trial 6. Note the amount of variation in both the independent variable and the dependent variable.

Solution. The means for all three examples are $(\overline{x},\overline{y})=(14,17)$. In the first example use Graphs →Legacy Dialogs → Scatter/Dot → Define Simple Scatter, move Trial 1 into the *X* Axis and Trial 2 into the *Y* Axis, hit OK. Next right-click anywhere on the graph and open the SPSS chart editor. Draw the best-fit line by clicking on the icon, see **Figure 6b**, and note the coefficient of determination. Since all the values were plotted at a single point, SPSS was unable to make a determination with regard to the coefficient of determination, thus it was left blank. In the second example move Trial 3 into *X* Axis and Trial 4 into the *Y* Axis as before and click OK. This time $R^2=1$, (R,R^2 can be used interchangeably with r,r^2), which tells us all the variation in the dependent variable is explained by all the variation in the independent variable. As the *x* variable changes, the *y* variable changes by a proportionate amount. There is 100% correlation between these two variables.

Example 6b.

Using the data set Anxiety 3, find the regression line for trial5, *x* value, and trial6, *y* value. Find the explained variation, unexplained variation, total variation, coefficient of variation, and the standard error of estimate. Show that the coefficient of determination equals the explained variation divided by the total variation.

Solution: Use Analyze → Regression → Linear, move trial5 into the independent window and trial6 into the dependent window, click OK. The coefficients of the regression line are in the coefficients output box under B, see **Figure 8a**. The constant coefficient, b_0, is 8.486 and the variable coefficient, b_1, is .603. The equation is $\hat{y}=8.486$ + .608 (*trial5*). The variations are in the output box labeled ANOVA; see **Figure 8b**. The explained variation is listed as Regression Sum of Squares and is 27.365. The unexplained variation is listed as Residual and is 14.635. The total variation is the sum of the explained and the unexplained and is 42. The coefficient of determination is $R^2=\dfrac{27.365}{42}=.652$, see **Figure 8c**.

Coefficients[a]

Model		Unstandardized Coefficients		Standardized Coefficients	t	Sig.
		B	Std. Error	Beta		
1	(Constant)	8.486	2.000		4.244	.002
	trial5	.608	.141	.807	4.324	.002

a. Dependent Variable: trial6

Figure 8a

ANOVA[a]

Model		Sum of Squares	df	Mean Square	F	Sig.
1	Regression	27.365	1	27.365	18.698	.002[b]
	Residual	14.635	10	1.464		
	Total	42.000	11			

a. Dependent Variable: trial6

b. Predictors: (Constant), trial5

Figure 8b

Model Summary

Model	R	R Square	Adjusted R Square	Std. Error of the Estimate
1	.807[a]	.652	.617	1.210

a. Predictors: (Constant), trial5

Figure 8c

Residuals

In Example 6b the point (14, 17) is on the line, but the other points are not all on the line; see **Figure 7c**. The value $(y - \hat{y})$ is called the residual, see **Figure 9a**, and it is the vertical distance the y value is from the projected $\hat{y}$ value. Residuals are the difference between the actual value y and the predicted value $\hat{y}$. The least-squares line is the line that minimizes the sum of the squares of the residuals. The residuals must be squared to find the best-fit line since the sum of the residuals is 0. The data values are (1, 4), (2, 24), (4, 8), and (5, 32). The best-fit line is $\hat{y} = 5 + 4x$ and the points on the line are (1, 9), (2, 13), (4, 21), and (5, 25).

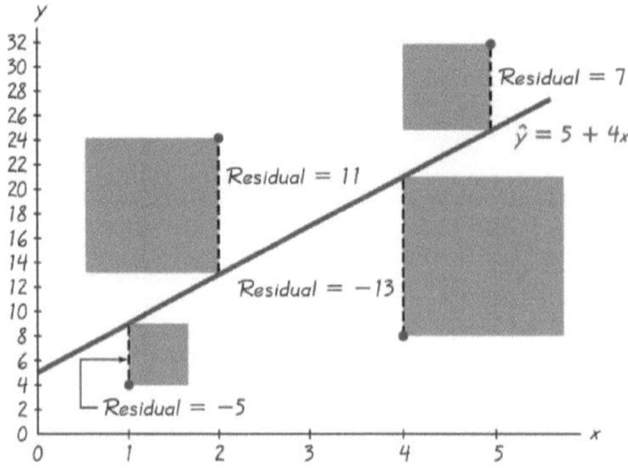

Figure 9a

The best-fit line is obtained by minimizing the sum of the squares of all the residuals. The deviations are shown in **Figure 9b**. ($^{-}$ = 9, $\hat{y}$ = 13, y = 19)

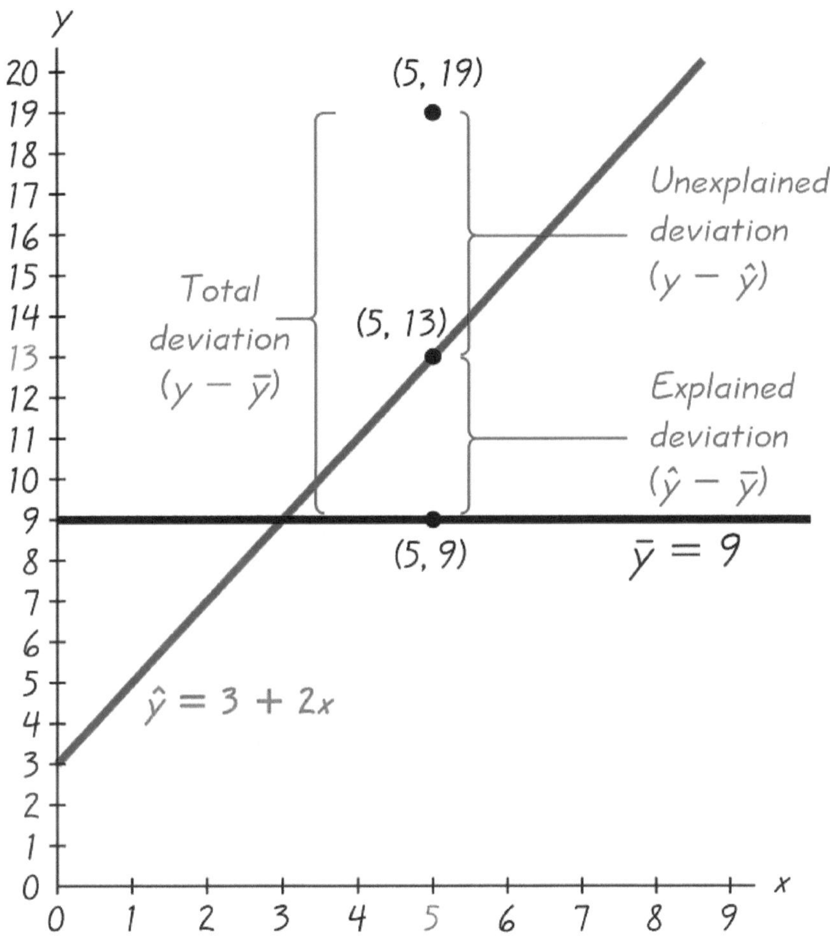

Figure 9b

The sum of the residuals in **Figure** 9a is 0. Thus, to avoid negative terms in our sum, we square all the deviations and add them up. Thus we have Total Variation = Explained Variation plus Unexplained Variation, or using symbols, $\sum(y-\bar{y})^2=\sum(\hat{y}-\bar{y})^2+\sum(y-\hat{y})^2$ and so $r^2=\dfrac{\sum(\hat{y}-\bar{y})^2}{\sum(y-\bar{y})^2}$, note this is a symbolic equation and not a mathematical equation. The correct mathematical equation is $r^2=\dfrac{\sum(y-\bar{y})^2-\sum(y-\hat{y})^2}{\sum(y-\bar{y})^2}=1-\dfrac{\sum(y-\hat{y})^2}{\sum(y-\bar{y})^2}$. The percentage of the explained variation is 1 minus the percentage of the unexplained variation. Therefore if we need to compute the coefficient of determination by hand, we can create two columns of differences, square them, add them up, find the ratio, and subtract from 1. The difference is then r^2. A more useful formula for computing r is

$r = \frac{n\sum xy - (\sum x)(\sum y)}{\sqrt{[n\sum x^2 - (\sum x^2)]\cdot[n\sum y^2 - (\sum y)^2]}}$. This is mathematically equivalent to the form given above, but it's easier to compute columns of squares rather than columns of squares of differences. The variable n is the number of data pairs and xy is the product of the x values and the y values. As long as we have the SPSS software, then SPSS will do all the computations for us, easing the pain of computations and giving us all the answers in the output.

Predicted Values

Once we have a regression line, we can make predictions of a dependent value for a given independent value.

Example 7.

Use the data set Anxiety 3 to find the regression line for Trial 3 and Trial 4 and find the predicted value if $x = 22$. Do the same for Trial 5 and Trial 6.

Solution. Use Analyze → Regression → Linear, move Trial 3 into the independent window and Trial 4 into the dependent window. SPSS should have the output as shown in **Figure 10a.**

Coefficients[a]

Model		Unstandardized Coefficients		Standardized Coefficients	t	Sig.
		B	Std. Error	Beta		
1	(Constant)	3.000	.000		.	.
	trial3	1.000	.000	1.000	.	.

a. Dependent Variable: trial4

Figure 10a

The constant term is 3 and the coefficient of the variable term, Trial 3, is 1. Thus the best-fit line has equation $(trial4) = 1 \cdot (trial3) + 3$. So if $x = 22$, then y should equal 25. SPSS can also find a predicted dependent value for a given independent value. Put 22 as the 13th data value in the variable Trial 3. Use Analyze → Regression → Linear, move Trial 3 into the independent window and Trial 4 into the dependent window, and click on Save. A new dialog box will open up; see **Figure 10b**.

Figure 10b

Check the box marked Unstandardized Predicted Values, and click continue and OK. In the SPSS Data Editor a new variable will appear and should have all the same values as for Trial 4, but also a value of 25 when Trial 3 has a value of 22. (25 = 1*22 + 3) See **Figure 10c Pre_1**.

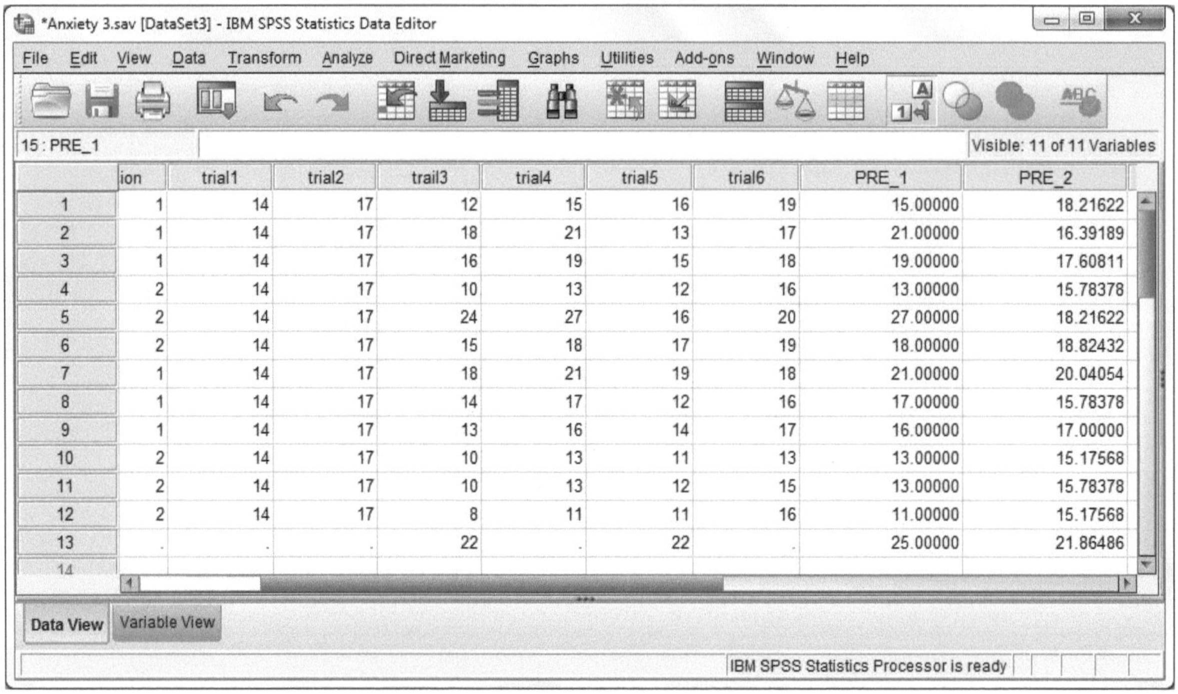

*Anxiety 3.sav [DataSet3] - IBM SPSS Statistics Data Editor

File Edit View Data Transform Analyze Direct Marketing Graphs Utilities Add-ons Window Help

15 : PRE_1 — Visible: 11 of 11 Variables

	ion	trial1	trial2	trail3	trial4	trial5	trial6	PRE_1	PRE_2
1	1	14	17	12	15	16	19	15.00000	18.21622
2	1	14	17	18	21	13	17	21.00000	16.39189
3	1	14	17	16	19	15	18	19.00000	17.60811
4	2	14	17	10	13	12	16	13.00000	15.78378
5	2	14	17	24	27	16	20	27.00000	18.21622
6	2	14	17	15	18	17	19	18.00000	18.82432
7	1	14	17	18	21	19	18	21.00000	20.04054
8	1	14	17	14	17	12	16	17.00000	15.78378
9	1	14	17	13	16	14	17	16.00000	17.00000
10	2	14	17	10	13	11	13	13.00000	15.17568
11	2	14	17	10	13	12	15	13.00000	15.78378
12	2	14	17	8	11	11	16	11.00000	15.17568
13	.	.	.	22	.	22	.	25.00000	21.86486
14									

Data View | Variable View

IBM SPSS Statistics Processor is ready

Figure 10c

Doing the same process for Trial 5 and Trial 6 gives $(trial6) = 8.486 + .608(trial5)$. The predicted value for 22 is 21.86486; see **Figure 10c Pre_2**. Note r and r^2 are given in the model summary output; see **Figure 10d**.

Model Summary[b]

Model	R	R Square	Adjusted R Square	Std. Error of the Estimate
1	.807[a]	.652	.617	1.210

a. Predictors: (Constant), trial5

b. Dependent Variable: trial6

Figure 10d

If the residuals were plotted with $\hat{y}$ as the independent variable, then their best-fit line would be horizontal at the value where residuals = 0; see **Figure 10e**. To create this graph first compute another variable called residuals, use Transform → Compute Variable, and put residuals in the target variable and Trial 6 – PRE_2 as the equation in numeric expression; see **Figure 10f**. (Pre_2 is the predicted variable with Trial 5 as the independent variable and Trial 6 as the dependent variable.) Next use Graphs → Legacy Dialog → Scatter/Dot → Define and put residual in the Y Axis and Unstandardized Predicted Values [PRE_2] in the X Axis. Click OK. Double click on the graph and click Add Fit Line at Total in the Chart Editor, then close the chart editor.

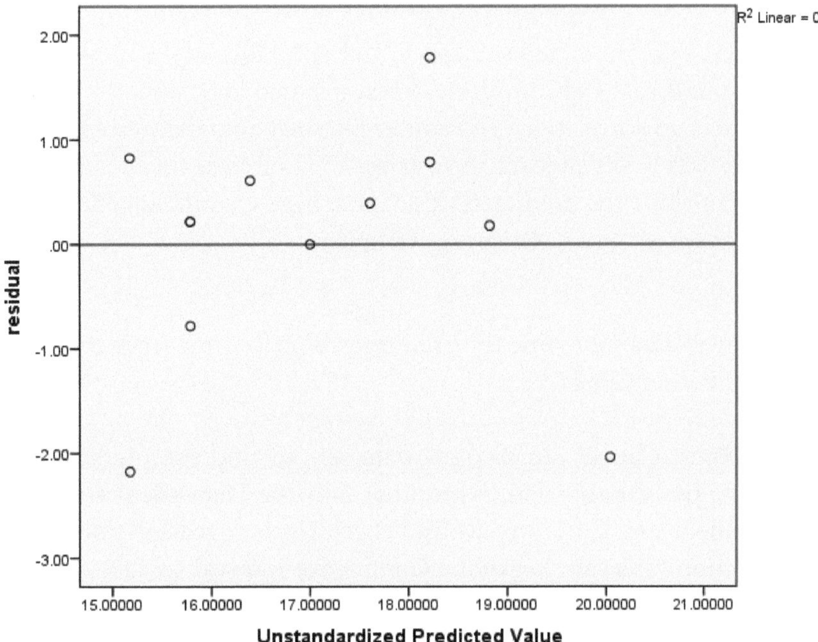

Figure 10e

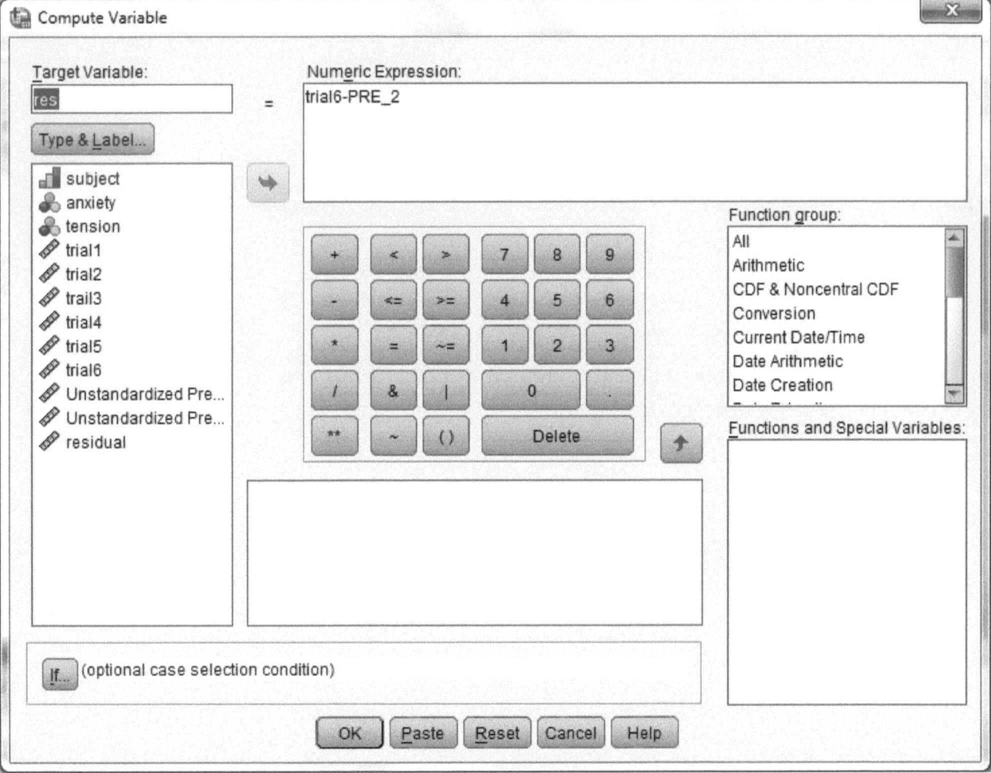

Figure 10f

Confidence Interval for Predicted Values

The 95% confidence interval for predicted values will give us an idea of the range for the true population predicted value. Recall the Breast Cancer Survival example had an r value of -.102, which is small, so we might conclude there is no correlation. The scatter plot had dots that were not grouped very closely together, but the p-value was 0, which leads us to reject the null hypothesis and conclude a correlation exists. This leads us to conclude the predicted value must have a confidence interval that is very large. Indeed that is the case, as we will see in our next example.

Example 8.

Use Breast Cancer Survival Data set to fine the predicted value for time (months) and a 95% confidence interval for the true population value.

Solution: Open the Brest Cancer Survival.sav data set. To find the predicted values use Analyze Regression Linear, move the variable Time (months) into the Dependent slot and the variable Age (years) into the Independent slot. Click on Save and check Unstandardized under Predicted Values and Individual under Prediction Intervals. Leave the Confidence Interval set at 95%; see **Figure 11a**. Click Continue and then OK. The output will be as in **Figure 11b**.

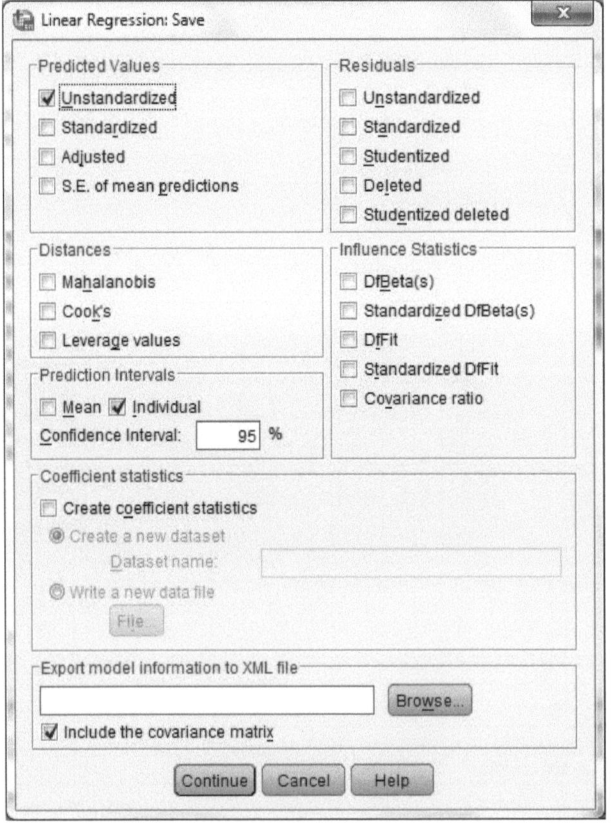

Figure 11a

*Breast Cancer Survial.sav [DataSet4] - IBM SPSS Statistics Data Editor

File Edit View Data Transform Analyze Direct Marketing Graphs Utilities Add-ons Window Help

63 : LICI_1 -10.6211185214487 Visible: 15 of 15 Variables

		pathsize	lnpos	histgrad	er	pr	status	pathscat	ln_yesno	time	pathscat2	PRE_1	LICI_1	UICI_1
1	60	99.00	0	3	Negative	Negative	Censored	99	No	9.47	.	46.13506	-11.75961	104.02972
2	79	99.00	0	Unknown	Unknown	Unknown	Censored	99	No	8.60	.	41.81450	-16.14740	99.77640
3	82	99.00	0	2	Unknown	Unknown	Censored	99	No	19.33	.	41.13231	-16.84909	99.11371
4	66	99.00	0	2	Positive	Positive	Censored	99	No	16.33	.	44.77067	-13.13471	102.67605
5	52	99.00	0	3	Unknown	Unknown	Censored	99	No	8.50	.	47.95424	-9.94126	105.84974
6	58	99.00	0	Unknown	Unknown	Unknown	Censored	99	No	9.40	.	46.58985	-11.30340	104.48311
7	50	99.00	0	2	Positive	Negative	Censored	99	No	17.67	.	48.40903	-9.48938	106.30744
8	83	99.00	0	3	Negative	Negative	Censored	99	No	9.30	.	40.90491	-17.08353	98.89335
9	46	99.00	17	Unknown	Unknown	Unknown	Censored	99	Yes	27.63	.	49.31862	-8.58885	107.22609
10	54	99.00	6	2	Positive	Positive	Censored	99	Yes	11.13	.	47.49944	-10.39423	105.39311
11	67	99.00	1	Unknown	Unknown	Unknown	Censored	99	Yes	11.07	.	44.54327	-13.36483	102.45138
12	54	99.00	0	3	Unknown	Unknown	Censored	99	No	7.10	.	47.49944	-10.39423	105.39311
13	63	99.00	0	Unknown	Unknown	Unknown	Censored	99	No	12.83	.	45.45286	-12.44594	103.35167
14	44	.10	1	Unknown	Unknown	Unknown	Censored	<= 2 cm	Yes	108.90	1.00	49.77342	-8.14020	107.68704
15	60	.15	0	1	Positive	Positive	Censored	<= 2 cm	No	16.37	1.00	46.13506	-11.75961	104.02972
16	49	.20	0	1	Unknown	Unknown	Censored	<= 2 cm	No	94.33	1.00	48.63643	-9.26384	106.53670
17	41	.20	0	1	Unknown	Unknown	Censored	<= 2 cm	No	87.50	1.00	50.45561	-7.46925	108.38047
18	39	.26	0	2	Unknown	Unknown	Censored	<= 2 cm	No	73.03	1.00	50.91041	-7.02330	108.84411
19	65	.30	0	Unknown	Unknown	Unknown	Censored	<= 2 cm	No	82.60	1.00	44.99807	-12.90485	102.90098
20	45	.30	0	1	Unknown	Unknown	Censored	<= 2 cm	No	96.40	1.00	49.54602	-8.36439	107.45643
21	57	.30	0	3	Unknown	Unknown	Censored	<= 2 cm	No	45.27	1.00	46.81725	-11.07570	104.71020
22	47	.30	0	2	Unknown	Unknown	Censored	<= 2 cm	No	42.23	1.00	49.09123	-8.81358	106.99603
23	77	.30	0	1	Positive	Positive	Censored	<= 2 cm	No	63.47	1.00	42.26930	-15.68094	100.21953

Data View Variable View

IBM SPSS Statistics Processor is ready

Figure 11b

We can see the first predicted value is 46.135 with a 95% confidence interval of [-11.7596, 104.0297], which is almost the entire range of all the data values. The results are similar for all the other predicted values. Thus, even though there is a strong correlation for the linear-regression line, the predicted value doesn't tell us much about the true population predicted value.

For our next example we'll use the classic ice cream–crime rates relationship. There is a relationship between ice cream sales and crime rates, but is there a causation between these two variables?

Example 9a.

For a small town the following table gives the amount of ice cream sold in gallons per month, x, and the crime rate for the month, y, excluding petty crimes and misdemeanors.

Mn	Jan	Feb	Mar	Apr	May	Jun	Jul	Aug	Sep	Oct	Nov	Dec
x	23	13	17	19	24	26	30	33	29	25	20	26
y	17	10	9	11	16	20	25	39	28	30	18	27

Test the linear correlation, use the eight-step process, and then in Example 9b, find the regression line and make a prediction for the number of crimes committed if 31 gallons of ice cream are sold.

Solution: Open the data set ice cream.sav. First we'll test if there is indeed a linear correlation. Use Analyze → Correlate → Bivariate, move the variables gallons and number into the variable window, and click OK. The Pearson Correlation Coefficient is .883 and the p-value is .000, which tells us it's significant. The eight-step process is as follows:

Step 1. $H_0 : \rho = 0$

Step 2. $H_1 : \rho \neq 0$

Step 3. $\alpha = .05$

Step 4. We will test the correlation coefficient using a two-tailed test since we're looking for a relationship, but necessarily a direct or inverse relationship. Assumptions are the data values for both variables come from normal populations.

Step 5. $r = .883$, p-value = .000

Step 6. Since we didn't calculate a t-value, we'll skip finding a critical t value. When we find the linear-regression equation, the t value is listed next to the correlation coefficient, which is under the column headed Beta, as in **Figure 12a**. The obtained t value is 5.964 and a critical t value with 10 degrees of freedom at a 5% level is $t = \pm 2.23$.

Step 7. The p-value, .000, is less than the significance level, .05, that is .000 < .05.

Step 8. We'll reject the null hypothesis; it appears there is a correlation between the two variables ice cream and crime rates.

Example 9b.

Find the regression line and make a prediction for the number of crimes committed if 31 gallons of ice cream are sold.

Solution: To find the equation of the best-fit line we'll use Analyze → Regression → Linear, move gallons into the independent variable and number into the dependent variable. The equation is $\hat{y} = -12.636 + 1.409x$; see **Figure 12a**. To find a predicted $\hat{y}$ value when x = 31, we can use this equation or we can input 31 as the 13th data value under gallons and use Analyze → Regression → Linear, move gallons into the independent variable and number into the dependent variable, click on Save, and check Unstandardized under Predicted Values; see **Figure 10b**. Click continue then OK. In the ice cream data editor we can find that the predicted number of crimes that will be committed in a month when 31 gallons of ice cream are sold is 31, see **Figure 12b**.

Coefficients[a]

Model		Unstandardized Coefficients		Standardized Coefficients	t	Sig.
		B	Std. Error	Beta		
1	(Constant)	-12.636	5.760		-2.194	.053
	gallons of ice cream sold	1.409	.236	.883	5.964	.000

a. Dependent Variable: number of crimes committed

Figure 12a

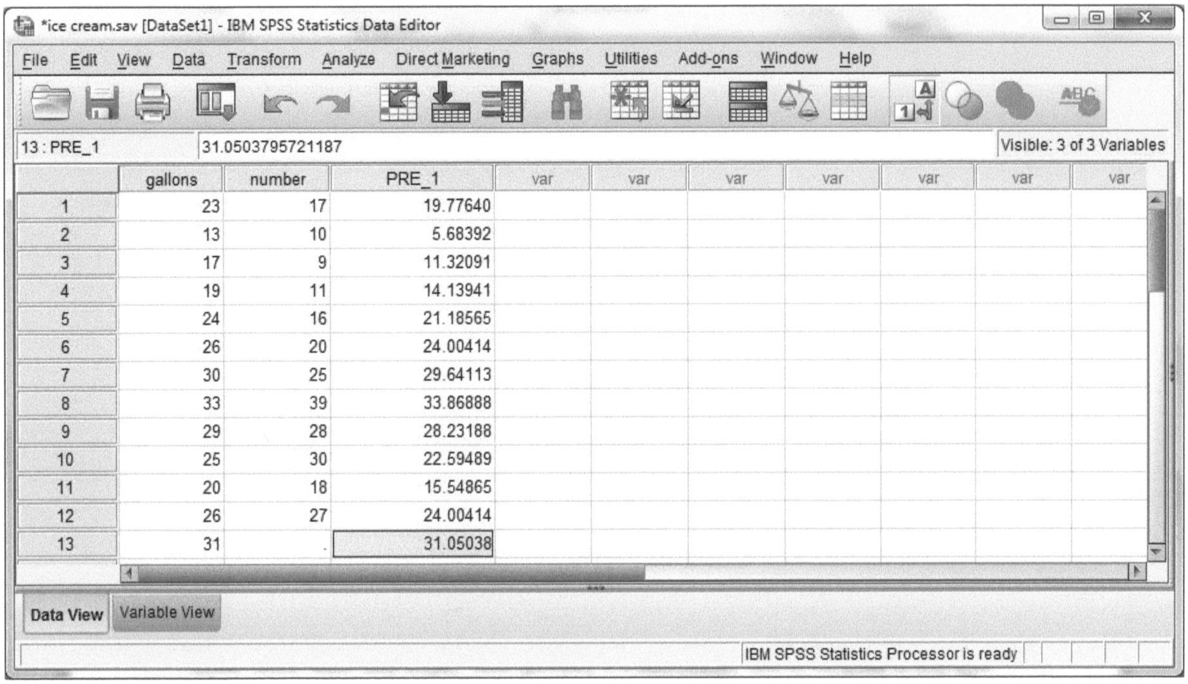

*ice cream.sav [DataSet1] - IBM SPSS Statistics Data Editor

File Edit View Data Transform Analyze Direct Marketing Graphs Utilities Add-ons Window Help

13 : PRE_1 31.0503795721187 Visible: 3 of 3 Variables

	gallons	number	PRE_1	var	var	var	var	var	var	var
1	23	17	19.77640							
2	13	10	5.68392							
3	17	9	11.32091							
4	19	11	14.13941							
5	24	16	21.18565							
6	26	20	24.00414							
7	30	25	29.64113							
8	33	39	33.86888							
9	29	28	28.23188							
10	25	30	22.59489							
11	20	18	15.54865							
12	26	27	24.00414							
13	31	.	31.05038							

Data View Variable View

IBM SPSS Statistics Processor is ready

Figure 12b

Even though the correlation coefficient is significant, it probably doesn't mean that selling ice cream causes crime rates to rise. There probably is a third reason, such as ice cream is sold more during the summer months when daylight hours are longer and kids aren't going to school. The fact that people have time with nothing to do might be a reason for more crime. The extra time and the summer heat is called a lurking variable, since summer is an opportunity for people to buy more ice cream and commit more crimes. Summer affects both variables.

Causality, and Confounding and Lurking Variables

Just because two variables have a strong correlation doesn't mean they're related. Remember, as statisticians we don't say why things are the way they are. We simply state it appears there's a relationship between two variables if they have a strong correlation. But we must always keep in mind that a third variable, called a lurking variable, may be affecting the outcome for our two variables; see **Figure 13a**. This is the situation for our two variables ice cream sales and crime rates, where summer is the lurking variable.

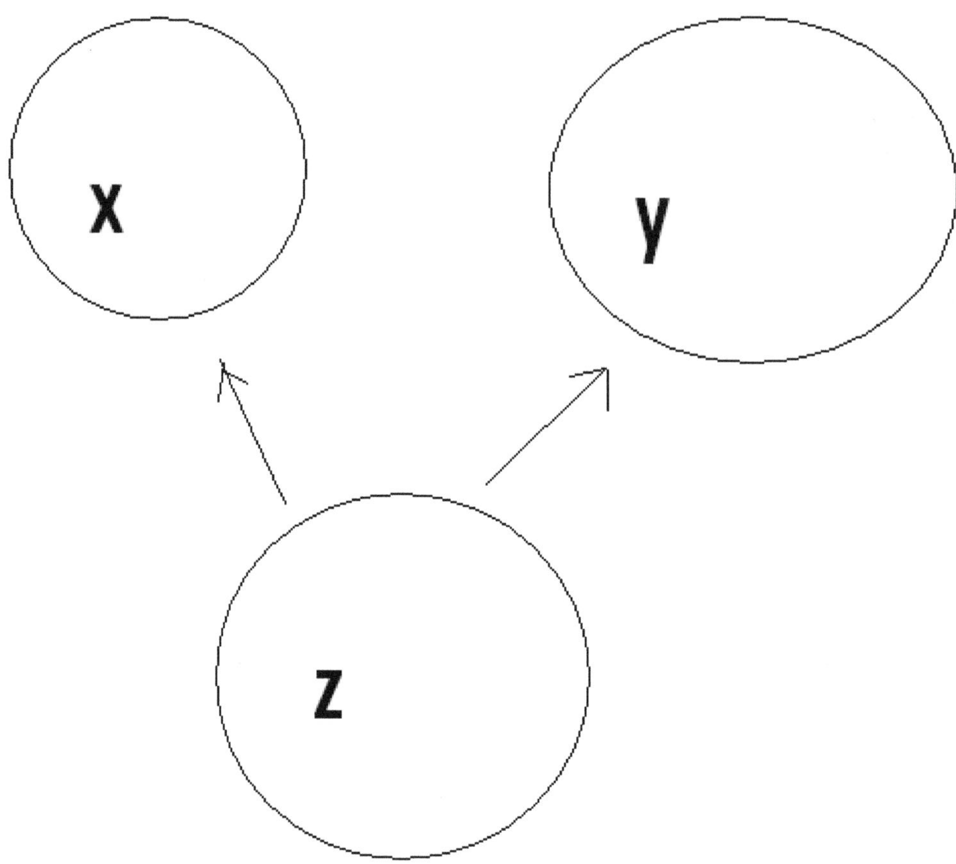

Figure 13a. Z is a lurking variable that affects both *x* and *y*. (There may or may not be an affect between *x* and *y*.)

A confounding variable, see **Figure 13b**, will affect the *y* variable but not the *x* variable. This is the situation of sons' heights and fathers' heights. The father's height will affect the son's height, but the mother's height will also affect the son's height. However, the mother's height cannot affect the father's height. In this case the confounding variable, *z*, can only affect the dependent variable, *y*, and not the independent variable, *x*.

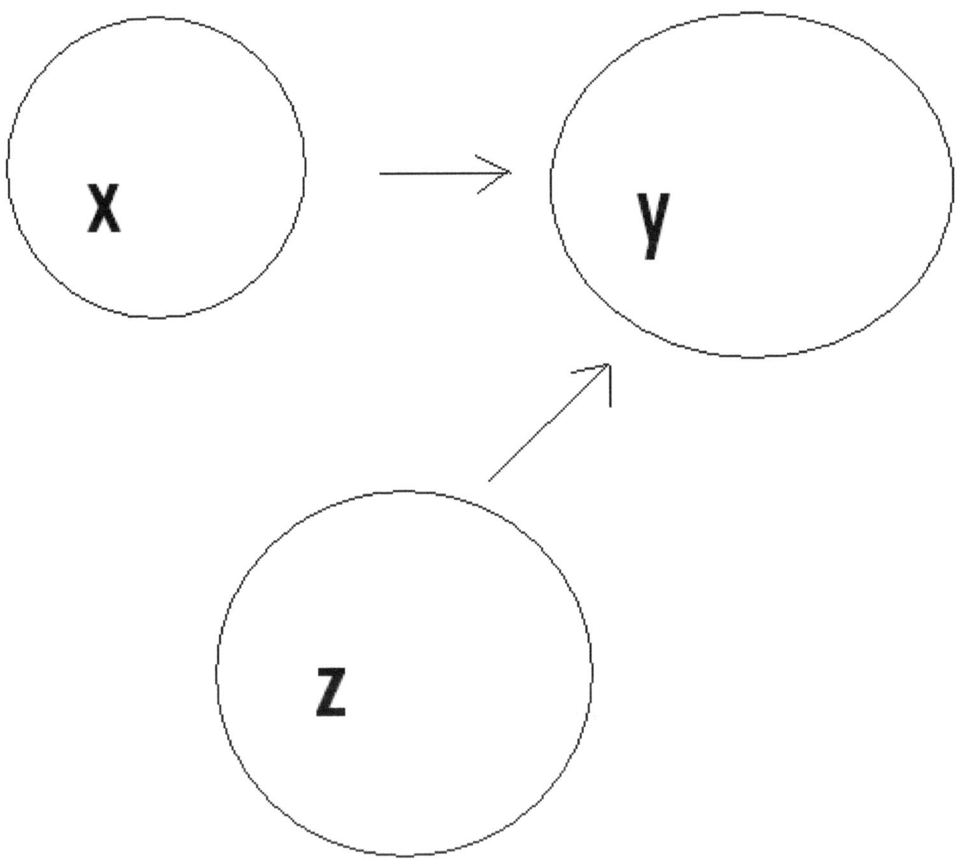

Figure 13b. Z is a confounding variable that affects *y* but not *x*. (However, *x* does affect *y*.)

Our final example, in this chapter, shows there is a relationship between Quality of Marriage and Parent–Child Relationship. The number and types of lurking and confounding variables for marriage and parent–child relationship are well beyond the scope of this text.

Example 10.

Open the data set marriage_rel.sav and test if the two variables Qua_mar and Qua_pcrel are related. Use the eight-step process for doing a test. Also find a linear regression model to predict parent–child relations based on Quality of Marriage and find a predicted Qua_pcrel value for a given Qua_mar value of 76.

Solution: Open the data set marriage_rel and use Analyze → Correlate → Bivariate, and move both variables Qua_mar and Qua_pcrel into the variable window. Click OK and the output should read as in **Figure 14a**.

Correlations

		quality of marriage	quality of parent–child relation
quality of marriage	Pearson Correlation	1	.393*
	Sig. (2-tailed)		.035
	N	29	29
quality of parent–child relation	Pearson Correlation	.393*	1
	Sig. (2-tailed)	.035	
	N	29	29

*Correlation is significant at the 0.05 level (2-tailed).

Figure 14a

Since we're testing the correlation coefficient, we will write down the eight-step process.

Step 1. $H_0 : \rho = 0$

Step 2. $H_1 : \rho \neq 0$

Step 3. $\alpha = .05$

Step 4. This is a two-tailed Pearson correlation coefficient test. Assumptions are the data values for both variables come from normal populations.

Step 5. The Pearson correlation coefficient is .393 and the obtained *p*-value is .035, which is significant at the 0.05 level for a two-tailed test. Using Analyze → Regression → Linear, moving Quality of Marriage into the independent variable and Quality of Parent–Child Relationship into the dependent variable and click OK; the obtained *t* value is 2.224, as shown in **Figure 14b**.

Step 6. Since $df = 27$ ($n - 2 = 29 - 2 = 27$), the critical *t* value for a two-tailed test at the .05 level is ± 2.05. We assume the reader by now can find a critical *t*-value.

Step 7. Comparing the obtained *t*-value 2.224 against the critical *t*-value, ± 2.05, we have $2.224 > 2.05$. Alternatively comparing the *p*-value, .035 against the alpha level, .05, we have $.035 < .05$.

Step 8. Decision. Since *t*-value is in the rejection region and the *p*-value is less than the alpha level, we reject the null hypothesis; there is a relationship between the variables Qua_mar and Qua_pcrel.

Coefficients[a]

Model		Unstandardized Coefficients		Standardized Coefficients	t	Sig.
		B	Std. Error	Beta		
1	(Constant)	13.091	12.589		1.040	.308
	quality of marriage	.355	.160	.393	2.224	.035

a. Dependent Variable: quality of parent–child relation

Figure 14b

The coefficients for the best-fit line are given in **Figure 14b**. The linear-regression line has equation $\hat{y} = 13.091 + .355x$. To find the predicted value for x = 76, we substitute 76 for x and solve for $\hat{y}$. $\hat{y} = 13.091 + .355 * 76 = 40.071$. Or put 76 as the 30th data value under Qua_mar and use Analyze → Regression → Linear, move Qua_mar into the independent window and Qua_pcrel into the dependent window, click on save, check Unstandardized in Predicted Values window, and click continue and OK. The predicted values will be in the changed data editor window. The predicted value is 40.1. The differences are due to rounding.

Let's review what we learned in this chapter.

1. The correlation coefficient is used to test if two variables have a relationship.
   i.) If the correlation coefficient is positive, then the relationship is direct.
   ii.) If the correlation coefficient is negative, then the relationship is indirect or inverse.
2. The t-value is the test statistic to test against a critical t-value obtained from a significance level, alpha level. The larger the data set, the more significant the statistic.
3. We always use an eight-step process when doing a test.
4. The null hypothesis will always be that there is no relationship between variables, rho = 0.
5. The alternative hypothesis can be either directional (that is, one of positive or negative direction), or non-directional (that is, the direction is not given).
6. The coefficient of determination is r^2, which tells us the percentage of variance in the dependent variable that can be explained by the variance in the independent variable.
7. The coefficient of alienation is $1 - r^2$ and is the unexplained variance.
8. The best-fit line is the line that best fits the data points in a scatter plot.
9. The equation of the best-fit line can be used to find predicted values.
10. Predicted values are indicated by $\hat{y}$ (y hat).
11. Residuals are the difference between the actual value and the predicted value, $(y - \hat{y})$.
12. A confounding variable is a third variable that may affect one of the variables in the relationship.
13. A lurking variable is a third variable that affects both the variables in a relationship.

Class Work 11 a&b

a) The following table gives son's heights corresponding to their father's heights. Draw a scatterplot and best fit line. Test, at the 5% level, if there is a correlation between son's heights and father's heights. b) Find the coefficient of determination and the coefficient of alienation. Find the equation for the best-fit line and find a predicted value for a son's height given a father's height of 70.5 inches. The data set is fathers_sons.sav

Sons	74	70	69	69	72	75	69	69	68	66	69	65
Fathers	71	71	68	70	74	74	73	68	65	67	71	66

Homework 11.

1. Compute and test the correlation coefficient for the average temperature difference (in °C) between the inside and outside of a house and the average daily gas consumption (in cu.ft.). chpt11problem1.sav

Temp. Diff.	10.5	11.5	11.4	12.5	13.5	15.0	15.2	15.6	16.5	17
Gas Consump.	70.1	83.0	82.7	80.0	75.5	78.5	81.1	86.5	105.5	86

2. 2. The following data give the number of sports injuries for a particular number of training hours. Test at the 5% level if there is a correlation between these two variables. chpt11problem2.sav

Training 12 13 22 12 11 31 27 31 8 16 14 26
36 26 15 11 16 10 15 16 22 24 26 31 12
24 33 21 12 36

Injuries 8 7 2 5 7 1 5 3 4 2 7 2
2 2 5 4 7 8 3 7 3 6 7 2 3
3 3 5 7 4

3. Is high school GPA a predictor for first-year college GPA? The following data give the high school GPA and the first-year college GPA of eleven students. Test if there is a correlation between these two variables. Use alpha = .05. chpt11problem3.sav

HS GPA	3.5	2.9	3.9	3.8	2.8	2.4	3.2	3.7	2.7	3.3	3.6
FYC GPA	3.3	2.8	3.6	3.5	3.0	2.0	2.9	3.4	3.0	3.4	3.6

4. The following table (*Statistical Abstract of the United States*) gives the U.S. population (in millions) for various years of the twentieth century. chpt11problem4.sav

1900	76.1	1920	106.5	1940	132.5	1960	180.0
1905	83.8	1925	115.8	1945	133.4	1965	193.5
1910	92.4	1930	123.1	1950	151.9	1970	204.0
1915	100.5	1935	127.3	1955	165.1	1975	215.5

1980	227.2	2000	281.4
1985	237.9	2005	298.2
1990	249.4	2010	308.7
1995	262.8		

Test at the 1% level if there is a correlation between the year and the population of the United States.

5. Find the t value that corresponds to the correlation coefficient in problem 1. Decide if the t value implies the correlation is direct or indirect. Also find the critical t value that corresponds to the test of the correlation coefficient.

6. Find the t value that corresponds to the correlation coefficient in problem 2. Decide if the t value implies the correlation is direct or indirect. Also find the critical t value that corresponds to the test of the correlation coefficient.

7. Find the t value that corresponds to the correlation coefficient in problem 3. Decide if the t value implies the correlation is direct or indirect. Also find the critical t value that corresponds to the test of the correlation coefficient.

8. Find the t value that corresponds to the correlation coefficient in problem 4. Decide if the t value implies the correlation is direct or indirect. Also find the critical t value that corresponds to the test of the correlation coefficient.

9. Find the coefficient of determination for the data in problem 1. Also find the coefficient of alienation and explain what they mean in context of the data set.

10. Find the coefficient of determination for the data in problem 2. Also find the coefficient of alienation and explain what they mean in context of the data set.

11. Find the coefficient of determination for the data in problem 3. Also find the coefficient of alienation and explain what they mean in context of the data set.

12. Find the coefficient of determination for the data in problem 4. Also find the coefficient of alienation and explain what they mean in context of the data set.

13. Find the equation of the best-fit line for the data in problem 1. From the equation find the predicted value, $\hat{y}$ = gas consumption, if x = temperature difference is 14. SPSS can also be used.

14. Find the equation of the best-fit line for the data in problem 2. From the equation find the predicted value, $\hat{y}$ = sports injuries, if x = training hours = 27. SPSS can also be used.

15. Find the equation of the best-fit line for the data in problem 3. From the equation find the predicted value, $\hat{y}$ = first year college GPA, if x = High School GPA = 3.75. SPSS can also be used.

16. Find the equation of the best-fit line for the data in problem 4. From the equation find the predicted value, $\hat{y}$ = U.S. population, if x = year = 2012. SPSS can also be used.

17. Explain what might be a lurking variable in the ice cream–crime number in Example 8.

18. Explain what might be a confounding variable for sons' heights in classwork problem.

19. Find the correlation coefficient of this data set: chpt11problem19.sav

Temp. Diff. 15.5 16.5 16.4 17.5 18.5 20.0 20.2 20.6 21.5 22
Gas Consump. 80.1 93.0 92.7 90.0 85.5 88.5 91.1 96.5 115.5 96

This data set is the same as in problem 1, but with 5 added to all *x* values and 10 added to all *y* values. Is the correlation coefficient the same as in problem 1? What does this imply?

20. Find the correlation coefficient of this data set: chpt11problem20.sav

Temp. Diff. 77.5 82.5 82 87.5 92.5 100 101 103 107.5 110
Gas Consump. 801 930 927 900 855 885 911 965 1155 960

This data set is the same as in problem 19, but with 5 multiplied to all *x* values and 10 multiplied to all *y* values. Is the correlation coefficient the same as in problem 1? What does this imply?

Just for fun: When a sack of gold was divided among 6 pirates, 1 piece was left over. In the ensuing struggle for that one piece, one pirate was shot dead. Now they were able to divide the gold evenly among the 5 remaining pirates and every pirate was happy. How many pieces of gold were in the bag? (Find the smallest such number.)

Chapter 12

Chi-Square Distribution

Objectives

After completing this chapter, a student should:

- ✓ Know the formula for calculating a Chi-Square value
- ✓ Understand the goodness-of-fit test
- ✓ Understand the independence of categories test
- ✓ Know the assumptions and degrees of freedom for a Chi-Square test

Chi-Square Tests

Chi-square tests are called nonparametric tests because we don't assume we have the parameters that we had for a normal distribution; that is, we don't assume we have the population mean and the population standard deviation. Furthermore, we may have a nominal variable (categorical) as the independent variable and a scale or ordinal variable (numerical) as the dependent variable. In these cases we use the chi-square test.

The chi-square test is not as powerful as a parametric test, since it might not reject a false null hypothesis—a type II error—and so should be used only when the parameters of a z test, t test, or F test cannot be met, or if the variables are ordinal or nominal. We also assume the data are randomly selected.

There are two types of chi-square tests: the goodness-of-fit test and the independence-of-categories test.

Goodness-of-Fit Test

A goodness-of-fit test is used to test if categories will fit a theoretical or proscribed distribution. For example, if a car manufacturer claims that 50% of their cars are red, 20% are green, and 30% are blue, then a sample of 100 cars can be used to test if those ratios are really accurate. The chi-square value is calculated using the formula $\chi^2 = \sum \frac{(O-E)^2}{E}$, where O is the observed value for each cell, E is the expected value for each cell, and the difference, (O – E), is squared to avoid adding negative values. SPSS will calculate the chi-square value by way of nonparametric tests.

Expected Value

If all categories are supposed to be equal, that is, each cell is supposed to have the same number of data values, then the expected value for each cell is the total number of data values divided by the number of cells. If each category has a different value, either theoretical or observed, then a different expected value has to be assigned to each cell. The expected value is the theoretical probability multiplied by the total number of values.

Example 1.

In 1860, Gregor Mendel experimented with genetics to determine how characteristics were passed on to following generations. Peas have two characteristics, smooth and wrinkled. Mendel guessed that smoothness was a dominant trait, thus the only way a pea can be wrinkled was if it inherited a wrinkled gene from both parent plants. Mendel assembled data from 7,324 second-generation hybrid pea plants whose parent plants were both SW. (S represents smooth trait, W represents wrinkled.) From these two genes we have four possible outcomes SS, SW, WS, and WW and the only way a pea can be wrinkled is to inherit WW. Therefore a quarter of the offspring should be wrinkled and three quarters should be smooth. Of the 7,324 second-generation hybrid pea plants, 5,474 of these were smooth and the rest (1,850) were wrinkled. (1) Use a chi-square test to see if the data fit the theoretical probabilities.

Solution: Open a new Data editor and put the values 1 and 2 in the first column and the values 5474 and 1850 in the second column. Name the first variable texture (1 for smooth, 2 for wrinkled), and the second variable freq (frequency); see **Figure 1a**. Change texture to a nominal variable and freq to a scale variable. Next weight the variable texture by freq. Go to Data → Weight Cases by and weight cases by freq. Next go to Analyze → Nonparametric Tests → Legacy Dialogs →Chi-square and select texture as the Test Variable. Under Expected Values type in .75 and.25, click Add; see **Figure 1b**.

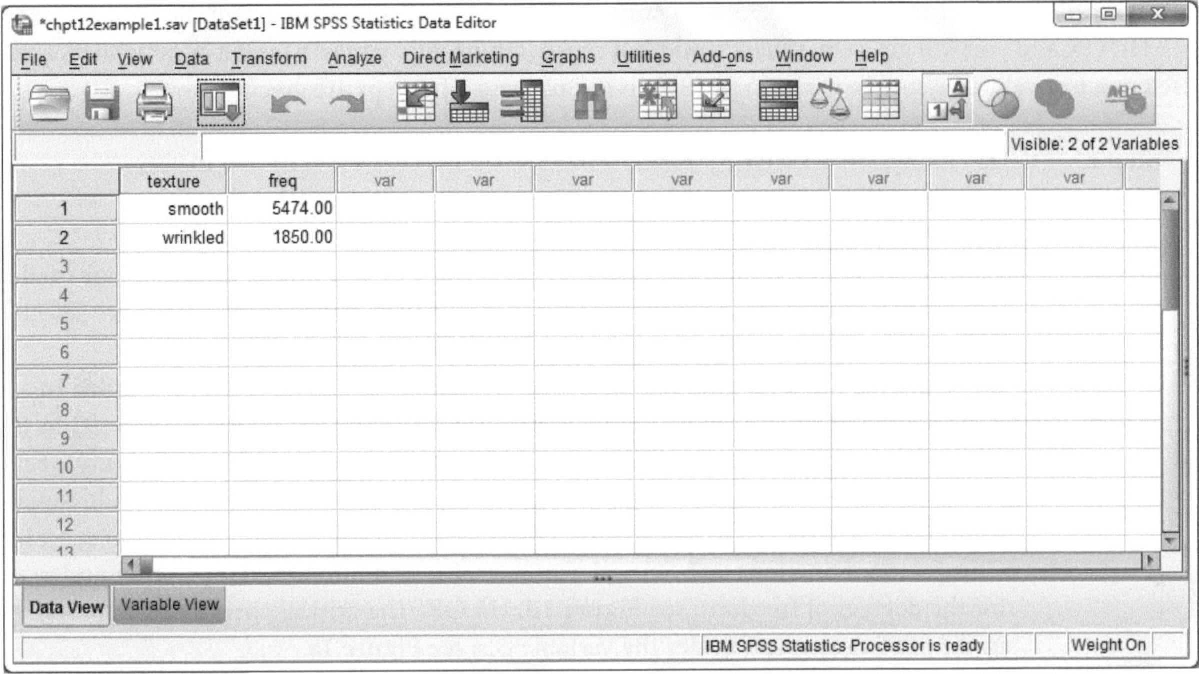

Figure 1a

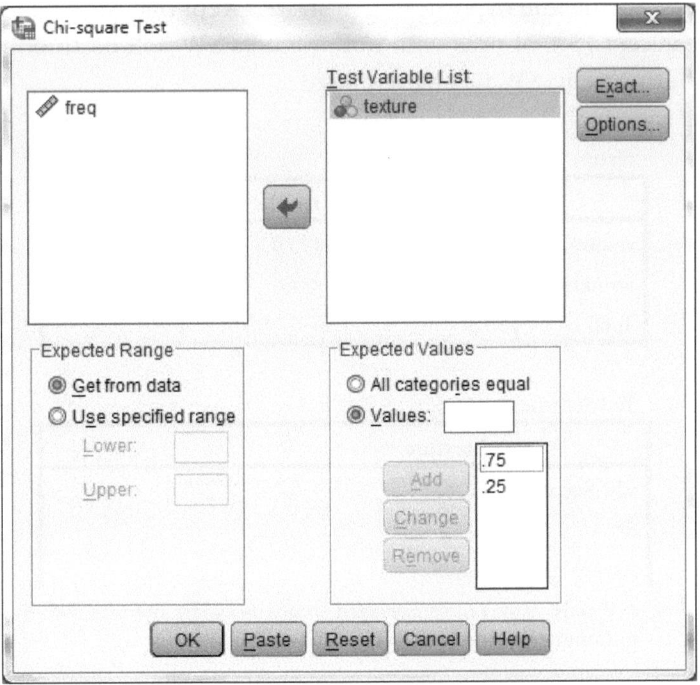

Figure 1b

SPSS will automatically associate .75 with smooth (1) in the variable texture and .25 with wrinkled (2). Hit OK and you will get a chi-square value, but more importantly you will have a *p*-value that can be used as a test value. Be sure to perform the 8-step test process. The steps are given below:

Step 1. $H_0 : P_{smooth} = .75, P_{wrinkled} = .25$.

Step 2. $H_1 : H_0$ is false.

Step 3. $\alpha = .05$.

Step 4. Use a Chi-Square goodness-of-fit test. The Chi-Square value is always a right-tailed value since chi-square is always positive. Assumptions are the data values are selected randomly and independently, and the cells have values greater than 5.

Step 5. The obtained Chi-Square value is .263, *df* = 1 and the *p*-value is .608; see **Figure 1c**.

Step 6. The critical chi-square value is 3.84, which was found by entering 1 – .05 = .95 in the next column of the SPSS data editor and calling it *a* for area (or *p* for probability). Then use Transform $\rightarrow$ Compute Variable, call the target variable *ccs* for critical chi-square value, click on Inverse DF (for Inverse Distribution Function), and double-click on Idf. Chisq under Functions and Several Variables. Move *a* into the first position and use 1 for the degrees of freedom; see **Figure 1d**. Hit OK. The critical chi-square value will be in the SPSS data editor under the variable ccs; see **Figure 1e**.

Step 7. Compare the obtained Chi-Square value against the critical Chi-Square value, .263 < 3.8414, alternatively compare the *p*-value against α, .608 > .05.

Step 8. We see the chi-square value does not lie in the rejection region, see **Figure 1f** (the shaded area is the rejection region), alternatively the *p*-value is greater than .05, thus we fail to reject the null hypothesis and so we accept that the proportions are correct. That is, we accept 75% of peas with both parents SW will be smooth and 25% of the peas with both parents SW will be wrinkled.

texture

	Observed N	Expected N	Residual
smooth	5474	5493.0	-19.0
wrinkled	1850	1831.0	19.0
Total	7324		

Test Statistics

	texture
Chi-Square	.263[a]
df	1
Asymp. Sig.	.608

a. 0 cells (0.0%) have expected frequencies less than 5. The minimum expected cell frequency is 1831.0.

Figure 1c

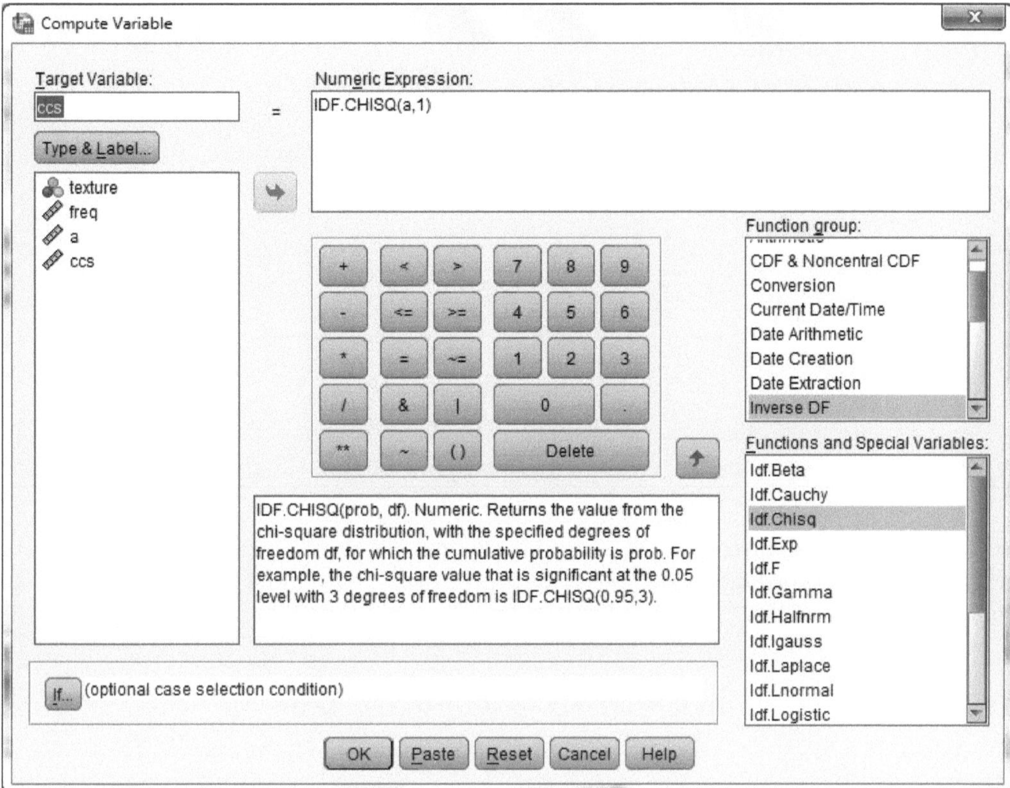

Figure 1d

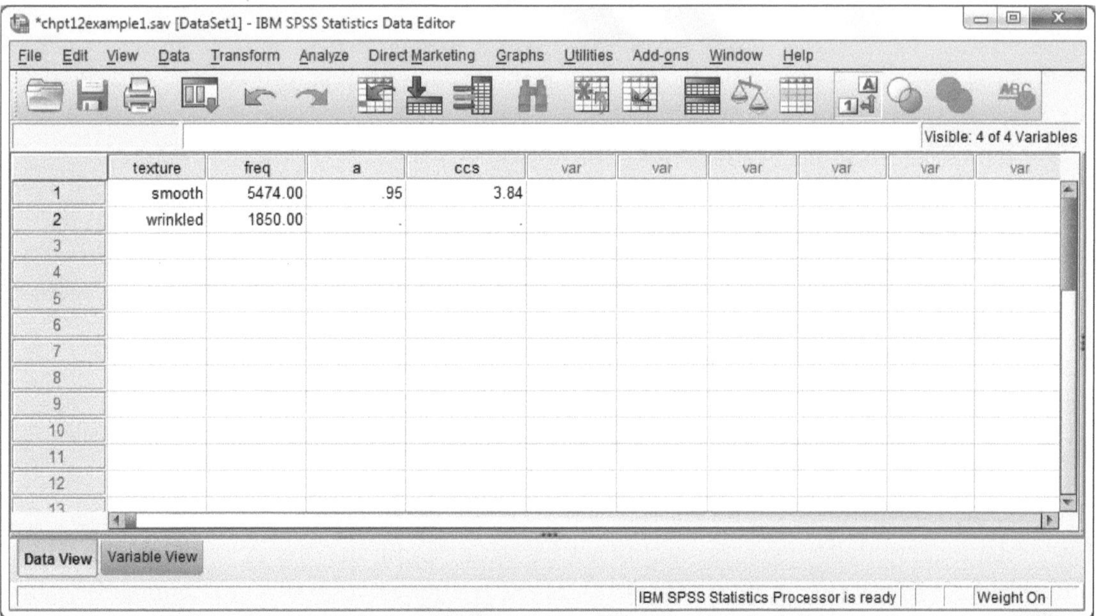

Figure 1e

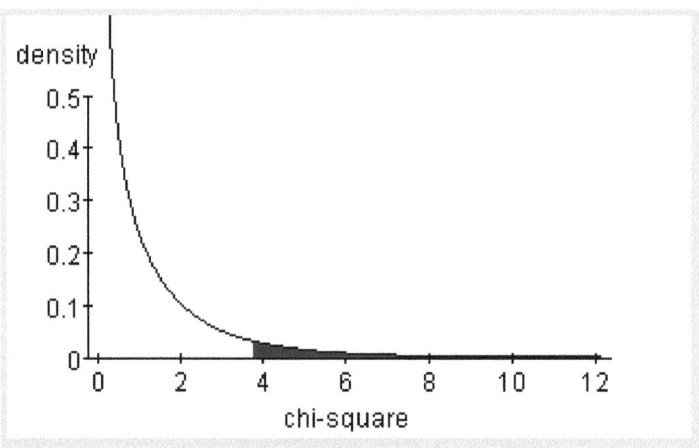

Figure 1f

We can also use the Chi-Square statistic to test if two qualitative or discrete variables are related. This is called Chi-Square Independence-of-Categories Test.

Independence-of-Categories Test

An independence-of-categories test is a test of numerical values in categories. If the categories are independent, then the numerical values should follow the independence rule given in Chapter 5. That is, if categories A and B are independent, then $P(A \cap B) = P(A) \cdot P(B)$. We can make a contingency table and calculate the numerical values for each cell and use the chi-square formula, $\chi^2 = \sum \frac{(O-E)^2}{E}$, where O is the observed value for each cell and E is the expected value for each cell and the difference $(O - E)$ is squared to avoid adding negative values, but why bother? SPSS will gladly do all the work for us. We just have to know how to process the procedure.

Example 2.

Open the Student data set (2). We would like to test if the two variables "own a car" (owncar) and "residency on campus" (res) are related. We would expect that a commuter is more likely to own a car than a resident.

Solution. Go to Analyze → Descriptive Statistics → Crosstabs and select "Do you own a car?," *owncar* as the Row, and "Resident or Commuter" *res* as the Column; see **Figure 2a**. Hit Ok and the contingency table will be given in the output, as in **Figure 2b**. The total number of data values is 219 (lower right corner). The number of data values that reflect "do not own a car" and "resident" is 6. The number of data values that reflect "do not own a car" and "commuter" is 72. The total of data values that reflect "do not own a car" is 78 (6 + 72 = 78). Likewise the second row is the number of data values for "do own a car" and "commuter" or "resident." The third row is the column total (the sum of the two numbers above the bottom number). The grand total of 219 is also the total of the column totals; that is, 219 is the sum of the two numbers immediately to the left.

To perform the chi-square test, go back to Crosstabs, click on Statistics, and choose Chi-Square. Also click on Cells and select Expected (keep Observed, the default, selected as well); see **Figure 2e**. The output contingency table gives the table for Resident or Commuter and whether they own a car or not with the observed and expected values in each cell; see **Figure 2c**. The observed values are the Count rows and the Expected Count rows are the expected values that are obtained by the equation $Expected = \frac{rowtotal \bullet columntotal}{grandtotal}$. The Chi-Square value is computed by the formula $\chi^2 = \sum \frac{(O-E)^2}{E}$. The chi-square value is given in the next table; see **Figure 2d**.

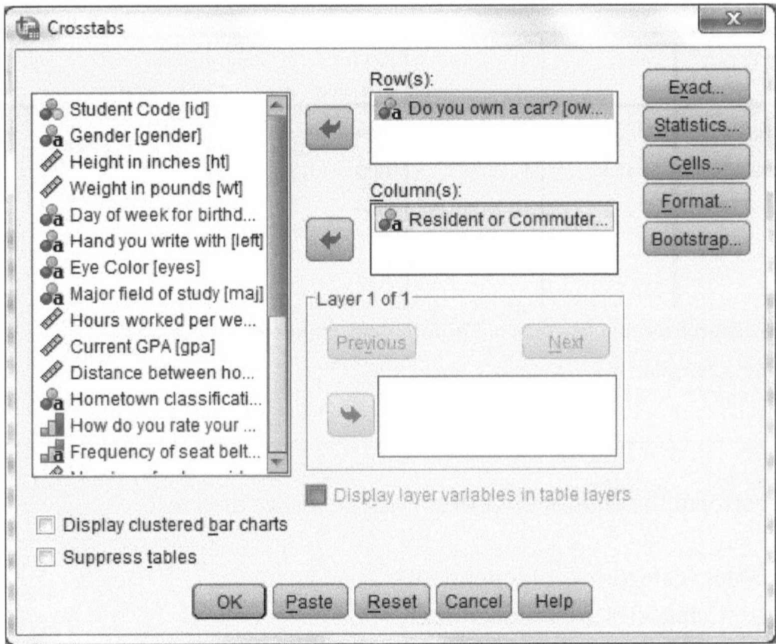

Figure 2a

Do you own a car? * Resident or Commuter? Cross-tabulation

Count

		Resident or Commuter?		Total
		C	R	
Do you own a car?	N	6	72	78
	Y	44	97	141
Total		50	169	219

Figure 2b

Do you own a car? * Resident or Commuter? Cross-tabulation

			Resident or Commuter?		Total
			C	R	
Do you own a car?	N	Count	6	72	78
		Expected Count	17.8	60.2	78.0
	Y	Count	44	97	141
		Expected Count	32.2	108.8	141.0
Total		Count	50	169	219
		Expected Count	50.0	169.0	219.0

Figure 2c

Chi-Square Tests

	Value	df	Asymp. Sig. (2-sided)	Exact Sig. (2-sided)	Exact Sig. (1-sided)
Pearson Chi-Square	15.759[a]	1	.000		
Continuity Correction[b]	14.453	1	.000		
Likelihood Ratio	17.952	1	.000		
Fisher's Exact Test				.000	.000
N of Valid Cases	219				

a. 0 cells (0.0%) have expected count less than 5. The minimum expected count is 17.81.

b. Computed only for a 2x2 table

Figure 2d

As always we'll perform the eight-step process whenever we do a test.

Are the two variables (categories) independent?

Step 1. H_0 : Categories are independent.

Step 2. H_1 : Categories are not independent.

Step 3. $\alpha = .05$.

Step 4. We will use the Chi-Square test of independence. The Chi-Square value is always a right-tailed value since chi-square is always positive. We assume values are randomly selected and categories are independent. The expected value for each cell must be 5 or greater.

Step 5. Use Analyze → Descriptive Statistics → Crosstabs, move *owncar* into Row(s) and *res* into Column(s), click the tab Statistics, and check Chi-Square and Continue. Also click the tab Cells and check Expected under Counts; see **Figure 2e**. Click Continue and OK. The Pearson Chi-Square = 15.759, df = 1 and the p-value = .000.

Step 6. Since $df = 1$, the critical chi-square value is 3.84; see Example 1, Step 6.

Step 7. Compare 15.759 > 3.84, which puts the chi-square value in the rejection region, alternatively .000 < .05, which says the p-value is significant.

Step 8. We fail to accept the null hypothesis. It appears that owning a car and being a resident or commuter are not independent variables.

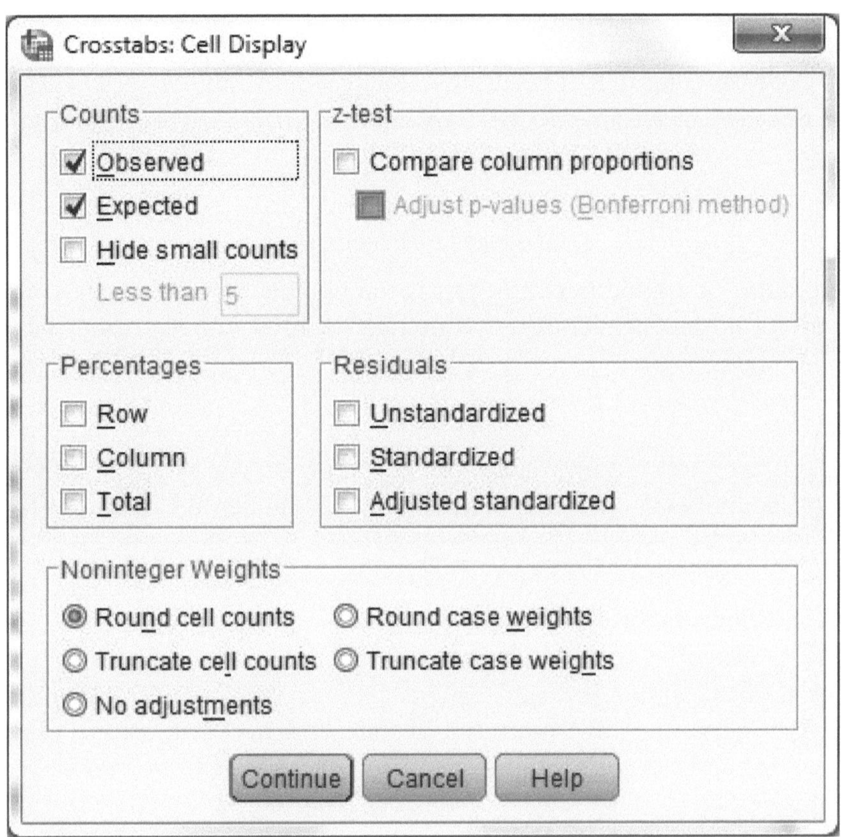

Figure 2e

Asymptotic Significance (2-sided)

The p-value associated with the Pearson Chi-Square value is .000 but under the heading "Asymp. Sig. (2-sided)." The meaning of this heading is that the significance level is determined by a function asymptotic to the normal distribution since the sample size is large. When the sample size is small the significance is determined by an exact function, binomial distribution, which can be computed but would take too long to compute if the sample size is large. To illustrate the asymptotic function is close to the normal curve let's compare the chi-square values for $df = 1$ against the squares of critical z values of a one-tailed test with half the significance level. They will be very close, such as the critical z value for a one-tailed test with significance level .05 is 1.645. $1.645^2 = 2.706025$. For $df = 1$ and a significance level of .10, the critical chi-square value is 2.7055. Likewise the z value for a two-tailed test at the 5% level is 1.96, so the critical chi-square value, 3.8414, should be about $1.96 \quad = 3.8416$, which it is; differences can be attributed to the way the values were calculated.

Minimum Expected Value

The Chi-Square test needs an expected value of at least a 5 in each cell. If that is not the case for a 2 x 2 contingency table, then the process will automatically combine cells together until the number in the cell is greater than 5 using the continuity correction value. If the contingency table is larger than 2 x 2, then SPSS will not calculate a continuity correction value. You need to make the changes in the data set if that is the case. Copy the column that needs to be changed into another variable, give it a new name, and replace those values that have an expected frequency less than 5 with one value larger.

Degrees of Freedom

The degrees of freedom for a goodness-of-fit test is the number of cells minus 1, that is, $n - 1$. The degrees of freedom for an independence test is number of rows minus 1 times number of columns minus 1, that is, $(R-1)\cdot(C-1)$.

Example 3a.

Using the data set Student, test if the variables (drive), "How do you rate your driving?", and Gender (gender) are independent of each other. Put drive in Row and gender in Column. What are the results?

Solution:

Step 1. H_0 : Categories are independent.

Step 2. H_1 : Categories are not independent.

Step 3. $\alpha = .05$.

Step 4. We will use the Chi-Square test of independence. We assume values are randomly selected and categories are independent. The expected value for each cell must be 5 or greater.

Step 5. Use Analyze → Descriptive Statistics → Crosstabs, move drive into Row(s) and gender into Column(s). Click on Statistics, check Chi-Square and click on Cells, check Expected. The Pearson Chi-Square value = 17.727, df = 2 and the *p*-value = .0; see **Figure 3a**.

Step 6. Since *df* = 2, the critical chi-square value = 5.991. [Open a new data editor and put .95 in the first column and call the variable *a*. Use Transform →Compute Variable, call the target variable *ccs* for critical chi-square value, click on Inverse DF (for Inverse Distribution Function) and double click on Idf.Chisq under Functions and Several Variables. Move *a* into the first position and use 2 for the degrees of freedom, click OK.]

Step 7. Compare the obtained Chi-square value against the critical Chi-square value, 17.727 > 5.9914, which puts the obtained value in the critical region of the distribution; see **Figure 3b**. Alternatively, the *p*-value is less than 5%.

Step 8. Our decision is to reject the null hypothesis. We have enough statistical evidence to suggest the categories are not independent.

Chi-Square Tests

	Value	df	Asymp. Sig. (2-sided)
Pearson Chi-Square	17.727[a]	2	.000
Likelihood Ratio	18.033	2	.000
N of Valid Cases	218		

2 cells (33.3%) have expected count less than 5. The minimum expected count is 3.49.

Figure 3a

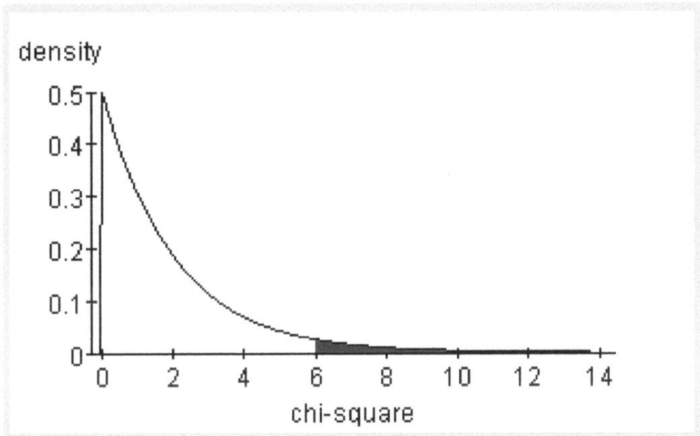

Figure 3b

Assumptions

When testing independence of categories, we make the following assumptions:

1. The data are drawn from random samples.
2. The expected value for each cell is the product of the row total times the column total divided by the grand total, which implies we're assuming the categories are independent.
3. The expected value for each cell must be greater than or equal to 5.

Example 3b.

Since some cells had less than 5 as the expected value, we should combine a few of the cells together. Combine the below-average values with the average value and rework the chi-square test.

Solution. Open the data set Student.sav. Use Transform →Recode into different Variables. Move the variable "How do you rate your driving?" (drive) into the Variables list. Under the Output variable assign the new variable a name such as new_drive and click change (you can leave label blank); see **Figure 3c**. Click old and new values. A new dialog box will open. Under old value, put in 1. Under new value put in 2, then click add. Also put in $2 \rightarrow 2$ and $3 \rightarrow 3$; see **Figure 3d**. Then click continue and OK. Redo the test, put new_drive in Row and gender in Column.

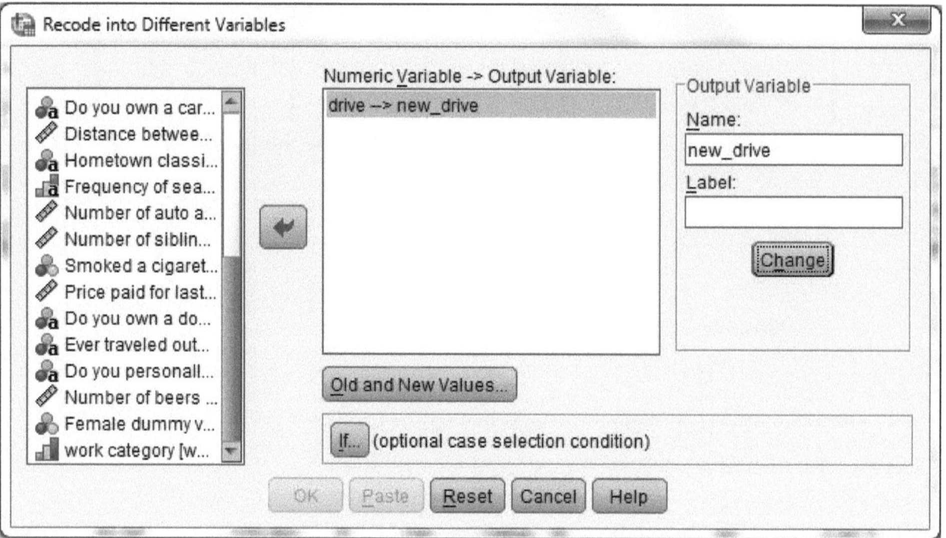

Figure 3c

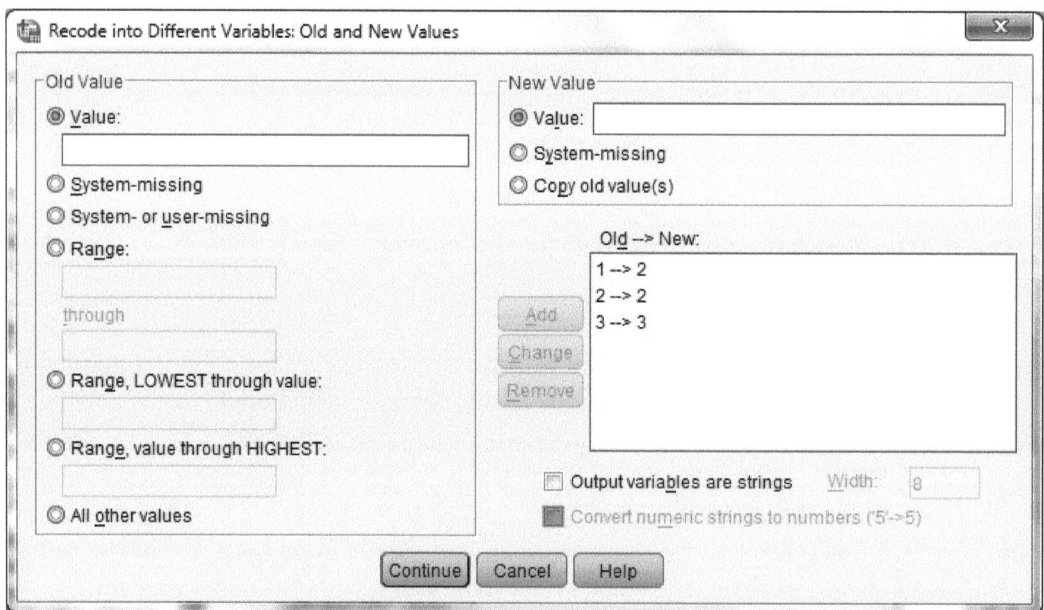

Figure 3d

Step 1. H_0 : Categories are independent.
Step 2. H_1 : Categories are not independent.

Step 3. $\alpha = .05$.

Step 4. We will use a Chi-Square test of Independence. We assume values are randomly selected and categories are independent. The expected value for each cell must be 5 or greater.

Step 5. The obtained chi-square value is 17.554 and the p-value is .000, which is close to the values found above. The reason is that the data set is large. If a data set is large and the test result is significant, then a few changes in the table won't affect the outcome. However, df changed from 2 to 1.

Step 6. The critical chi-square value for df = 1 and a cumulative area of .95 is 3.84; see **Figure 1e**.

Step 7. Compare the obtained Chi-Square value against the critical Chi-Square value, 17.554 > 3.84. Alternatively, compare the p-value against α .000 < .05.

Step 8. We reject the null hypothesis. It appears the categories "How do you rate your driving?" and Gender are not independent.

If the sample set is large, greater than 50, then a few cells with expected frequency less than 5 will not have an effect upon the outcome of the test.

Example 4.

Test if the following categories are each equally distributed (a goodness-of-fit test).

Preference for School Voucher

For	Maybe	Against	Total
23	17	50	90

Since there are three categories and the total is 90, then the expected value for each cell is 30.

Solution: Assuming each category is equally likely, the expected data value for each cell is 30, thus the contingency table should look like this:

Preference for School Voucher

For	Maybe	Against	Total
23	17	50	90
30	30	30	90

In the top row are the observed values and in the bottom row are the expected values. The computed chi-square value is $\chi^2 = \frac{(23-30)^2}{30} + \frac{(17-30)^2}{30} + \frac{(50-30)^2}{30} = 20.6$, which is certainly much larger than the critical value of 5.991 (df = 2). However, we'll do the eight steps using SPSS. Open a new SPSS data editor and put 1, 2, 3 into the first column and label the variable voucher. In variable view change the values 1 to For, 2 to Maybe, and 3 to Against. Change the variable to Nominal. In the next column put 23, 17, 50 and label the variable freq. Set the freq variable to scale. Go to Data Weight Cases, and under Weight Cases By, put the variable freq, click OK. Now we're ready to perform the test.

Step 1. H_0 : All categories are equally distributed.

Step 2. H_1 : All categories are not equally distributed.

Step 3. $\alpha = .05$.

Step 4. We'll do a chi-square goodness-of-fit test. Assumptions are the data values are selected randomly and independently, and the cells have values greater than 5.

Step 5. Use Analyze → Nonparametric Tests → Legacy Dialogs → Chi-square and move voucher into Test Variable List. Under Expected Values we should have "All categories equal" checked. Click OK. The Chi-Square value is $\chi^2 = 20.6$, df = 2, and p-value =.0; see **Figure 4**.

Step 6. Since $df = 2$, the critical chi-square value at 5% is 5.991; see Step 6 in Example 3a.

Step 7. Compare our obtained Chi-Square value against the critical Chi-Square value gives the inequality 20.6 > 5.991, alternatively comparing the p-value against α gives the inequality .000 < .05.

Step 8. We reject the null hypothesis. It appears the categories "for," "maybe," and "against" are not equally distributed.

Test Statistics

	voucher
Chi-Square	20.600[a]
df	2
Asymp. Sig.	.000

a. 0 cells (0.0%) have expected frequencies less than 5. The minimum expected cell frequency is 30.0.

Figure 4

Let's review what we learned in this chapter.

1. A chi-square test is a nonparametric test used to test categories or discrete variables, which are nominal or ordinal values.
2. There are two types of chi-square tests.
   i.) Goodness-of-fit test
   ii.) Independence-of-categories test
3. The formula for a chi-square test is $\chi^2 = \sum \frac{(O-E)^2}{E}$.
4. The expected value for a goodness-of-fit test depends on whether there is a theoretical or a proscribed distribution for each category, and if all categories have an equal distribution. If all categories have an equal distribution, then the expected value for each cell is the total number of data values divided by the number of cells.
5. The expected value for the cells of independence-of-categories test is

$$Expected = \frac{rowtotal \bullet columntotal}{grandtotal}.$$

6. The minimum expected value for each cell must be greater than or equal to 5.
7. The degrees of freedom is $k - 1$ for a goodness of fit test, k is the number of categories and for an independence-of-categories test, the degree of freedom is $(R-1)\cdot(C-1)$, where R is the number of rows and C is the number of columns in a contingency table.
8. The total frequency for a goodness-of-fit test should be large, greater than 50.

Class Work 12

Take your bag of M&M's and test the hypothesis that the frequency distribution is not the same as the manufacturer's proposed frequency distribution. This is a goodness-of-fit test. To test if all the categories are the same, leave Expected values in the default mode "All categories equal." Use alpha = .05. Do not eat the M&M's before you count the number of each color. There needs to be enough M&M's so that the expected value of each cell is at least 5.

Manufacturer's Proposed Frequency Distribution
Brown = .30
Red = .20
Yellow = .20
Green = .10
Blue = .10
Orange = .10

Homework 12

1. Out of 100 cars surveyed, 43 were red, 10 were green, and 47 were blue. Do these numbers fit the manufacturer's claim that 50% of its cars are red, 20% are green, and 30% are blue? Use a 5% level of significance.

2. Use the data set Student to test if seat-belt usage is independent of gender.

3. A community college claims only 10% of their hourly instructors are dissatisfied with their treatment as sub-par employees, while 50% of the hourly faculty are completely satisfied being treated as expendable and worthless employees and the rest are partially satisfied. A recent survey of hourly faculty found these results: chpt12problem3.sav

Hourly Faculty	Completely Satisfied	Partially Satisfied	Dissatisfied
Number of Responses	120	230	17

Each hourly employee in the survey chose only one of the three categories. Do these results support the claim of the community college that hourly faculty are satisfied with their treatment as employees? Test against 50% completely satisfied, 40% partially satisfied and 10% dissatisfied.

4. Another survey at a different school found the following results for the satisfaction level of their employees. chpt12problem4.sav

Employees	Faculty	Administrators	Clerical	Support Staff
Satisfied	82	38	34	41
Not Satisfied	18	22	11	17

Do these results support the claim that satisfaction level is different for different job classifications?

This is called a chi-square test of Homogeneity of Proportions, but it is the same as a test of independence of categories, except it is several samples with one variable rather than one sample with two variables as in problem #2.

5. The number π is irrational, which means the decimal representation of π does not follow a pattern. The digits 0 through 9 will appear in non-discernible patterns and so will act like a random-number generator. The frequencies for the first hundred digits are given in the following table. Test at the 5% level if the digits are uniformly distributed. chpt12problem5.sav

Digit	0	1	2	3	4	5	6	7	8	9
Frequency	8	8	12	11	10	8	9	8	12	14

6. Based on results from a study done by AT&T and the Automobile Association of America, the following table gives a frequency of drivers who had an accident while using a cell phone and those who did not use a cell phone. Test at the 5% level if cell phone use is independent of accidents. chpt12problem6.sav

	Had Accident Last Year	Had No Accident Last Year
Cell Phone User	23	282
Not Cell Phone User	46	407

Just for fun. Use the data set Salem to test if the variables proparri and accuser are independent. What does this result say about the Salem Witch trials in 1692? Proparri: indicator variable identifying persons who supported Rev. Parris in 1695 records (1 = supporter); Accuser: indicator variable identifying accusers and their families (1 = accuser); Defend: indicator variable identifying accused witches and their defenders (1 = defender).

Alternative classwork for chapter 12 (in case everyone doesn't have a bag of m&m's.)

In a recent m&m study the following frequencies were observed for each color:

Color	Brown	Red	Yellow	Green	Orange	Blue
Frequency	84	90	130	123	101	96

Since the manufacture no longer claims the colors fit a predetermined percentage, then we'll test if all categories are equal.

References

(1) Johann Gregor Mendel article from Muskingum University
(2) SPSS data sets

Chapter 13
Other Non-Parametric Tests

Objectives

After completing this chapter, a student should:

- ✓ Understand when to use nonparametric tests
- ✓ Be able to use the Wilcoxon Signed Rand Test
- ✓ Be able to use the Mann-Whitney U Test
- ✓ Be able to use the Kruskal-Wallis H Test
- ✓ Be able to use Spearman's Rank Order Correlation

Sometimes we have data that do not follow any type of distribution, such as normal, uniform, binomial, etc., but we can still run tests, check hypotheses, and form conclusions. These types of tests are called nonparametric and are not as robust as parametric tests since we don't assume that any parameters, such as the mean, median, standard deviation, or proportion are given. In fact, in some of these nonparametric tests, we ignore the numerical values entirely and simply work with the sign of the number, + or –, or the rank of the data value. Appropriately these tests are called the sign test and the rank test.

Wilcoxon Signed Rank Test

Recall a paired samples test of dependent means that we used in Chapter 8 had a null hypothesis that the mean of the difference was zero and the alternative hypothesis that the mean of the difference was not zero, or was positive or was negative. In a similar fashion the Wilcoxon Signed Rank Test is the nonparametric version of the paired samples test.

Example 1.

Use the Wilcoxon Signed Rank Test on chpt8example1.sav to see if the conclusions are the same as in Chapter 8, Example 1.

Solution: Open the data set chpt8example1.sav, use Analyze → Non-Parametric → Legacy Dialogs → 2 Related Samples, and move the variables before and after into the Test Pairs List. Check Wilcoxon as the Test Type, also click on Exact and check Exact, leave the time limit as 5 min—see **Figure 1a** (exact is based on the binomial probability)—and click OK. The output should be as in **Figure 1b**. The z value is -2.207 and the p-value is .027 for asymptotic significance (similar to normal distribution) and .031 for exact significance. Since both of these 2-tailed values are less than .05 then the test is significant. Nonparametric test are usually two-sided, since we don't have a lot of assumptions to work with, which is why the p-value is given as two-sided, but in this case we want to show there are more positive signs than negative signs (the nonparametric tests do not calculate before – after, but after – before). The exact significance values have two-tailed and one-tailed results. Here's the eight-step process: The test variable, W, is the minimum absolute value of the sum of the signed ranks of differences between the two related variables, not counting differences equal to 0. To find the sum, find the difference of the two variables and then rank the absolute value of the differences with the rank of ties being averaged. Then multiply the rank with the sign of the difference and find the sum of the negative ranks and the sum of the positive ranks.

The test variable, W, is the minimum absolute value of the sums. The test statistic is $\frac{n(n \quad)}{\quad}$,

where n is the number of nonzero differences and $\sigma_W = \sqrt{\frac{n(n+1)(2n+1)}{24} - \frac{t^3 - t}{48}}$; t is the number of ties in a group of tied ranks. In this example, to find the ranks, use Transform → Compute Variable, call the target variable d for difference and under numeric expression put after – before (see **Figure 1c**) and click OK. The differences are 2, 0, 4, 3, -2, -1, 5, 3, 3, 2. We drop the difference of 0 and rank the absolute value of the rest; in ties, each value gets the same rank. Multiply the rank with the sign of the value.

Difference	-1	-2	2	2	3	3	3	4	5
Absolute Value	1	2	2	2	3	3	3	4	5
Rank	1	3	3	3	6	6	6	8	9
Signed Rank	-1	-3	3	3	6	6	6	8	9

The number of ranked values is 9 and the sum of the negative ranks is -4, while the sum of the positive ranks is 41. We choose $W = 4$, the minimum of the absolute values of -4 and 41. The median value is the average rank, which is 22.5 (we expect both the positive ranks and the negative ranks to have a sum of 22.5). There are two groups of three tied ranks, so t = 3 and we need to compute $\frac{t^3 - t}{48}$ twice, $\frac{3^3 - 3}{48} \times 2 = 1$. Therefore the standard deviation, which is the average of the sum of squares of the ranks, is adjusted by subtracting 1 from the variance, $\sigma_w = \sqrt{\frac{9 \times 10 \times 19}{24} - 1} = \sqrt{70.25}$. So $z = \frac{4 - 22.5}{\sqrt{70.25}} = -2.207$.

Step 1. H_0 : median difference between the pairs is 0
Step 2. H_1 : median difference between the pairs is not 0
Step 3. $\alpha = .05$
Step 4. Type of test: Wilcoxon Signed Ranked Test, we assume that data are paired andcome from the same population, each pair is chosen randomly and independently, and data need not be normal.
Step 5. The obtained values are z = -2.207 with a p-value of .027. The exact p-value for a one-tailed test is .016
Step 6. The critical z value for a two-tailed test is ± 1.96.
Step 7. Compare: -2.207 < -1.96, which says the obtained value is in the rejection region, alternatively .027 < .05, or using the exact value for a one-tailed test, .016 < .05.
Step 8. Decision: Reject the null hypothesis: there is a difference in means. It appears the psychologist's self-esteem courses had a positive effect on the inmates.

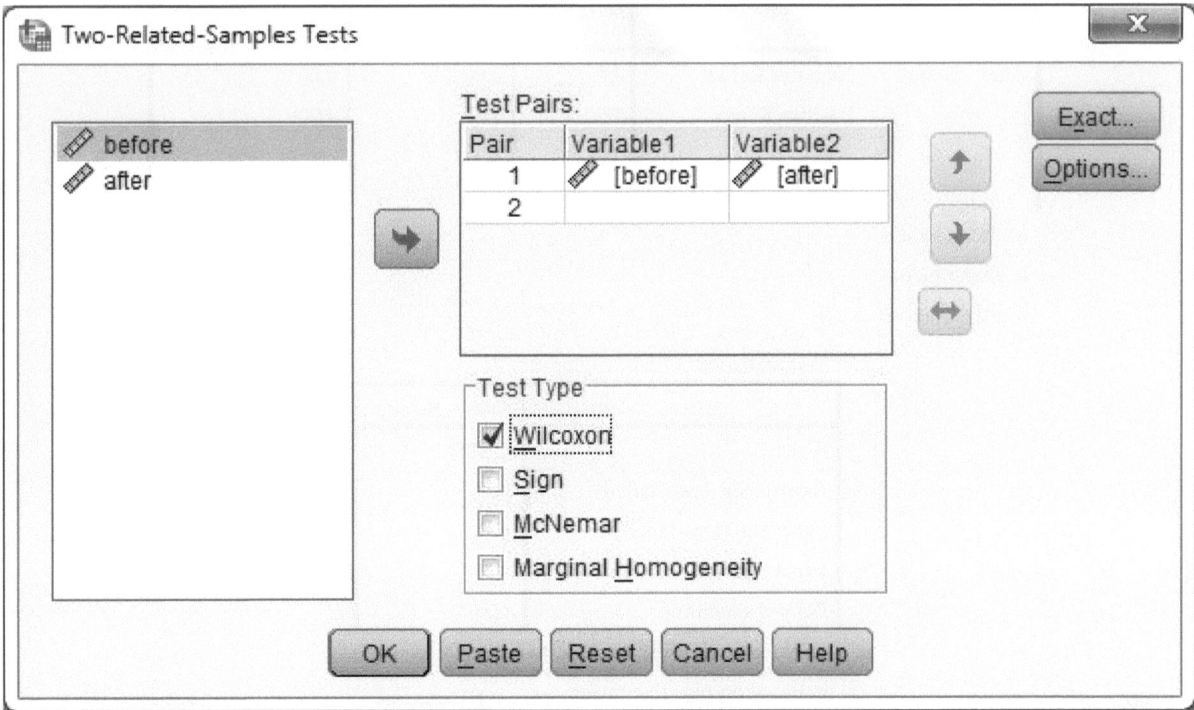

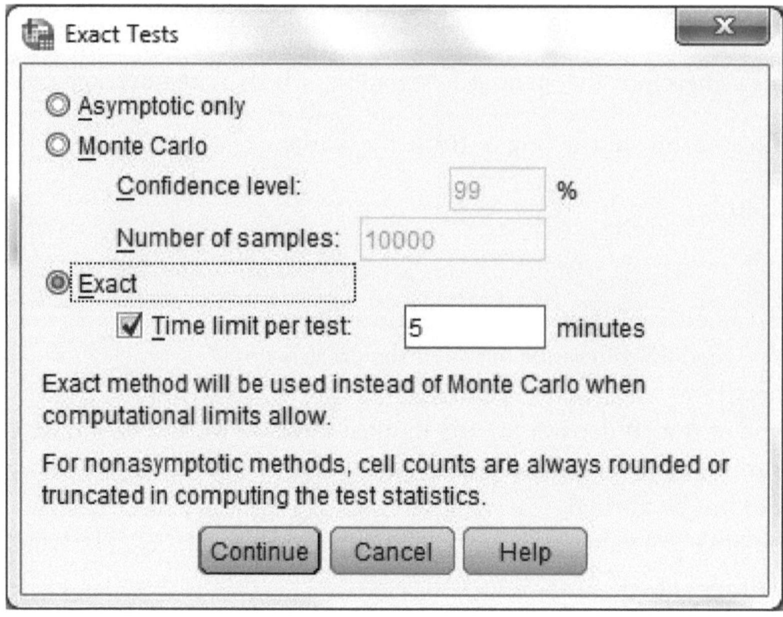

Figure 1a

Wilcoxon Signed Rank Test

Ranks

		N	Mean Rank	Sum of Ranks
After – Before	Negative Ranks	2[a]	2.00	4.00
	Positive Ranks	7[b]	5.86	41.00
	Ties	1[c]		
	Total	10		

a. after < before

b. after > before

c. after = before

Test Statistics[a]

	After – Before
Z	-2.207[b]
Asymp. Sig. (two-tailed)	.027
Exact Sig. (two-tailed)	.031
Exact Sig. (one-tailed)	.016
Point Probability	.006

a. Wilcoxon Signed Rank Test

b. Based on negative ranks

Figure 1b

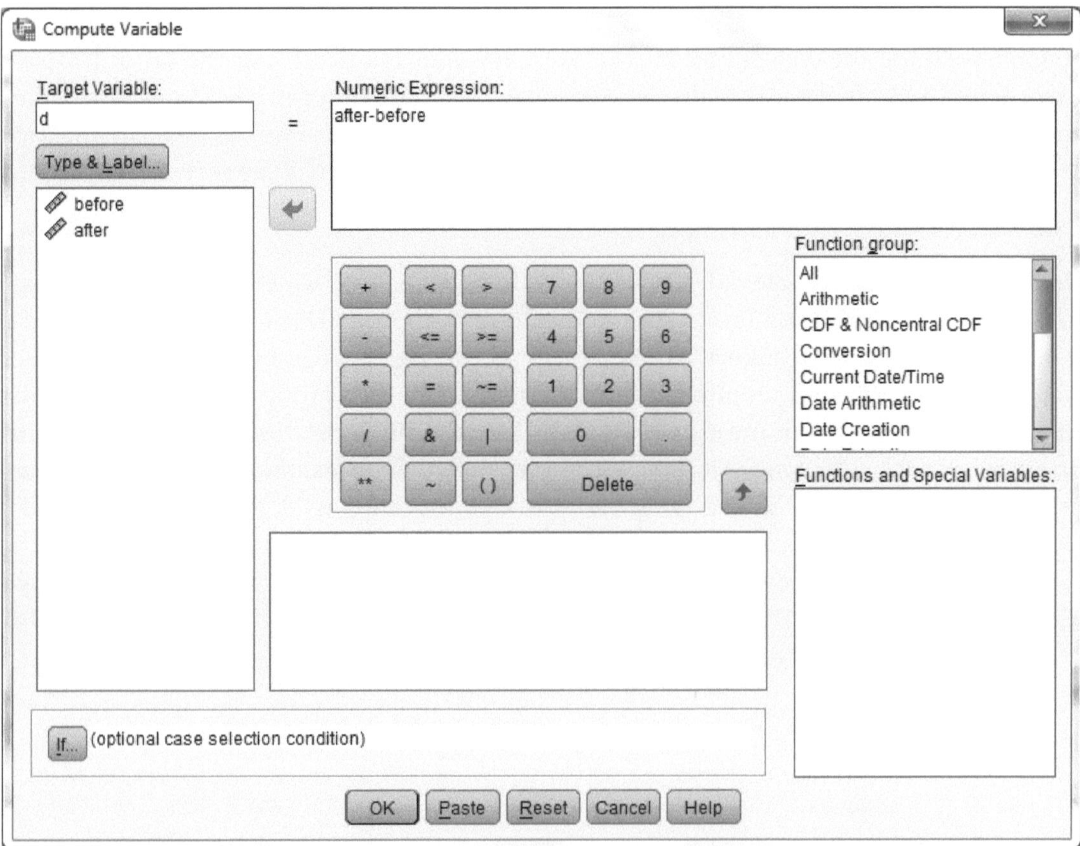

Figure 1c

Assumptions

1. Data are paired and come from the same population.
2. Each pair is chosen randomly and is independent.
3. The data are measured on an interval scale (ordinal is not sufficient because we take differences) but need not be normal.

The Mann-Whitney U Test

The nonparametric version of the independent samples T-test is the Mann-Whitney U Test. The **assumptions** are that the two groups are independent of each other, the data values are ordinal or scale, and both groups have an equal distribution. In the test we rank all the data as one variable and then test if the sums of the ranks are equal. If there is a difference in means between the two groups, then the data from one group should be clustered on one side of the ranking and the data from the other group should be clustered on the other side of the ranking. If there is no difference between the two groups, then the data from both groups should be evenly mixed throughout the ranked data.

Example 2.

As an example we'll use the data set Salem (1). We'll consider tax data from the set of people who support Rev. Parris and tax data from the set of people who don't support Rev. Parris. Since the two histograms are not normal, they're skewed, yet have the same shape; hence we can compare their means using the Mann-Whitney U Test. Test if the mean tax paid by the ProParris supporters was the same as the mean tax by the non-ProParris supporters.

Solution: Open the data set Salem. Use Graphs ⟶ Legacy Dialogs ⟶ Histogram. Move Tax paid (pounds) into the variable box and ProParris into the Rows box. Check Display normal curve and click OK. The output should be as in **Figure 2a**. As we can see the two histograms are slightly skewed but of the same shape. The third assumption is satisfied; that is, the two groups have an equal distribution. To analyze the difference in means we'll use the Mann-Whitney U Test as follows: Use Analyze ⟶ Nonparametric tests ⟶ Legacy Dialogs ⟶2 Independent Samples, move Tax paid into the Test Variables List and proparri into the Grouping Variable. Define groups as "0, not a supporter" and "1, a supporter"; see **Figure 2b**. Click continue and OK. The output should read as in **Figure 2c**.

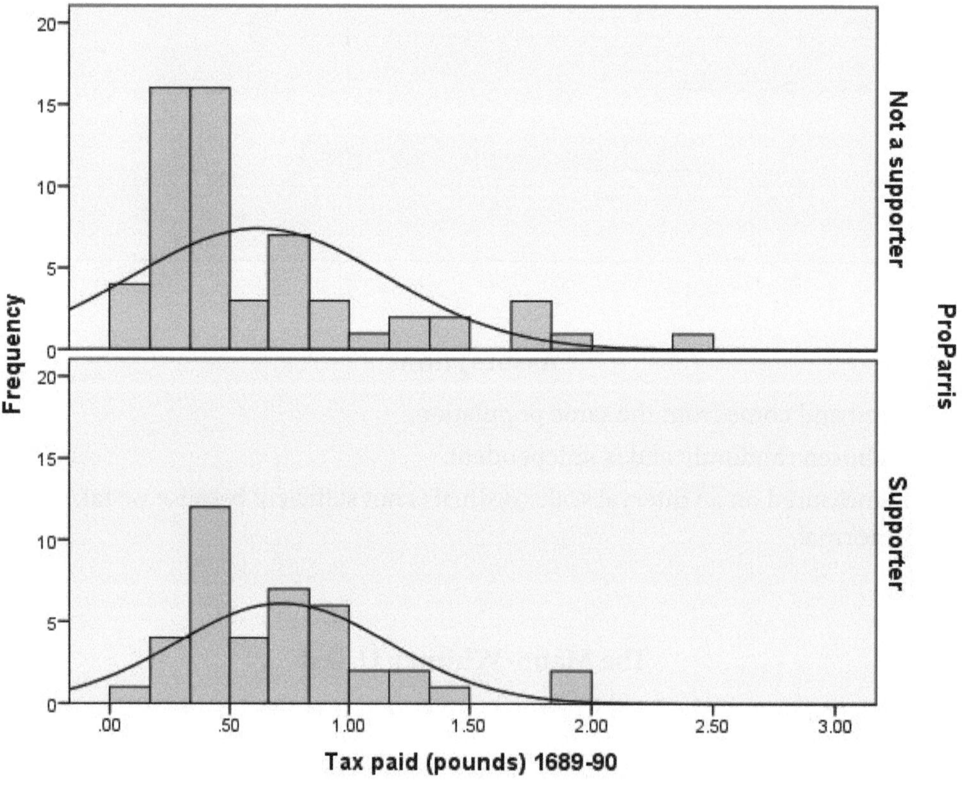

Figure 2a

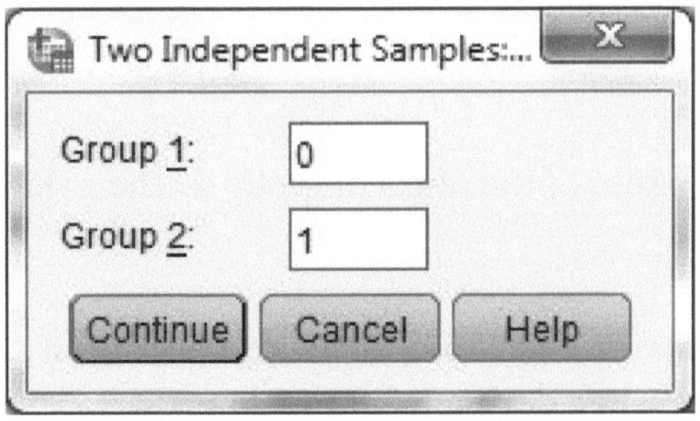

Two Independent Samples:...

Group 1: 0

Group 2: 1

Continue Cancel Help

Figure 2b

Ranks

	ProParris	N	Mean Rank	Sum of Ranks
Tax Paid (Pounds) 1689–90	Not a Supporter	59	45.67	2694.50
	Supporter	41	57.45	2355.50
	Total	100		

Test Statistics[a]

	Tax Paid (Pounds) 1689–90
Mann-Whitney U	924.500
Wilcoxon W	2694.500
Z	-2.004
Asymp. Sig. (two-tailed)	.045

a. Grouping Variable: ProParris

Figure 2c

The *p*-value is .045, which leads us to reject the null hypothesis and state that there appears to be a difference in the amount of tax paid by supporters of Rev. Parris and nonsupporters of Rev. Parris.

The eight-step process is as follows:

Step 1. H_0: the two groups have the same sum of rankings

Step 2. H_1: the two groups do not have the same sum of rankings

Step 3. $\alpha = .05$

Step 4. We'll use the Mann-Whitney U Test. We'll assume both groups are independent of each other, the data values are scale, and both groups have the same distribution.

Step 5. The Mann-Whitney U statistic is 924.5 and is the smaller of U_1 or U_2. $U_1 = R_1 - \frac{n_1(n_1+1)}{2}$, where n_1 is the sample size for group 1 and R_1 is the sum of the ranks of group 1. Likewise $U_2 = R_2 - \frac{n_2(n_2+1)}{2}$, where n_2 is the sample size for group

2 and R_2 is the sum of the ranks of group 2. The z statistic is $z = \dfrac{U - \frac{n_1 n_2}{2}}{\sqrt{\frac{n_1 n_2 (n_1 + n_2 + 1)}{12}}}$ and is approximately equal to -2.004 with a p-value of .045. Note: there is an adjustment for the variance when dealing with ties.

Step 6. The critical z value for a two-tailed test at a 5% significance level is ± 1.96.

Step 7. Since -2.004 < -1.96, we see the obtained z value is in the rejection region, alternatively the p-value of .045 < .05.

Step 8. Our decision is to reject the null hypothesis; there is a difference in the mean tax paid by these two groups. By looking at the mean tax paid by these two groups, it appears that supporters of Rev. Parris paid more tax than nonsupporters of Rev Parris.

Kruskal-Wallis H Test

When we ran tests using one-way ANOVA, we assumed all samples came from normal populations with equal variances. If that is not the case, then we can use a nonparametric test called the Kruskal-Wallis H Test. Also the Kruskal-Wallis Test can be used with data at the ordinal level of measurement, such as ranked data. The Kruskal-Wallis Test has the following **assumptions**:

1. We have more than two random samples.
2. Our null hypothesis is that the means from all the populations are equal.
3. Each sample has at least five data values.
4. The variances still have to be somewhat equal and each group has to have identical shape and scale.

Example 3.

Check if the time intervals for Old Faithful Geyser in Yellowstone National Park have changed over the years from 1951 to 1996.

Solution: Open the data set chpt13example3.sav. We'll first check to see if the distributions are somewhat equal. Use Graphs → Legacy Dialogs → Histogram, move Time_int into the Variable dialog box and Year into the Panel by Rows dialog box. Check Display normal curve (see **Figure 3a**) and click OK. The output should be as in **Figure 3b**. Since we don't have normal distributions for all the panels, yet they are somewhat similar, we'll use the Kruskal-Wallis Test. We can verify that the data set does satisfy the Homogeneity of Variance Test (use ANOVA) and thus we can test if the means are equal in all panels. If it passes this test, then the population means for the four different years are the same.

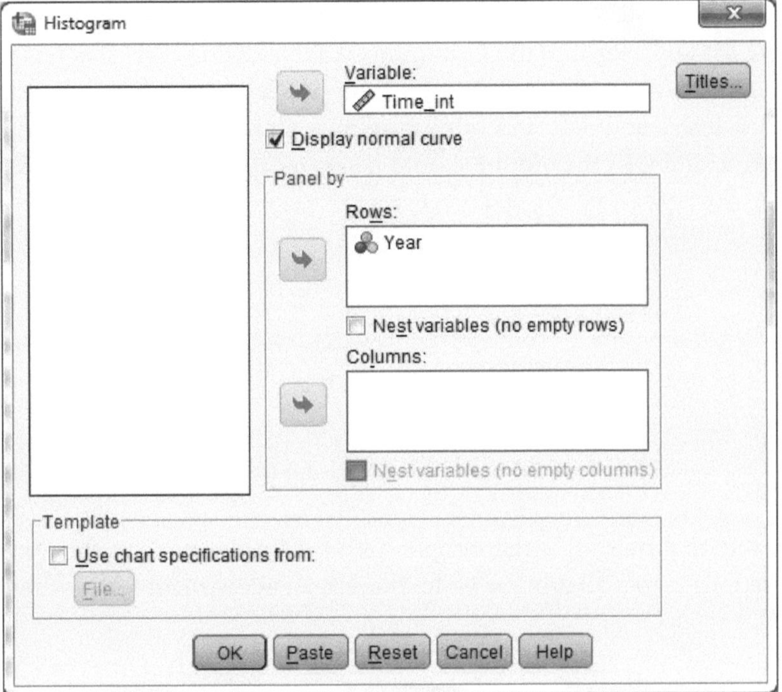

Figure 3a

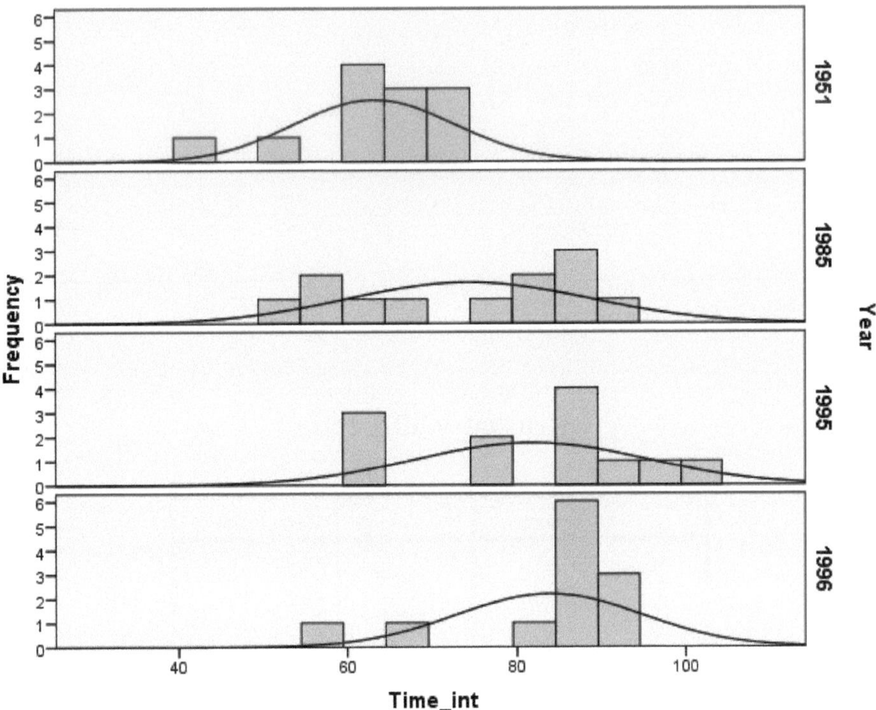

Figure 3b

Use Analyze → Nonparametric Tests → Legacy Dialogs → K independent samples, move Time_int into the Test variable list and Year into the Grouping Variable dialog box. This test will ask us to define the groups by range instead of individual groups. Our smallest group number was 1 and our largest group number was 4. Put 1 and 4 in the minimum and maximum boxes of the range window; see **Figure 3c**. Hit continue and OK; the output should be as in **Figure 3d**. The test statistic is a chi-square statistic because the formula is $H = \frac{12}{n(n+1)}\sum_{i=1}^{k} n_i\left(\overline{R_i} - \overline{\overline{R}}\right)^2$, which is the basic format for the chi-square statistic, i.e. ($\chi^2 = \frac{(observed - expected)^2}{expected}$). (This formula for *H* is a simplification of a more complicated formula that clearly illustrates the χ^2 equation.) $\overline{R_i} = \frac{R_i}{n_i}$ and $\overline{\overline{R}} = \frac{\sum_{i=1}^{k} R_i}{\sum_{i=1}^{k} n_i}$, where R_i is the ranked sum for the *i*th group, n_i is the sample size of the *i*th group, *k* is the number of groups, and *n* is the total observations across all groups. Note: there is an adjustment for *H* when dealing with ties.

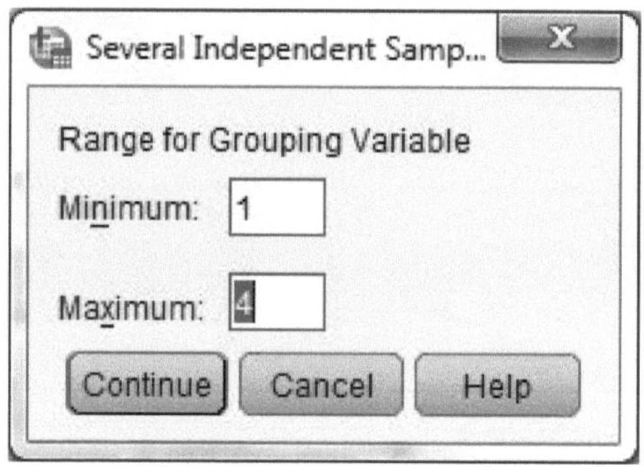

Figure 3c

Kruskal-Wallis Test

Ranks

	Year	N	Mean Rank
Time_int	1951	12	13.08
	1985	12	22.13
	1995	12	29.88
	1996	12	32.92
	Total	48	

Test Statistics[a,b]

	Time_int
Chi-Square	14.479
df	3
Asymp. Sig.	.002

a. Kruskal-Wallis Test

b. Grouping Variable: Year

Figure 3d

Whenever we do a test, we list the eight-step process.

Step 1. H_0: The mean time intervals for the four years are the same.

Step 2. H_1: At least one population mean from the four years is different.

Step 3. $\alpha = .05$

Step 4. Since the samples are not necessarily from normal distributions, we'll use the nonparametric equivalent of one-way ANOVA, the Kruskal-Wallis H Test.

Step 5. The obtained chi-square value is 14.479 with a p-value of .002.

Step 6. The critical chi-square value with three degrees of freedom is 7.8147 (in general, the degrees of freedom equals $k - 1$, where k is the number of groups).

Step 7. Compare 14.479 > 7.8147, alternatively .002 < .05.

Step 8. We see from the above that the obtained value is in the critical region; hence we reject the null hypothesis in favor of the alternative hypothesis. It appears the population time intervals between eruptions of Old Faithful is not the same.

One of the reasons nonparametric tests are used is that it's usually easier to compute the test statistic if working the problem by hand instead of using a computer. Since this text is based on the assumption that everything is being done with a computer, that reason is no longer valid. However, it is nice to know where the theory came from and how to use it if one day we don't have access to our computers.

Spearman's Rank Order Correlation

Our last nonparametric test is the Spearman's Rank Order Correlation Test, which is used to test if there is a linear relationship between two ordinal variables. The parametric equivalent is the Pearson Correlation Coefficient, which tests if there is a relationship between two scale variables.

Example 4.

Two groups of people measure the teaching ability of five hourly instructors and rank the instructors on a scale of 1 to 5, from best to worst. The two groups are students and full-time faculty. We want to measure if the two groups rank the instructors with the same ranking; that is, do the students and full-time faculty agree as to who is a good instructor. The data are in chpt13example4.sav.

Solution: Open the data set chpt13example4.sav. Use Analyze → Correlate → Bivariate, move the two variables "students" and "faculty" into the Variables dialog box, deselect Pearson, the default option, and select Spearman since we're using ranked data. Perform the eight-step test process as given below. The formula to calculate the obtained value is $r_s = 1 - \frac{6\sum d^2}{n(n^2-1)}$, where d is the difference in the corresponding ranks and n is the number of ranks. If tied ranks exist, then Pearson's correlation coefficient, $\rho = \frac{\sum_{i=1}^{n}(x_i - \bar{x})(y_i - \bar{y})}{\sqrt{\sum_{i=1}^{n}(x_i - \bar{x})^2 \sum_{i=1}^{n}(y_i - \bar{y})^2}}$, between ranks should be used for the calculation. All the tied values must be assigned the same rank, which is the average of their positions in the ascending order of values.

Step 1. $\rho_s = 0$ We use ρ_s for the Spearman rho and ρ for the Pearson rho.
Step 2. $\rho_s \neq 0$ We could also use the symbols > or <, but usually nonparametric tests are two-tailed.
Step 3. $\alpha = .05$
Step 4. Since we are using ordinal data, we'll use the nonparametric equivalent of the correlation coefficient, Spearman's Rank Test.
Step 5. $r_s = 0$ Spearman's rank correlation coefficient is 0 and the significance is 1.
Step 6. $r_{sc} = 1$ The critical value was obtained from a table, as SPSS does not have an inverse function to find a critical Spearman's rank correlation coefficient. To find a table on the Internet, just search for "critical values for Spearman's rank correlation coefficient."
Step 7. 0 < 1 or using p-values 1 > 0.
Step 8. Conclusion Students and faculty are not in agreement as to whom is a better instructor. There is no correlation between the rankings.

We want to emphasize once again that a statistical test does not give any clue as to a cause-and-effect relationship. All we can say is a relationship exist or it doesn't exist. Causes may be infinite in number. For example, students may like instructors who are younger and speak their language, while faculty may prefer instructors to dress sharp and be clear in their presentations at all times—or it could be the other way around. As they say in law, a person should not be convicted on circumstantial evidence alone, because there is an infinite number of ways to arrive at a set of circumstances. Likewise in statistics, the number of causalities that might affect a relationship between two variables is infinitely large.

Let's review what we've learned in the chapter.

1. We use the nonparametric test when our data aren't scale data or if the normal parameters of the population aren't satisfied.
2. The Wilcoxon Signed Rank Test is the nonparametric equivalent of the paired samples t test.
3. The Mann-Whitney U Test is the nonparametric equivalent of the independent samples t test.
4. The Kruskal-Wallis Test is the nonparametric equivalent of one-way ANOVA.

5. The Spearman's Rank Order Correlation Test is the nonparametric equivalent of the Pearson Correlation Coefficient Test.

Class Work 13

In Chapter 11 we tested the correlation between the heights of a father and son and we found a positive correlation. In this class work, we'll test if there is a correlation between father's height and son's height, but we'll use their ranks. Open the data set fathers_sons.sav and rank the fathers' heights along what their corresponding sons' heights rank. Then use the Spearman's Rank Order Correlation Test to test if there is a relationship between father's height and son's height. chpt13classwork.sav

Homework 13

1. The following are the critical-thinking scores of students before and after studying abstract reasoning. Test at the 5% level using the nonparametric Wilcoxon Signed Rank Test that their critical-thinking skills improved after studying abstract reasoning. chpt8problem1.sav

Before	72	75	69	70	71	82	34	59	61	75	88	90
After	72	76	75	80	76	85	45	66	72	74	88	93

2. Identical preschool twins are separated and randomly selected to go to one of two rooms. One room has alphabet coloring books, preschool readers, and alphabet puzzles. The other room has non-educational toys. Both groups spend two hours a day in their respective rooms for six months, after which a record is kept as to when they first start to read a primary text. Their ages, in months, are given in the following table. Is there evidence, using the nonparametric Wilcoxon Signed Rank Test, to suggest that alphabet books will help a child read sooner? Test at the 5% level. chpt8problem4

Basic Toys	70	71	69	63	64	62	63	75	65	52	64	60
Alphabet Toys	69	70	68	61	65	58	62	71	60	53	61	58

3. Open the data set GSS94 (2), and test if the number of hours watching television is different for men and women. Use the Mann-Whitney U Test for nonparametric statistics.

4. Open the data set GSS94, and test if the respondents' income is different for men and women. Use the Mann-Whitney U Test for nonparametric statistics.

5. Test, at the 5% level, if these three groups have the same mean. Use the nonparametric Kruskal-Wallis Test. chpt9problem5.sav

Group A	Group B	Group C
52	78	24
31	93	38
61	88	23
59	77	50
62		39
		90

6. Test, at the 1% level if these four treatments have the same mean. Use the nonparametric Kruskal-Wallis Test. chpt9problem6.sav

Treatment 1	3	2	3	4	
Treatment 2	5	6	7	5	6
Treatment 3	1	7	5	4	6
Treatment 4	2	2	1		

7. Compute and test the correlation coefficient for the average temperature difference (in °C) between the inside and outside of a house and the average daily gas consumption (in cu.ft.). Open the data set chpt11problem1.sav and rank the data. Then use the Spearman's Rank Order Correlation Test to test if there is a relationship between Temp. Diff. and Gas Consump. (chpt13problem7.sav)

Temp. Diff.	10.5	11.5	11.4	12.5	13.5	15.0	15.2	15.6	16.5	17
Gas Consump.	70.1	83.0	82.7	80.0	75.5	78.5	81.1	86.5	105.5	86

8. Is high school GPA a predictor for first-year college GPA? The following data give the high school GPA and the first-year college GPA of eleven students. Test if there is a correlation between these two variables. Use alpha = .05. Open the data set chpt11problem3.sav and rank the data. Then use the Spearman's Rank Order Correlation Test to test if there is a relationship between high school GPA and first-year college GPA. (chpt13problem8)

HS GPA	3.5	2.9	3.9	3.8	2.8	2.4	3.2	3.7	2.7	3.3	3.6
FYC GPA	3.3	2.8	3.6	3.5	3.0	2.0	2.9	3.4	3.0	3.4	3.6

Just for fun. The Barber of Seville Paradox. The barber of Seville shaves every man in Seville who doesn't shave himself. That is, if the man shaves himself, then the barber doesn't shave him. But if the man doesn't shave himself, then the barber must shave him. Now the question is, "Who shaves the barber?" And of course the barber is a man. Explain the paradox.

Reference

(1) The Salem Witchcraft site
(2) Data for Sociology

Chapter 14

Control Charts

Objectives

After completing this chapter, a student should:

- ✓ Be able to create a Control Chart and understand the output
- ✓ Be able to create a Pareto Chart and understand the output
- ✓ Know the three criteria to determine if a process is out of control

A coffee dispenser is supposed to dispense the same amount of coffee for every cup. The cup size allows for some variation in the amount of coffee dispensed; sometimes the amount is a little more, sometimes a little less, but the variation remains in a tolerable range. Over time, however, the variation will exceed the limitations of the cup, the coffee dispenser puts in very little coffee or it overflows the cup. At that point we say the coffee dispenser is out of control and needs to be recalibrated. But what we really would like to know is when the machine will start to be out of control before customers start complaining. A control chart will help us answer this question.

Control Charts

A control chart gives the average value of means, standard deviations or ranges of sets of data values; see **figures 1c**, **1d**, and **1f**. The solid line in the middle is the average value and the dotted lines above and below the solid line are three standard errors above and below the average. As long as all the points are within the dashed lines and the points are randomly distributed, that is the points do not have a run of 8 or more points on one side of the mean line, then we say the process is in control.

Example 1.

The following chart gives the amount of coffee dispensed from a coffee dispenser over an eight-day period. One sample was drawn every hour for eight hours. The amount of coffee dispensed is supposed to be 12 ounces. Test if the mean, range, and/or standard deviation are out of control. chpt14example1.sav

Days/Hours	1	2	3	4	5	6	7	8
Day 1	12.0	11.6	13.0	11.2	12.1	12.2	12.0	12.3
Day 2	12.0	11.2	10.9	12.3	12.1	12.0	12.6	11.5
Day 3	11.5	12.1	12.6	11.8	11.7	12.0	12.3	11.4
Day 4	12.6	11.1	12.0	12.2	10.9	13.1	11.2	11.1
Day 5	10.9	13.1	12.2	11.6	13.2	10.8	11.9	12.1
Day 6	10.9	10.8	13.1	12.9	11.0	12.5	13.1	11.2
Day 7	10.5	10.6	13.1	13.2	10.9	11.1	13.0	13.1
Day 8	10.6	13.1	10.6	13.2	10.7	13.2	13.0	10.8

Solution: Input the data into a new SPSS data editor. Use days as one variable and amount as the second variable. Input the data as we did for an ANOVA test or open the data set chpt14example1.sav. Use Analyze → Quality Control → Control Charts. Leave Variable Charts, X-bar, R, s as the default mode for control charts and click on Define; see **Figure 1a**. Move the variable amount into the Process Measurement dialogue box and days into the Subgroups defined by dialogue box. We have two choices for charts: X-bar using range and X-bar using standard deviation. In both cases we'll have two graphs, one for X-bar and the other for range or standard deviation. Since we were interested in variance, we'll select the X-bar using standard deviation; see **Figure 1b**. Hit OK and the output should be as in **figures 1c** and **1d**.

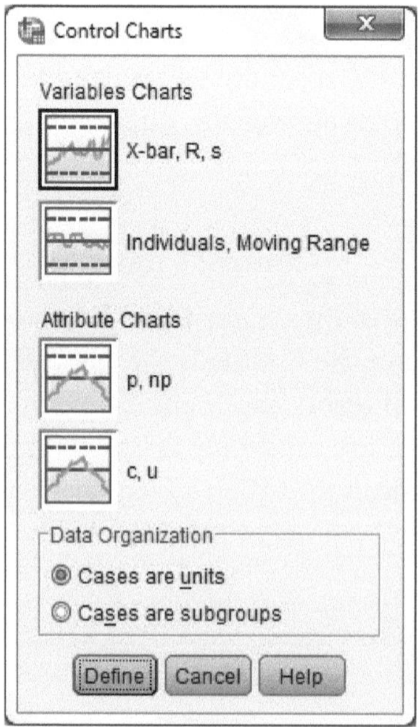

Figure 1a

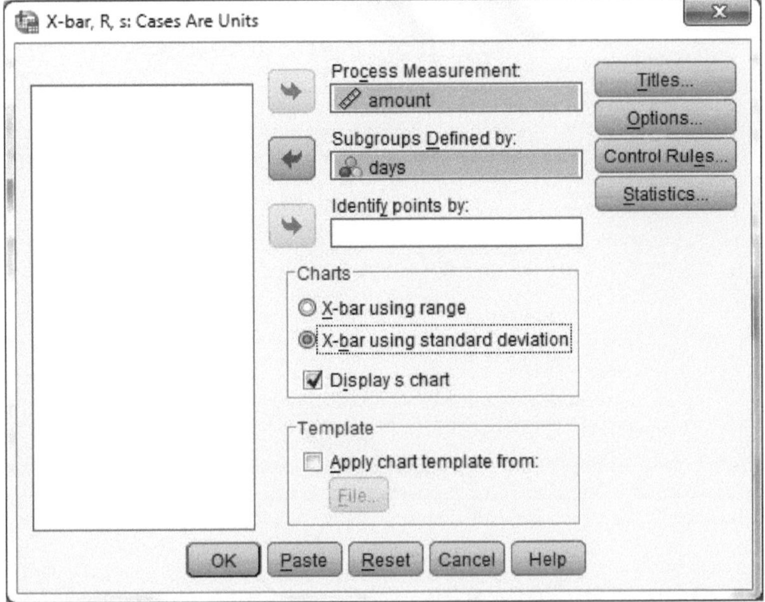

Figure 1b

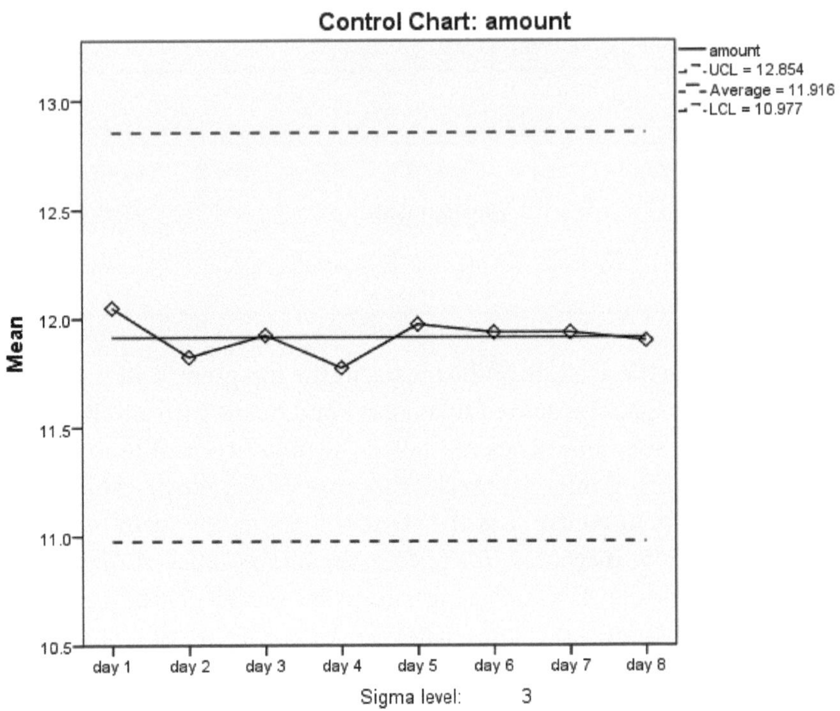

Figure 1c

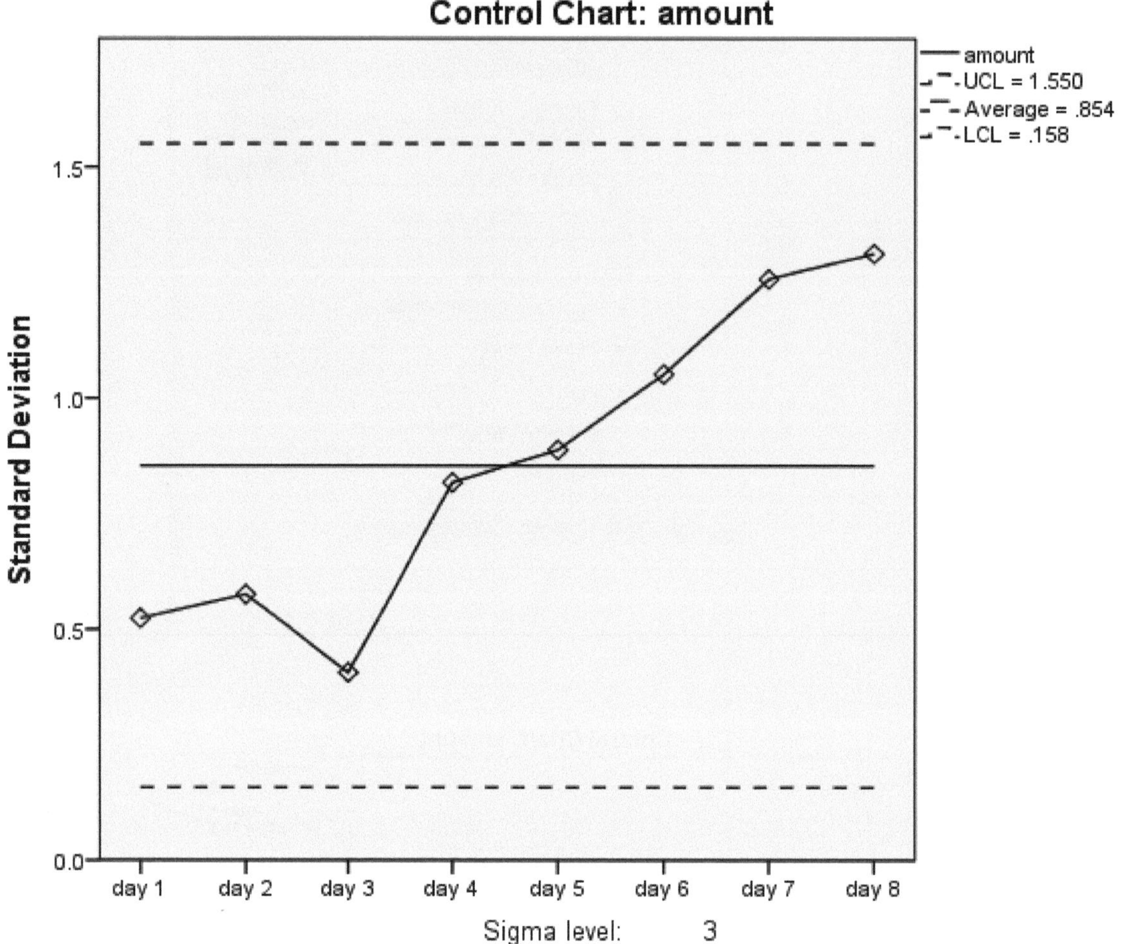

Figure 1d

The solid line in the middle is the average of the means in the top graph and the average of the standard deviations in the bottom graph. The dashed lines above and below the solid line are the values that are three standard errors off the solid line. They are called the upper control limit, UCL, and the lower control limit, LCL. We say a process is out of control if there are values above or below the dashed lines. In this case the means are clearly in control but the standard deviations seem to be increasing as the days progress. It may be if corrective measures aren't taken, the standard deviation will go out of control.

Redo the process but this time check X-bar using range. The output should be as in **figures 1e** and **1f**. This time we see the range is under control if we allow an average range of 2.113 ounces. (Remember range is maximum value minus minimum value.) The cup may not be able to accommodate a variance of an ounce in either direction, since too little is not a full cup and too much is an overflowed cup.

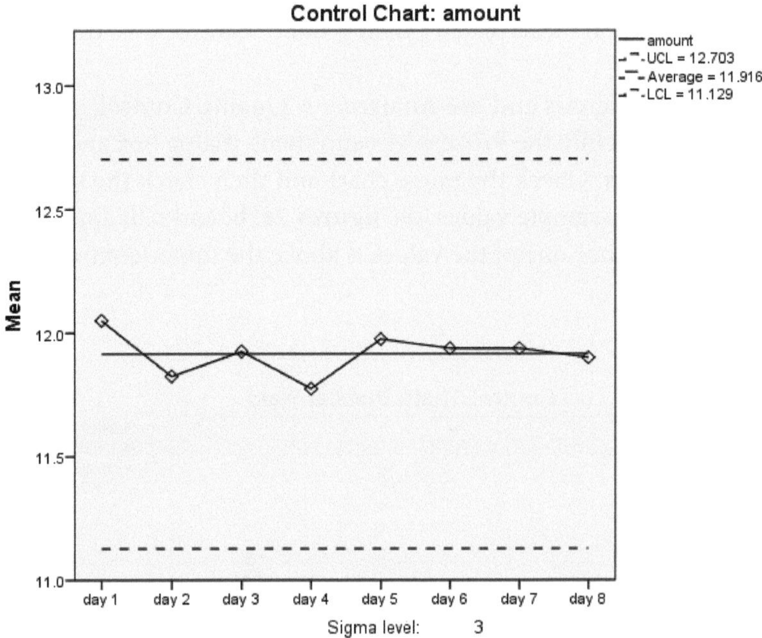

Figure 1e

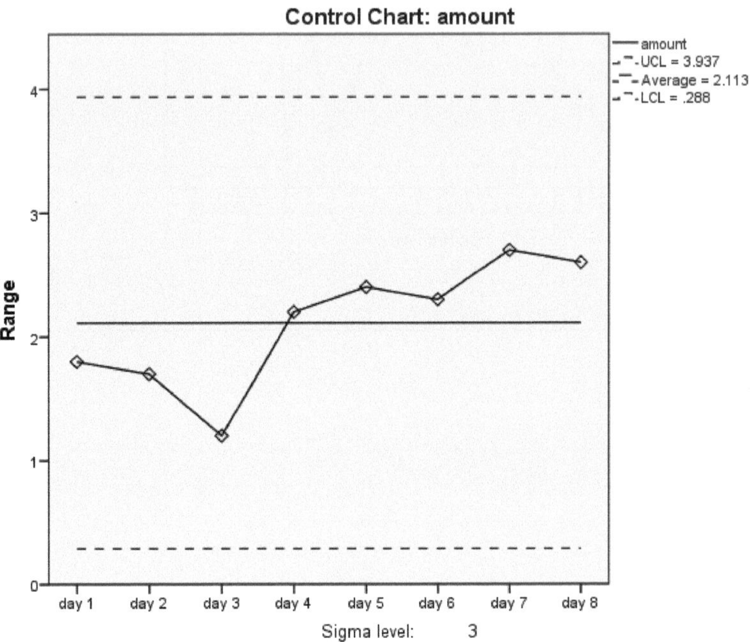

Figure 1f

Example 2.

Use the data set Pheresis (1) to see if mean platelet yield is out of control with different sample numbers.

Solution: Open the data set Pheresis and use Analyze → Quality Control → Control Charts, click on Define and move Platelet Yield into the Process Measurement: dialog box and Sample Number in the Subgroups Defined by: dialog box. Check the range chart and then check the standard deviation chart and look at the distribution of the sample values; see **figures 2a**, **b**, and **c**. It appears the mean values of Platelet Yield are out of control since one of the values is above the upper control limit.

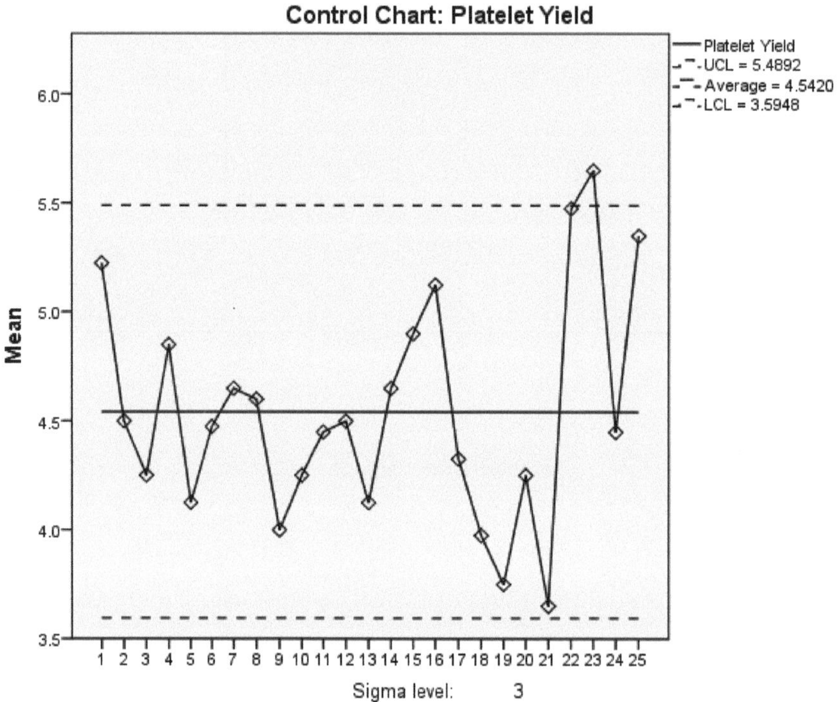

Figure 2a

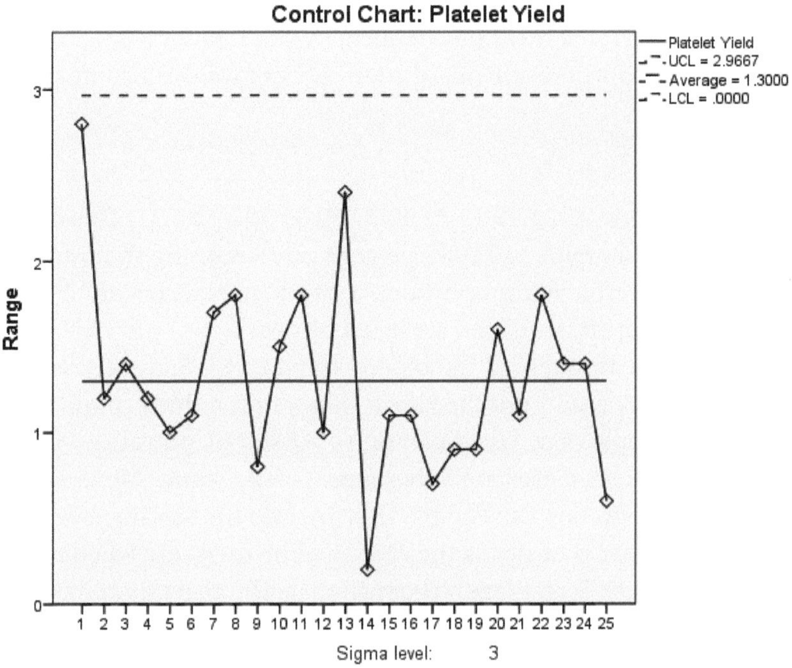

Figure 2b

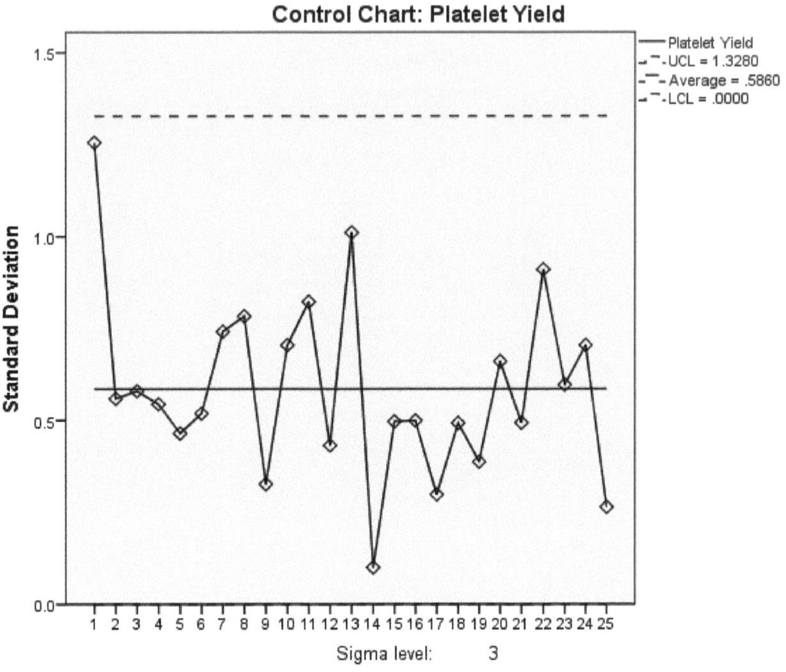

Figure 2c

Control Charts for Proportions

Sometimes we need a control chart that measures frequencies of a qualitative variable in a data set. That is, we may need to find if the proportions are out of control. Here we use a control chart for p, where p represents proportion.

Example 3.

Consider these frequency numbers: 25 24 22 25 27 30 31 30 33 32 33 32 31. This data set is taken from 13 consecutive years, randomly sampling 100,000 persons and counting the number who died from respiratory tract infections. (2) Perform a proportions control run and decide if the process is out of statistical control. (Criteria for statistical control are listed below.)

Solution: Input the numbers 1 through 13 in column 1 and call it Sample, put 100,000 in the next column in each cell and label it size, and put the above values in the third column and call it number, or open the data set chpt14example3.sav. Use Analyze → Quality Control → Control Charts, click on Attribute Charts p, np and check Cases are Subgroups. Click Define. Move number into Number Nonconforming, Sample into Subgroups Labeled by: and size into the variable box labeled Sample Size; see **Figure 3a**. You can check either p or np, as the chart will be the same whether the data values are measured in frequency (number) or percentages (proportions). The chart should be as in **Figure 3b**. We can see there is a definite trend upwards and we have a run of eight values above the center line. Hence we can conclude the process is out of control and infectious respiratory diseases may be on the rise.

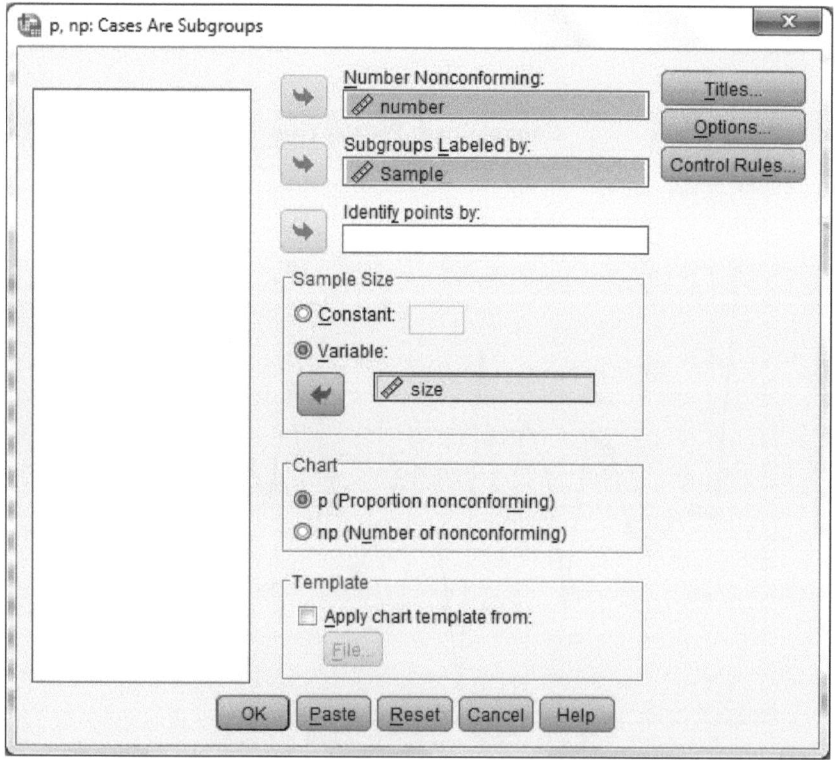

Figure 3a

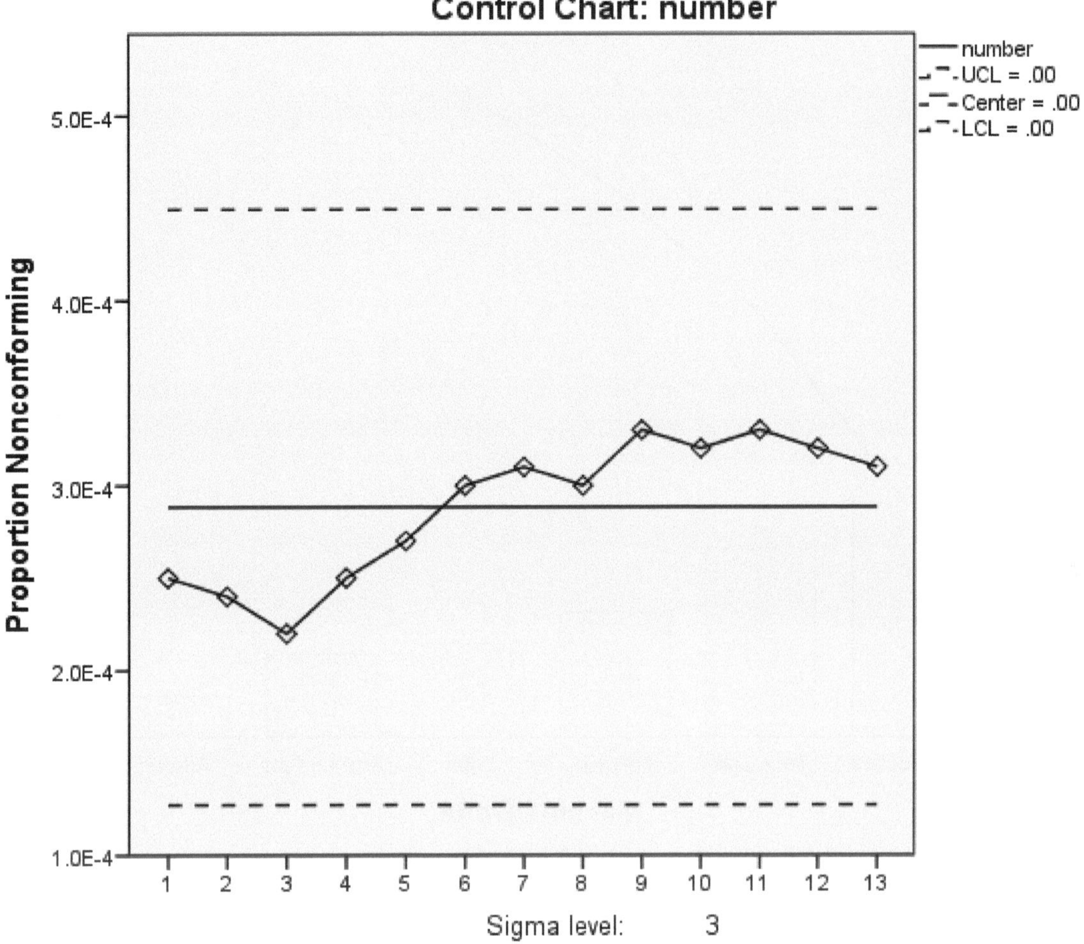

Figure 3b

Pareto Charts

A Pareto chart distributes the data frequencies from highest to lowest. It is the same as the bar chart but with the tallest bar listed first and the smallest bar listed last. It also has a cumulative relative frequency curve, as shown in **Figure 4**.

Example 4.

Use a Pareto chart to see which Apheresis machine is used most frequently.

Solution: Open the data set Pheresis and use Analyze → Quality Control → Pareto Chart → Define Simple and leave Counts or sums for groups of cases checked. Move Machine into the Category

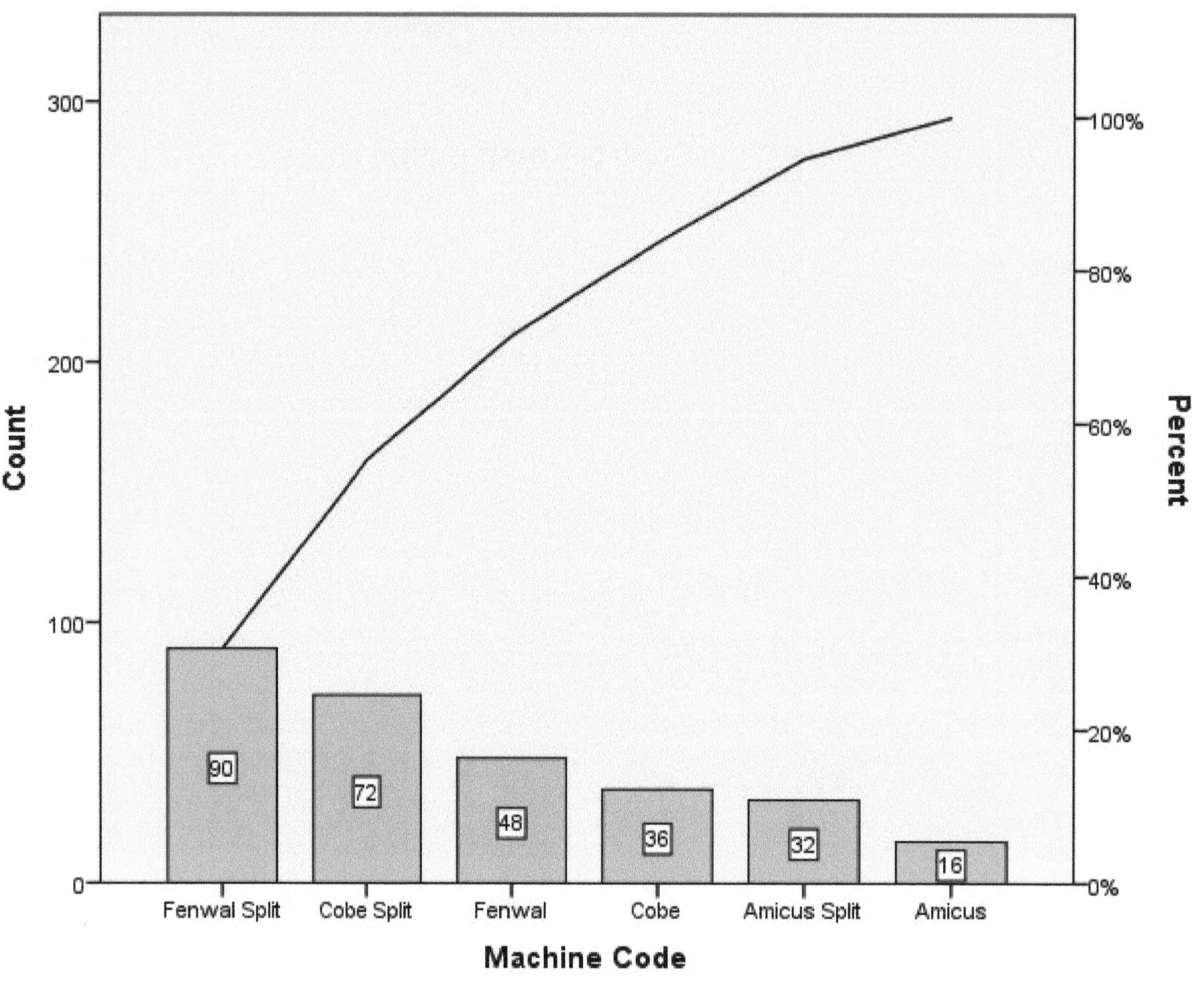

Figure 4

Criteria to Decide if a Process is Out of Statistical Control

1. There is a definite pattern, trend, or cycle that is obviously not random (such as the coffee dispenser having an upward trend in variation).
2. There is a point lying beyond the upper or lower control limits.
3. Run-of-Eight Rule: There are eight consecutive points all above or all below the center line. (If a process is statistically stable, then there's a 50% chance an arbitrary value will be above or below the center line, hence a run of eight values above or below the center line has a probability of $\frac{1}{256}$ of occurring.)

Let's review what we've learned in the chapter.

1. Control Charts

   i.) Control charts for the mean
   ii.) Control charts for the standard deviation
   iii.) Control charts for the range
   iv.) Control charts for proportions
2. Pareto Charts; bar charts with the tallest bar first and then successively smaller sizes.
3. Criteria to determine if a statistical process is out of control.

Class Work 14

Use the data set London1 to see if the mean, range, and standard deviation of carbon monoxide are out of control. (Select Cases are Subgroups.) This data set contains a subset of the hourly measurements of carbon monoxide (CO) for the year 1996. Also use the data set London2 (twenty-four observations per day, as opposed to six) to see if the results are the same.

Homework 14

1. The data set chpt14probelm1 represents a run of 25 days measuring the weight that seven individual aluminum cans can support before being crushed. Each day the company that produces aluminum cans took seven cans at random and measured the weight at which the cans were crushed. Their question is do the cans they produce have widely differing levels of crush resistance. Run a control test to see if there are any criteria that would say the can production is out of control. Check Cases are Subgroups and use Subgroups labeled by days. All seven samples go into the Samples window.

2. Old Faithful Geyser in Yellowstone was monitored for 25 years to record intervals between eruptions. Six intervals were recorded each day with the results given in chpt14problem2.sav. Construct control charts for the mean, X-bar, and the standard deviation, s. Are the time intervals getting out of control? Does this mean tourists will have to preplan their daily schedules in order to see Old Faithful erupt?

3. The Federal Reserve Bank controls the economy of the United States, but only the US Treasury has the authority to print money and issue coins. The weight of a quarter is supposed to be 5.670 grams, but any weight between 5.443 and 5.897 grams is acceptable. A new machine was just added to the quarter minting production line and a coin was randomly selected every twelve minutes for twenty hours. The weight of the coins can be found in chpt14problem3.sav. Is the new machine out of control? Check the three variables mean, range, and standard deviation to see if any one indicates the machine is out of control.

4. Persons with sleep apnea often use a CPAP machine to help control their apnea. (Sleep apnea is a momentary but frequent stoppage of breathing during sleep leading to all kinds of problems, not the least of which is sleep deprivation, which can lead to falling asleep while driving.) The CPAP blows air into the lungs and needs to be set to a certain level of force as determined by a sleep

study. The following data are for a person whose level is supposed to be set to 8. Check the three variables mean, range, and standard deviation to see if any one indicates the machine is out of control. chpt14problem4

5. The following data are based on "Trends in Infectious Disease Mortality in the United States During the 20th Century," by Armstrong et al, Journal of the American Medical Association, Vol. 275, No. 3. In thirteen consecutive years 100,000 children, aged 0 to 4 years, were randomly selected and the number who died from infectious disease was recorded. Do these numbers indicate a problem that needs to be addressed? Number who died: 30 29 29 27 23 25 25 23 24 25 25 24 23 chpt14problem5.sav

6. The following data are based on "Trends in Infectious Disease Mortality in the United States During the 20th Century," by Armstrong et al, Journal of the American Medical Association, Vol. 275, No. 3. In thirteen consecutive years 100,000 adults, 65 years or older, were randomly selected and the number who died from infectious disease was recorded. Do these numbers indicate a problem that needs to be addressed? Number who died: 270 264 250 278 302 334 348 347 377 357 362 351 343 chpt14problem6.sav

7. In twenty consecutive years, 1,000 adults were randomly selected and surveyed. The numbers given below are those who were victims of crime (based on data from the US Department of Justice, Bureau of Justice Statistics). Do these numbers indicate crime is getting out of hand?
28 33 26 29 30 33 39 21 26 28 31 35 27 25 30 37 28 31 30 26 chpt14problem7.sav

8. Out of twenty different classes using the new text *Statistics with Power: Using SPSS*, twenty students were asked if they thought this text was the greatest book, other than the *Holy Scriptures* and *The Book on Laws*, they've ever read. The number of students who responded affirmatively is given below. Do the data give any indication the book is out of control? 20 19 18 17 20 20 18 19 16 17 17 20 19 20 19 18 16 17 16 20 chpt14problem8.sav

9. Using the data set from problem 8 and a Pareto chart, determine which frequency is the highest with regard to the number of students who like the book.

Just for fun: Pick any seven numbers from the grid below such that no two numbers are in the same row or same column and add them up. Their sum should be 42. Why?

2	3	5	4	6	8	7
1	2	4	3	5	7	6
0	1	3	2	4	6	5
3	4	6	5	7	9	8
4	5	7	6	8	10	9
5	6	8	7	9	11	10
6	7	9	8	10	12	11

References

(1) Statlib Data sets

(2) Data are from "Trends in Infectious Disease Mortality in the United States During the 20th Century," by Armstrong et al, Journal of the American Medical Association, Vol. 275, No. 3.

Chapter 15

Two-Way ANOVA and Two-Way MANOVA

Objectives

After completing this chapter, a student should be able to:

- ✓ Determine when and how to use Two-Way ANOVA
- ✓ Determine the null and alternative hypothesis for Two-Way ANOVA
- ✓ Find the degrees of freedom for Two-Way ANOVA
- ✓ Use a Boxplot to draw conclusions in Two-Way ANOVA
- ✓ Determine when and how to use Two-Way MANOVA
- ✓ Determine the null and alternative hypothesis for Two-Way MANOVA
- ✓ Find the degrees of freedom for Two-Way MANOVA

Two-Way ANOVA

Factors

If we have two factors involved in a study, such as radio stations and age of participants or systolic blood pressure and gender, then we want to test three hypotheses. We want to test the hypothesis related to each factor and then tests if there is any difference between the factors, that is do the factors help contribute to the mean values of the treatment?

Example 1.

Open the data file labeled BP (blood pressure). We will test if there is a difference in the means of systolic blood pressure during mental arithmetic in our first factor: gender. We will test if there is a difference in the means of systolic blood pressure during mental arithmetic in our second factor: individuals with parental hypertension and individuals without parental hypertension. And we will test if one group has a higher mean in one factor while the other group has a lower mean in that factor and vice versa. That is do the factors make a difference? Is one gender more likely to have higher blood pressure with or without parental hypertension?

Solution: Use Analyze → General Linear Model → Univariate. Put systolic bp mental arithmetic (sbpma) as the dependent variable and subject sex (sex) and parental hypertension (PH) as the fixed factors, **see figure 1a**. Click options, descriptive statistics and homogeneity tests. Click continue and OK. The output should be as in **figure 1b**. Ignore the first two rows in the output and note the *p*-value in the last three rows. The *p*-value for sex is .000, which is significant and for parental hypertension the *p*-value is .003 which is also significant, but for crossing the factors, sex * ph, the *p*-value is .241 which is not significant. Since we only have two groups in each factor we don't need to perform a post hoc test because we know where the differences lie, they lie between the two groups, but let's plot the groups in interactive boxplot to see how the means are different. Use Graphs → Legacy Dialogs → Boxplot, click on Clustered → Define (leave Summaries for groups of cases checked) move systolic bp mental arithmetic into the Variable dialog box and move subject sex into the Category Axis and parental hypertension into the Define Clusters by dialog box. The Define Clusters Boxplot window should look like **figure 1c**. Hit OK and the output should look like **figure 1d**. Note that in both panels the medians go up to the right for both box plots. This indicates that both females and males have higher systolic blood pressure at ph+ then at ph-. Interchanging the category axis and the define clusters will give the same results but from a different perspective on the distribution of the data.

The boxplots indicate there is a difference in means for both subject sex and parental hypertension (the change from ph- to ph+ is not as obvious but the sample size is large which implies any change is actually significant) but the direction of the change is the same for both factors. This indicates the factors, sex and ph, do not affect the outcome of the mean values of the treatment sbpma. If, however the medians were reversed in the panels, see **figure 2c**, then the factors would contribute to the mean value of the treatment.

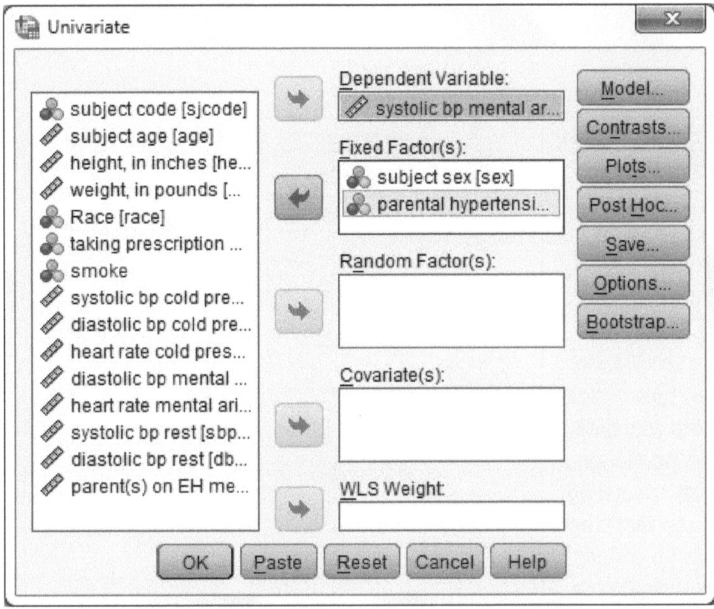

Figure 1a

Tests of Between-Subjects Effects

Dependent Variable: systolic bp mental arithmetic

Source	Type III Sum of Squares	df	Mean Square	F	Sig.
Corrected Model	7286.555[a]	3	2428.852	15.541	.000
Intercept	2917864.694	1	2917864.694	18669.573	.000
sex	5704.665	1	5704.665	36.501	.000
ph	1439.365	1	1439.365	9.210	.003
sex * ph	215.986	1	215.986	1.382	.241
Error	27194.434	174	156.290		
Total	2949638.444	178			
Corrected Total	34480.989	177			

a. R Squared = .211 (Adjusted R Squared = .198)

Figure 1b

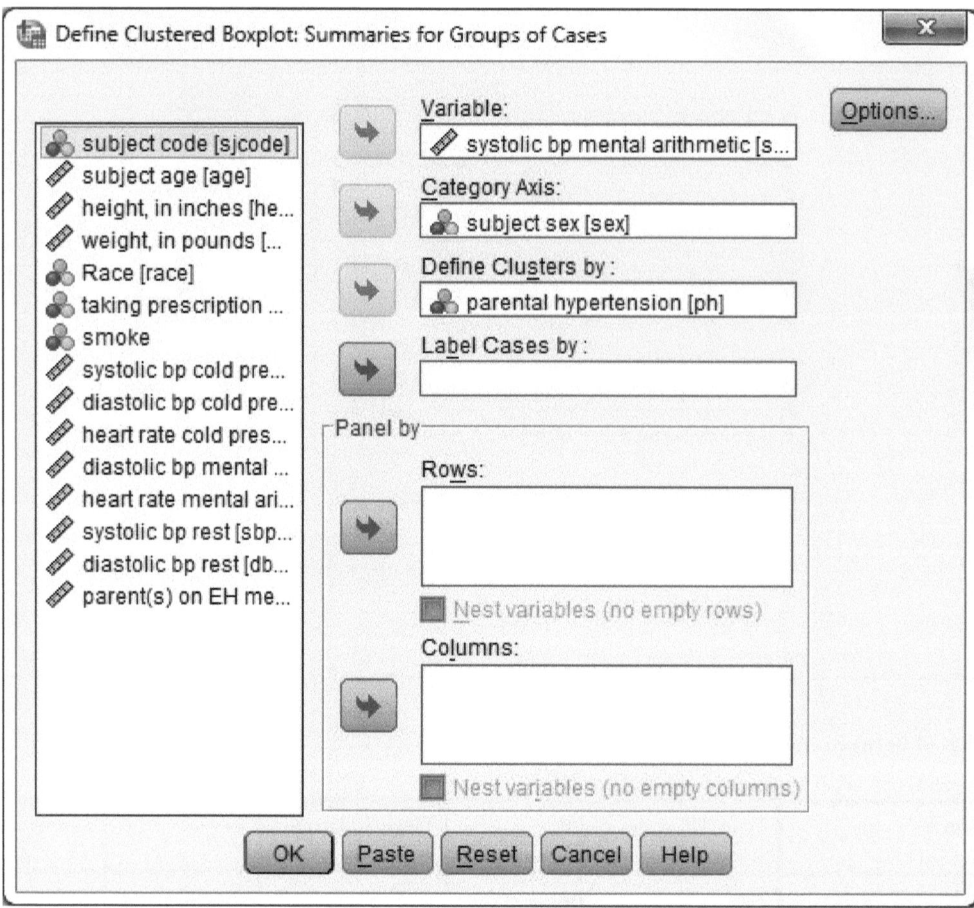
Define Clustered Boxplot: Summaries for Groups of Cases
subject code [sjcode]
subject age [age]
height, in inches [he...
weight, in pounds [...
Race [race]
taking prescription ...
smoke
systolic bp cold pre...
diastolic bp cold pre...
heart rate cold pres...
diastolic bp mental ...
heart rate mental ari...
systolic bp rest [sbp...
diastolic bp rest [db...
parent(s) on EH me...
Variable:
systolic bp mental arithmetic [s...
Category Axis:
subject sex [sex]
Define Clusters by :
parental hypertension [ph]
Label Cases by :
Panel by
Rows:
Nest variables (no empty rows)
Columns:
Nest variables (no empty columns)
Options...
OK
Paste
Reset
Cancel
Help

Figure 1c

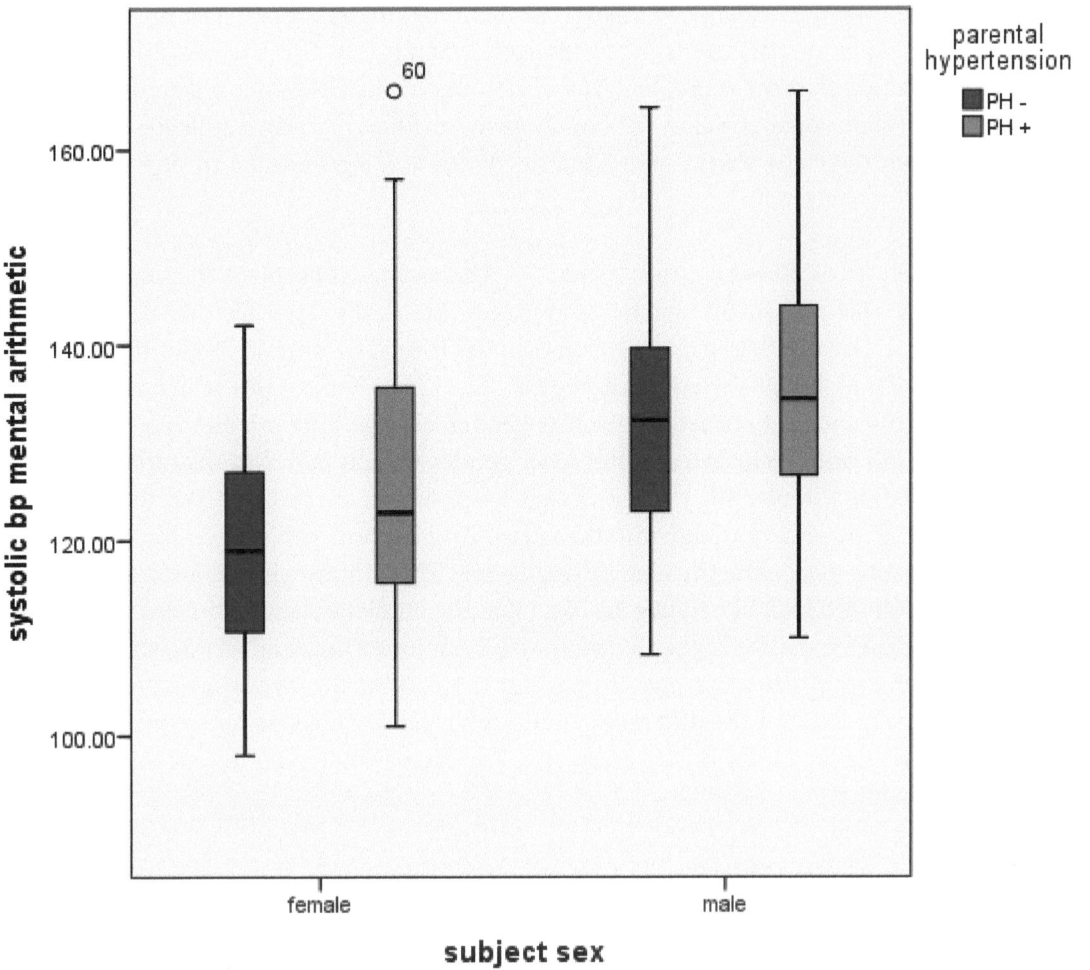

Figure 1d

Example 2.

Open the data file group. This set of data comes from a study conducted by Dr. Bonnie Klentz, who was interested in understanding the dynamics of individuals working in small groups. More specifically, she wanted to investigate whether people can accurately perceive fellow group members' output in a task. Two variables she thought might be important were the size of the work group and the gender of the participant. These two independent variables made up the four groups in the study: all-male groups of 2, all-male groups of 3, all-female groups of 2, and all-female groups of 3.

Participants were asked to find as many words as they could in a matrix of 16 letters (such as the Boggle game). Each time they found a word, participants wrote it on a slip of paper and put it in a box in the center of the table. Participants were told that the total number of words found by each group would be counted at the end of the study.

Our test will be on the participants estimation of their coworkers output (i.e., how many words do you think your coworkers found?) over the factors gender and group size. The accuracy of this estimation was calculated as a difference score - the actual number of slips of paper put in the box minus the estimated number of slips of paper put in the box. A positive difference score indicates that participants were underestimating their coworkers' performance. Is there a significance in any of the factors or between factors?

Solution: Use Analyze → General Linear Model → Univariate. Put subject's perception of coworker as the dependent variable and subject gender and group size as the fixed factors. Be sure to click on options and check Descriptive Statistics and Homogenity Tests. The output should be as in **figure 2a**. (Note Levene's Test of Equality of Variance is accepted since the p-value is .061 which is larger than .05.) We see the only significance is between the factors gender*grpsize. The *p*-value is .037. This says one gender will do better in one group size and the other gender will do better in the other group size. We can see the differences in Graphs → Legacy Dialog → Boxplot, check Clustered then Define. Move subject's perception of coworker variable into the variable dialog box, subject gender into the Category Axis, and group size into the Define Clusters by dialog box. The window should look like **figure 2b**. Hit ok and the box plots should look like **figure 2c**. Note that the medians change direction for the different group sizes. The median for male is higher in group size 2 but lower in group size 3, whereas the median for Female is lower in group size 2 but higher in group size 3. Thus the Group size affects the medians, and by transitivity the means, for the different genders. One gender does better in one group size while the other gender does better in the other group size. The factors make a difference, thus they are not independent of each other.

Tests of Between-Subjects Effects

Dependent Variable: subject's perception of co-worker

Source	Type III Sum of Squares	df	Mean Square	F	Sig.
Corrected Model	608.646[a]	3	202.882	3.572	.018
Intercept	261.750	1	261.750	4.609	.035
gender	83.519	1	83.519	1.471	.229
grpsize	24.554	1	24.554	.432	.513
gender * grpsize	256.569	1	256.569	4.518	.037
Error	4032.101	71	56.790		
Total	5240.000	75			
Corrected Total	4640.747	74			

a. R Squared = .131 (Adjusted R Squared = .094)

Figure 2a

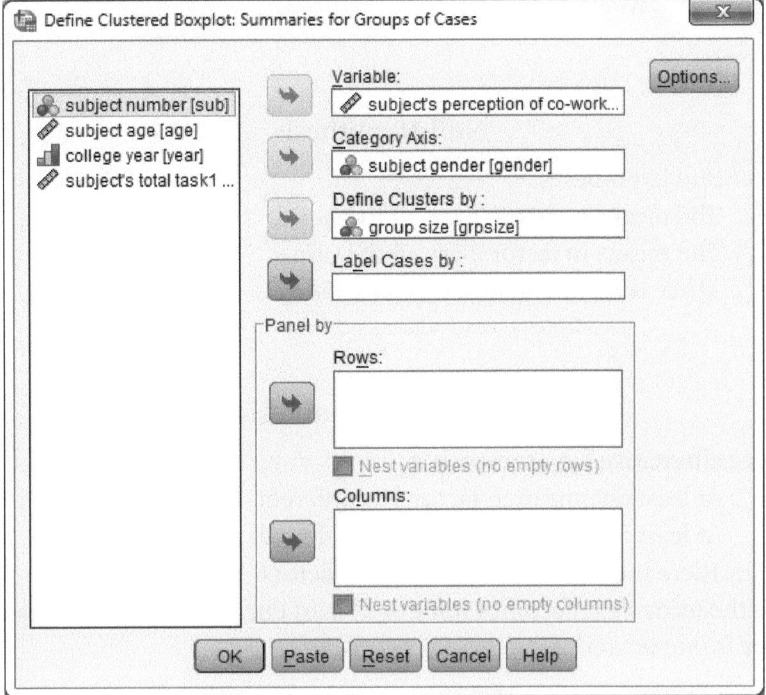

Figure 2b

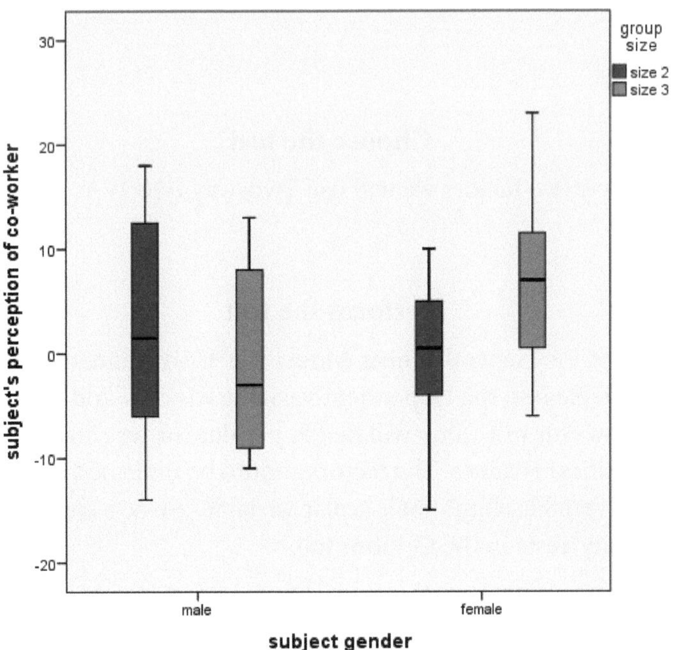

Figure 2c

This is the basics of Two-Way ANOVA. Let's list the steps.

Null Hypothesis

We will have three null hypotheses

Step 1. H_0 : The means in factor 1 are all the same.
H_0 : The means in factor 2 are all the same.
H_0 : There is no interaction between the factors.

Alterative Hypothesis

We will have three alternative hypotheses

Step 2. H_1 : At least one mean in factor 1 is different.
H_1 : At least one mean in factor 2 is different.
H_1 : There is interaction between the factors.

When we negate the term *all*, it becomes *some* or *at least one*, also when we negate *there is no interaction* it becomes *there is interaction.*

Significance Level

Step 3. $\alpha = .05$

Choose the test

Step 4. Since we have two factors we will use Two-way ANOVA.

Perform the test

Step 5. Use Analyze → General Linear Model → Univariate, move the dependent variable, the test variable into the Dependent variable window and the factors into the fixed factors window. Our test value will be the *p*-value, or we can use the *F* value and compare it against critical *F* values. The factors should be indicated as a nominal variable and the dependent variable should be a scalar variable. Always check Descriptive Statistics and Homogeneity Tests in the Options tab.

Critical F-Value

Step 6. The degrees of freedom for each factor is the number of groups minus one, k - 1, and the degrees of freedom for the treatment, interaction between the two factors, is the product of the two degrees of freedom for each factor. If factor one has $k_1 - 1$ degrees of freedom and factor two has $k_2 - 1$ degrees of freedom then the degrees of freedom for the interaction is $(k_1 - 1)\cdot(k_2 - 1)$. The degrees of freedom for the numerator is the sum of all the degrees of freedom of the two factors and the degrees of freedom of the interaction. The degrees of freedom for the denominator is the error *df*, which is total number of data values minus 1, the corrected degrees of freedom, and then minus the degrees of freedom of the numerator. To find the critical *F* value from SPSS place 1 minus the significance level in a new SPSS data editor and call the variable *cp* for cumulative probability. Use Transform → Compute Variable, call the target variable *cf* for critical *F* value and click on Inverse DF in the function group. Double click on IDF.F, move *cp* into the first position and then df num and df den into the second two positions. Hit OK and the critical *F* value will appear in the column headed by *cf*.

Compare

Step 7. Compare the *p*-value with the significance level. If the *p*-value is greater than the significance level we accept the null hypothesis. If the *p*-value is less than the significance level we fail to accept the null hypothesis. Alternatively we can compare the obtained *F*-value with a critical *F*-value found in step 6. If the obtained *F*-value is less than the critical *F*-value we accept the null hypothesis, but if the obtained *F*-value is greater than the critical *F*-value we fail to accept the null hypothesis. Remember the critical region for an *F*-test is always in the right tail region of the distribution, **see figure 3**.

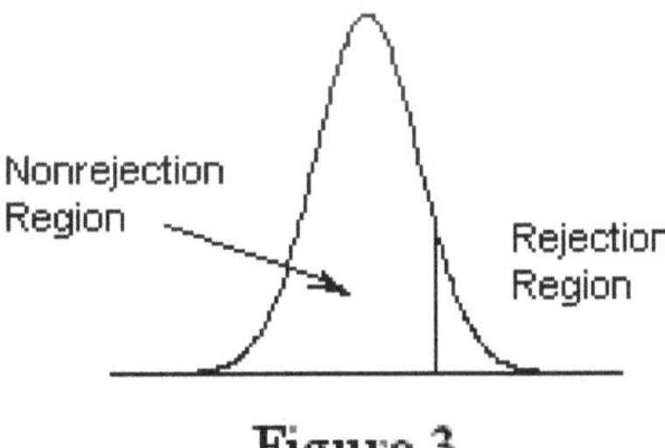

Figure 3

Decision

Step 8. From the comparison above make a decision based on the results of step 7. Be sure to write the answer in complete sentences in context of the situation for the problem, if given.

Post Hoc Test

Step 9. If there is a difference among the groups in the factors then we would like to know where the differences lie. To do this we use the Tukey Post Hoc Test. This test is only performed if we reject the null hypothesis for the factors. We do not perform this test if we reject the null hypothesis for the interaction between factors. If there are only two groups in each factor then the Post Hoc test is not necessary to determine where the difference lies since it can only be between those two groups.

Assumptions

The following assumptions are required for a two-way ANOVA test.

1. The populations from which the samples were obtained must be normally or approximately normally distributed.
2. The samples must be independent.
3. The variances of the populations must be equal.
4. The groups must have the same sample size.

Example 3.

The following data values are the number of germinated seeds using three types of corn seeds and five types of fertilizer. There are two values in each cell. Use Two-Way ANOVA on the data set to see if there is a difference in the mean number of germinated seeds or if the factors affect the rate of germination. Alpha = 5% chpt15example3.sav

	Fert I	Fert II	Fert III	Fert IV	Fert V
Seed A	105, 111	95, 100	94, 107	105, 103	100, 102
Seed B	110, 112	97, 99	100, 101	108, 112	108, 1104
Seed C	94, 97	86, 87	98, 99	96, 103	92, 99

Solution: The eight steps are as follows:

Step 1. $H_0 : \mu_{seedA} = \mu_{seedB} = \mu_{seedC}$ The means of all the seed types are the same.
$H_0 : \mu_{fertI} = \mu_{fertII} = \mu_{fertIII} = \mu_{fertIV} = \mu_{fertV}$ The means of the fertilizers are the same.
H_0 : The factors are independent, one factor will not affect the other factor.

Step 2. H_1 : At least one mean of the seed types is different.
H_1 : At least one mean of the fertilizers is different.
H_1 : The factors are dependent.

Step 3. $\alpha = .05$

Step 4. We will use two way ANOVA because we have two factors, seed type and fertilizer. We assume the samples are random and independent, the dependent variable has a somewhat normal distribution and the distribution on the dependent scores have an equal variance.

Step 5. Following the procedure outlined above the results show the obtained *F* value for the first hypothesis is 20.466 with a *p*-value of 0. The obtained *F* value for the second hypothesis is 8.946 with a *p*-value of .001. The obtained *F* value for the third hypothesis is 1.320 with a *p*-value of .306. **See figure 4a.**

Tests of Between-Subjects Effects

Dependent Variable: score

Source	Type III Sum of Squares	df	Mean Square	F	Sig.
Corrected Model	1128.800[a]	14	80.629	6.234	.001
Intercept	304819.200	1	304819.200	23568.495	.000
seed	529.400	2	264.700	20.466	.000
fert	462.800	4	115.700	8.946	.001
seed * fert	136.600	8	17.075	1.320	.306
Error	194.000	15	12.933		
Total	306142.000	30			
Corrected Total	1322.800	29			

a. R Squared = .853 (Adjusted R Squared = .716)

Figure 4a

Step 6. Since $n = 30$ and the number of groups in the first factor, seeds, is 3 and the number of groups in the second factor, fertilizer, is 5, the degrees of freedom for the first output, seed, is 2, df for the second output, fert, is 4 and the degrees of freedom for the interaction is $2 \times 4 = 8$ The corrected df is 29 so df denominator is 29 – (2 + 4 + 8) = 15. Hence, using 2, 4 and 8 for df1 and 15 for df2 we find the critical *F*-values to be:

$F_1 = 3.68$

$F_2 = 3.06$

$F_3 = 2.64$

Step 7. The obtained *F* values for hypothesis 1 and 2 are greater than the critical *F* values, and their *p*-values are less than .05, so we reject the null hypothesis for 1 and 2. The obtained *F* value for the interaction, 1.320 is less than the critical *F* value, 2.64. Also the *p*-value, .306 is greater than .05, so we accept the null hypothesis for the interaction between the two factors.

Step 8. The means between the different types of seeds is different, the means of the different fertilizers is different but the interaction between the seeds and fertilizers is not different. The different seeds perform at about the same level regardless of the fertilizer.

Step 9. The Turkey Post-Hoc test will tell where the differences lie. For seeds the difference is between seed C and seeds A, B. For fertilizers, the differences lie between Fert II and Fert I, IV and V.

When we use the boxplots we can see how the boxes for seed B is lower than the boxes for seeds A and C. Use Graphs → Legacy Dialogs → Boxplot → Clustered → Define. Move score into Variable, fert into Category Axis and seed into Define Clusters by:. Hit Ok and the output should be as in **figure 4b**. Notice how the boxplots have the same relative position with respect to each other in the different fertilizers with exception for Fert III.

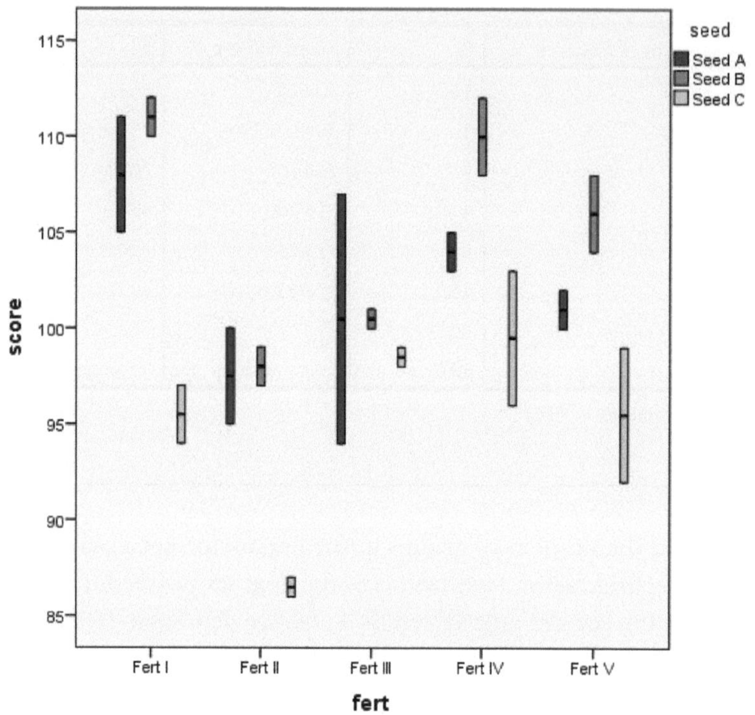

Figure 4b

Two-Way MANOVA

In two-way MANOVA we use two factors that have several groups in each factor and more than one dependent variable. We are testing to see if the factors interact in the effect on the dependent variables. Thus we are testing three hypotheses:

$H_0 1$: The different dependent variable vectors will not differ in factor 1.
$H_0 2$: The different dependent variable vectors will not differ in factor 2.
$H_0 3$: Factor 1 and factor 2 will not interact in the effect on the dependent variables.

Example 4.

Random samples of employees from different occupations were questioned about their income and job satisfaction.

Position	Level Education	High School	Bachelor	Master	Doctorate
Worker	Income	25,30,26,25,24	33,26,38,32,29,28,29	24	
	Job Satisfaction	3,4,5,3,4	2,1,4,4,3,2,2	1	
Skilled Worker	Income	30,50,34,37,36,43,31	36,58,48	42,51	60,35,42,35
	Job Satisfaction	2,5,5,5,5,5,5	2,5,4	3,2	5,5,4,3
Manager	Income	53,56,48	40,50,41,52	45,54,90,56,96,48	65,75,64,78,65
	Job Satisfaction	5,5,4	3,2,5,5	4,3,3,5,3,1	4,4,5,3,3
Director	Income	105		80,105,76,96,57,95	100,110,120,106,98,109
	Job Satisfaction	5		5,4,5,5,5,3	4,3,5,4,3,5

The two independent factors are Highest level of Education (high School, Bachelor degree, Master degree and Doctoral degree) and Position (Worker, Skilled Worker, Manager and Director) and the two dependent variables are Income (24 to 120 thousand) and Job Satisfaction (1 to 5, 1 being lowest and 5 being highest). Test, at the 5% level, the mean vectors for the dependent variables are the same and the factors do not interact in the effect of the dependent variables. Chpt15example4

Solution: We'll first look at the boxplots to get a visual representation of the data. This will help us better understand the test results. Use Graphs → Legacy Dialogs → Boxplot → Clustered → Define. Move Income into Variable, Level_Education into Category Axis and Position into Define Clusters by:. The graph should look as in **figure 5a**. Note that the skilled worker boxplots are all about the same level and the director boxplots are much higher on the graph. This implies there is a difference in the income for skilled workers and the income for directors.

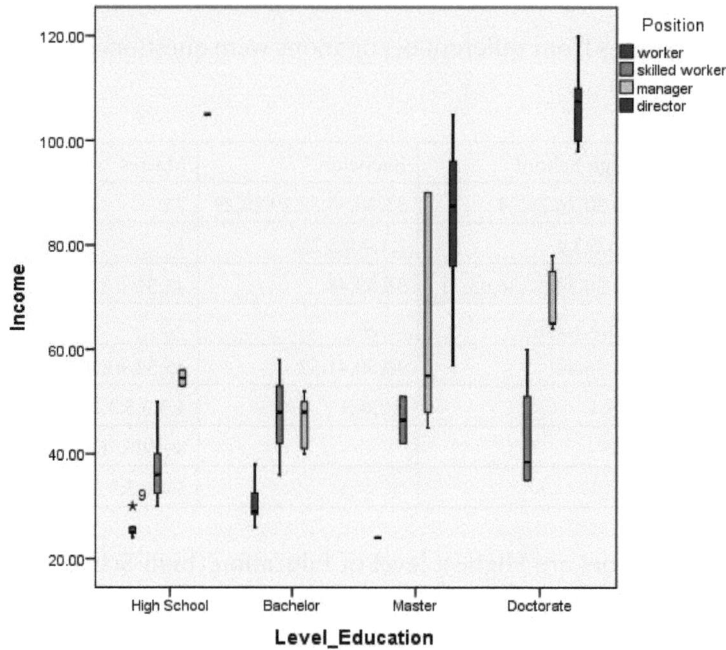

Figure 5a

Repeat the same process as above but replace Income with Job_Satisfaction in the Variable dialog box. The boxplots should be as in figure 5b. Note that workers with a Bachelor degree are less satisfied with their job than workers with a High School degree.

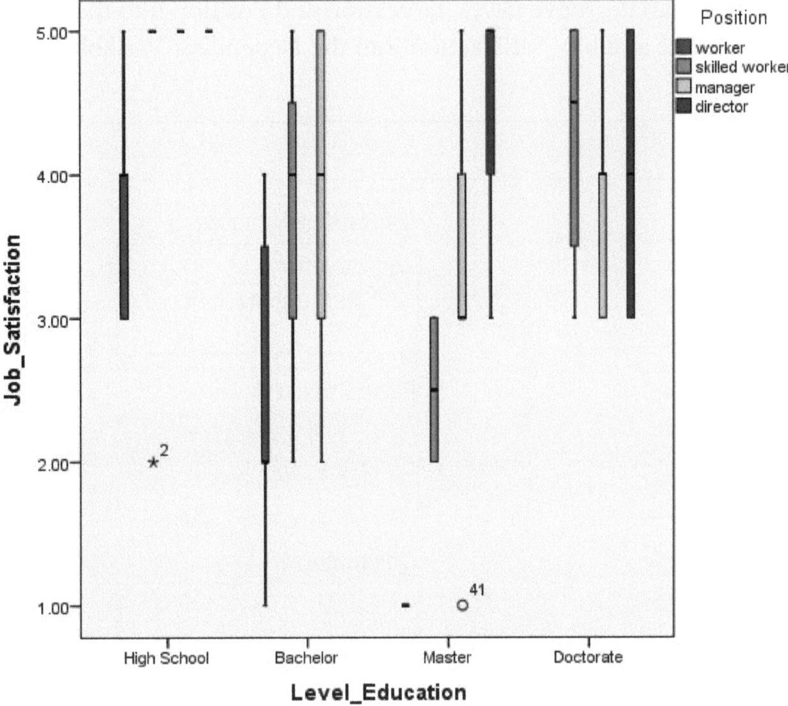

Figure 5b

Now we'll run the eight step process.

Step 1. H_0 : The population mean vectors of the dependent variables are equal in level of education.

H_0 : The population mean vectors of the dependent variables are equal in position.

H_0 : The factor Level of Education is independent of the factor Position.

Step 2. H_1 : At least one mean vector is different in levels of education.

H_1 : At least one mean vector is different in levels of position.

H_1 : The factors are dependent.

Step 3. $\alpha = .05$

Step 4. This test is two-way MANOVA. We have two factors, they are 1) Level of Education and 2) Position and two dependent variables and they are 1) Income and 2) Job Satisfaction. We assume the scores in each sample are random and independent of each other, the scores on all dependent variables follow a multivariate normal distribution in each group, the population covariance matrices for the dependent variables in each group are equal and the relationship among all pairs of dependent variables for each cell in the data matrix are linear.

Step 5. We need to find the *F*-value for Income and Job Satisfaction in both Level of Education, Position and their interaction and also the *p*-values. We also need to find the degrees of freedom for each *F*-value and the *p*-values for the test of equality of variance. Finally we need to check if our assumptions are satisfied. Use Analyze → General Linear Model

→ Multivariate, move Level_Education and Position into the Fixed Factors dialog box and Income and Job_Satisfaction into the Dependent Variables window, see **figure 5c**.

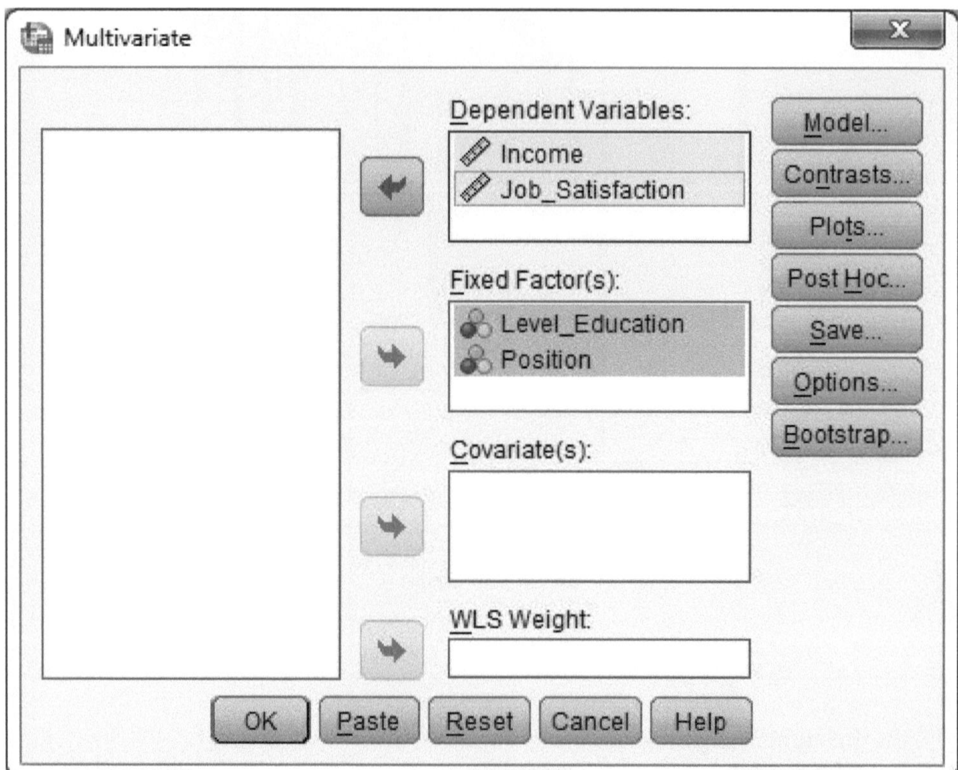

Figure 5c

Click on Options and check Descriptive Statistics and Homogenity Tests, see **figure 5d**. Hit continue.

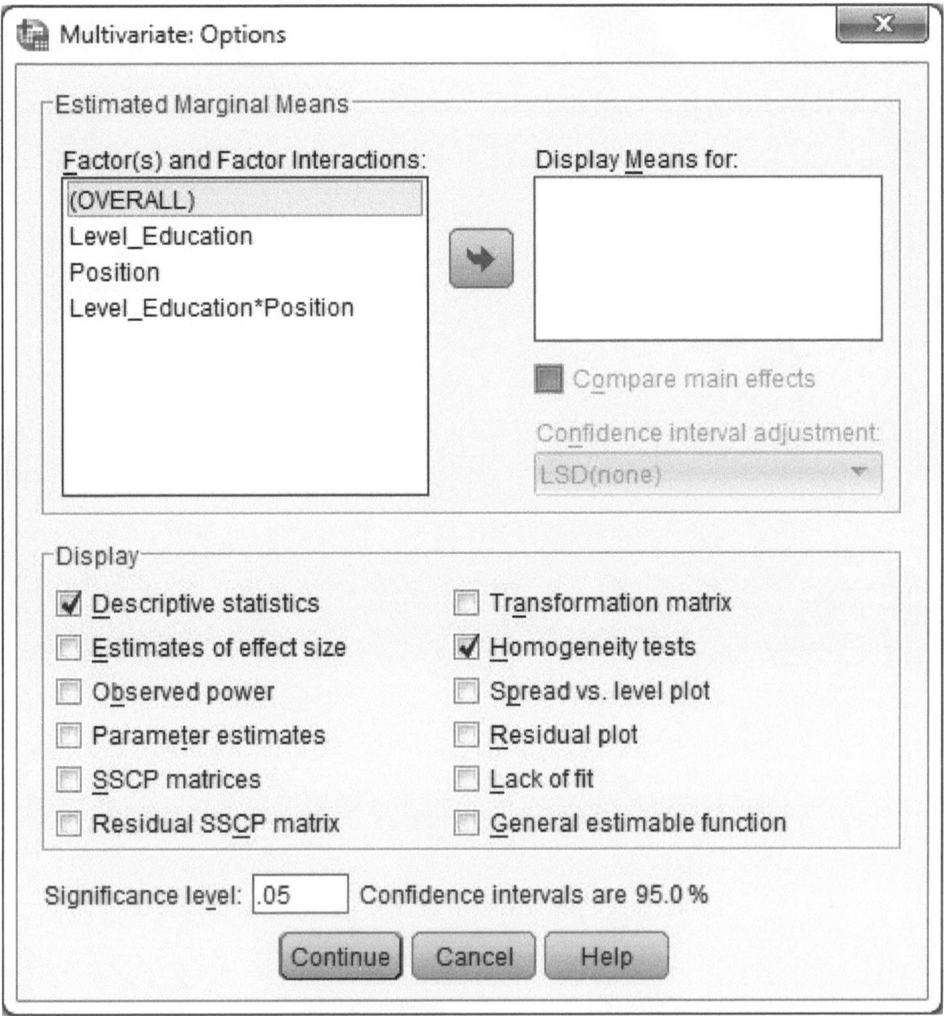

Figure 5d

Also click on Post Hoc tests, move Level_Education and Position into the Post Hoc Tests for: window and check Tukey, see **figure 5e**. Click Continue and OK.

The output will have a lot of tables. The first table is Between-Subjects Factors and it tells us how many subjects were in each group in each factor. The second table is the Descriptive Statistics table and it gives the mean and standard deviation for each group in each factor. The third table is Box's Test of Equality of Covariance Matrices, see **figure 5f**. The significance level is .028 which is less than .05 and thus we will use Pillai's Trace for the *F*-value and the significance level. When Box's Test of Equality is not significant we use Wilks' Lambda, however in most cases these two *F*-values are either both significant or both not significant. The next table is the Multivariate Tests. These will give us the *F*-values and *p*-values that we need along with df1 and df2, see **figure 5g**. The *F*-value of Pillia's Trace for Level_Education is 3.011 and the *p*-value is .010. The *F*-value of Pillia's Trace for Position is 11.094 and the *p*-value is 0. The *F*-value of

Pillia's Trace for interaction of Level_Education and Position is 1.590 and the *p*-value is .097. Thus the *F*-value for the factors is significant but for their interaction it is not significant.

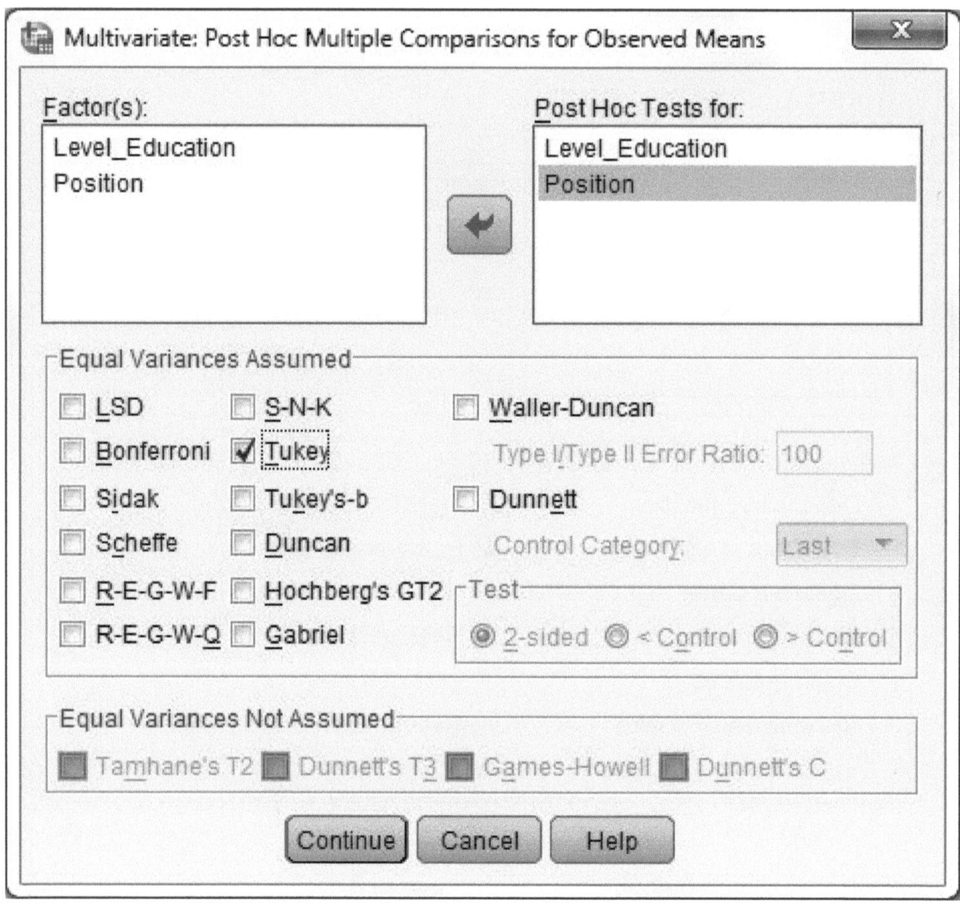

Figure 5e

Box's Test of Equality of Covariance Matrices[a]

Box's M	54.514
F	1.589
df1	27
df2	2140.013
Sig.	.028

Figure 5f

Tests the null hypothesis that the observed covariance matrices of the dependent variables are equal across groups. a. Design: Intercept + Level_Education + Position Level_Education + Position

Multivariate Tests[a]

Effect		Value	F	Hypothesis df	Error df	Sig.
Intercept	Pillai's Trace	.969	706.188[b]	2.000	45.000	.000
	Wilks' Lambda	.031	706.188[b]	2.000	45.000	.000
	Hotelling's Trace	31.386	706.188[b]	2.000	45.000	.000
	Roy's Largest Root	31.386	706.188[b]	2.000	45.000	.000
Level_Education	Pillai's Trace	.328	3.011	6.000	92.000	.010
	Wilks' Lambda	.695	2.991[b]	6.000	90.000	.010
	Hotelling's Trace	.405	2.970	6.000	88.000	.011
	Roy's Largest Root	.289	4.427[c]	3.000	46.000	.008
Position	Pillai's Trace	.840	11.094	6.000	92.000	.000
	Wilks' Lambda	.225	16.618[b]	6.000	90.000	.000
	Hotelling's Trace	3.156	23.143	6.000	88.000	.000
	Roy's Largest Root	3.062	46.951[c]	3.000	46.000	.000
Level_Education * Position	Pillai's Trace	.390	1.590	14.000	92.000	.097
	Wilks' Lambda	.632	1.658[b]	14.000	90.000	.079
	Hotelling's Trace	.548	1.722	14.000	88.000	.065
	Roy's Largest Root	.476	3.128[c]	7.000	46.000	.009

a. Design: Intercept + Level_Education + Position + Level_Education * Position

b. Exact statistic

c. The statistic is an upper bound on F that yields a lower bound on the significance level.

Figure 5g

The assumption the relationship among all pairs of dependent variables for each cell in the data matrix is linear can be checked by making a few scatter graphs for some of the cells. We'll use the Bachelor, Skilled Worker cell. Instead of inputting the values into a new data editor we'll use (chpt15eample4) Data → Select Cases, check 'If condition is satisfied' and click 'If', see **figure 5h**. In the new window move Level_Education into the condition window and set = 2 (2 is the value for Bachelor). Click the ampersand sign (&) and move Position into the condition window and set = 2 (2 is the value for Skilled Worker), see **figure 5i**. Click Continue and OK. Now we can check the linearity of the two dependent variables Income and Job_Satisfaction as we did in chapter 11. **Figure 5j** is a scatter-dot graph with the best fit line and the Coefficient of Determination. For this particular cell the two dependent variables do form a linear relationship. In a similar manner the linearity of all the cells can be checked. To go back to all cases, use Data → Select Cases and check "All cases". MANOVA also does tests equal to ANOVA for each factor. This is the 'Tests of Between-Subjects Effects'. First we need to check the equality of variance, see **figure 5k**. Income is significant but Job_Satisfaction is not significant, however the Post Hoc tests will indicate if there is a difference between groups even if equality of variance is significant. The test between subjects (two-way ANOVA) is given in **figure 5l** and the post hoc tests are in **figure 5m.**

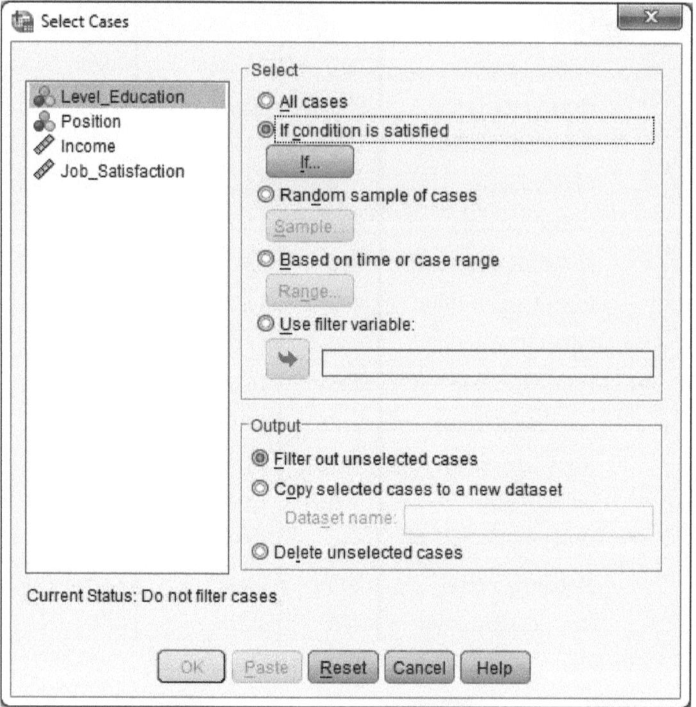

Figure 5h

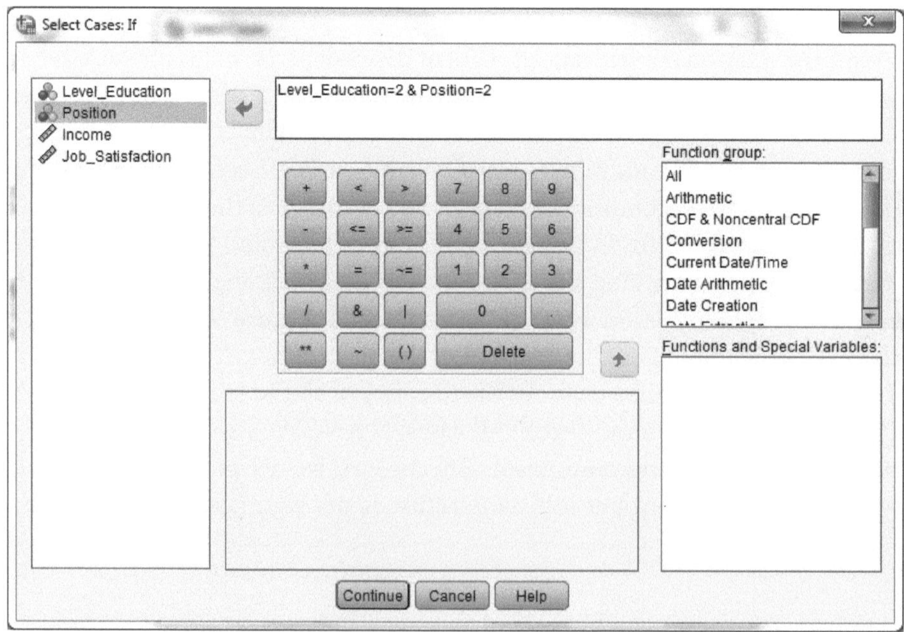

Figure 5i

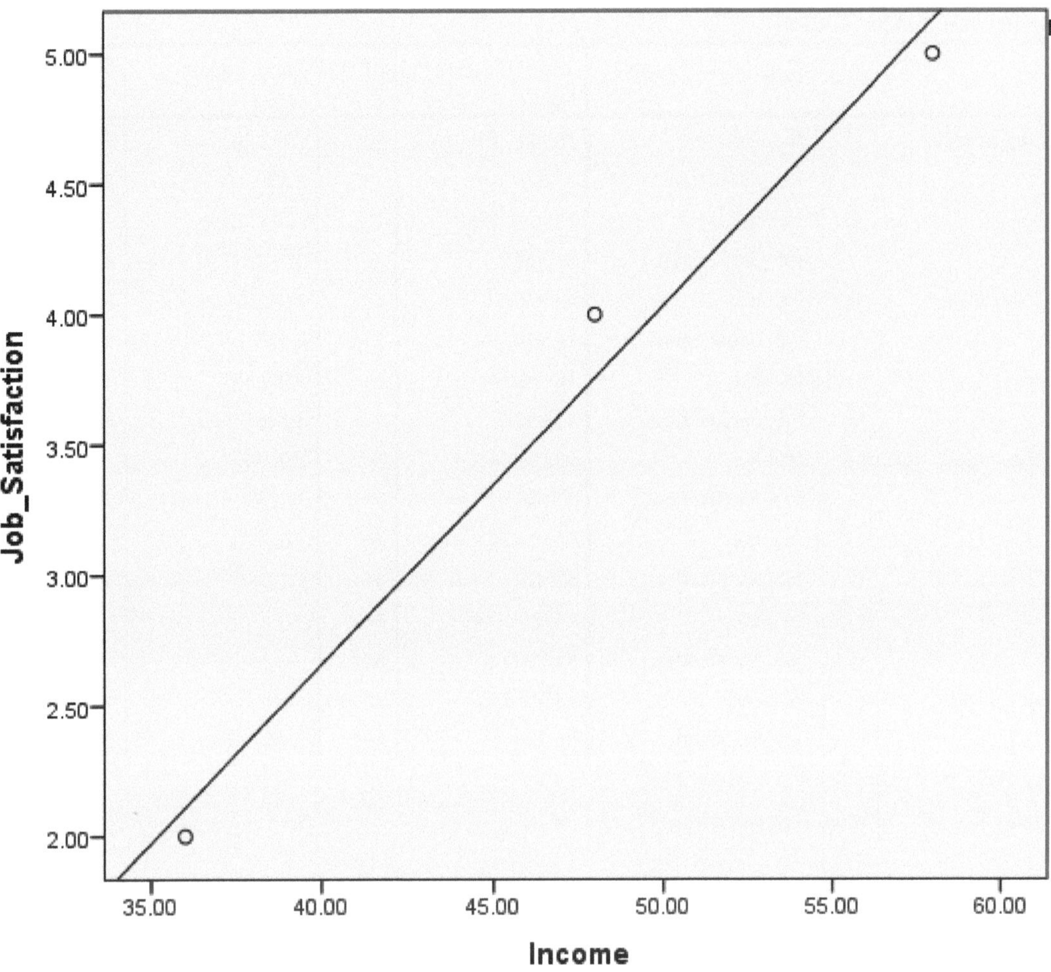

Figure 5j

Levene's Test of Equality of Error Variances[a]

	F	df1	df2	Sig.
Income	4.426	13	46	.000
Job_Satisfaction	.822	13	46	.635

Tests the null hypothesis that the error variance of the dependent variable is equal across groups.

a. Design: Intercept + Level_Education + Position + Level_Education * Position

Figure 5k

Tests of Between-Subjects Effects

Source	Dependent Variable	Type III Sum of Squares	df	Mean Square	F	Sig.
Corrected Model	Income	38380.710[a]	13	2952.362	25.295	.000
	Job_Satisfaction	36.655[b]	13	2.820	2.490	.012
Intercept	Income	129631.996	1	129631.996	1110.644	.000
	Job_Satisfaction	535.255	1	535.255	472.781	.000
Level_Education	Income	653.300	3	217.767	1.866	.149
	Job_Satisfaction	14.994	3	4.998	4.415	.008
Position	Income	16198.161	3	5399.387	46.260	.000
	Job_Satisfaction	13.037	3	4.346	3.838	.016
Level_Education * Position	Income	2021.926	7	288.847	2.475	.030
	Job_Satisfaction	7.104	7	1.015	.896	.517
Error	Income	5369.024	46	116.718		
	Job_Satisfaction	52.079	46	1.132		
Total	Income	232358.000	60			
	Job_Satisfaction	940.000	60			
Corrected Total	Income	43749.733	59			
	Job_Satisfaction	88.733	59			

a. R Squared = .877 (Adjusted R Squared = .843)

b. R Squared = .413 (Adjusted R Squared = .247)

Figure 5I

Multiple Comparisons

Tukey HSD

Dependent Variable	(I) Level_ Education	(J) Level_ Education	Mean Difference (I-J)	Std. Error	Sig.	95% Confidence Interval	
						Lower Bound	Upper Bound
Income	High School	Bachelor	1.1333	3.94492	.992	-9.3818	11.6485
		Master	-26.9333*	3.94492	.000	-37.4485	-16.4182
		Doctorate	-37.1333*	3.94492	.000	-47.6485	-26.6182
	Bachelor	High School	-1.1333	3.94492	.992	-11.6485	9.3818
		Master	-28.0667*	3.94492	.000	-38.5818	-17.5515
		Doctorate	-38.2667	3.94492	.000	-48.7818	-27.7515
	Master	High School	26.9333*	3.94492	.000	16.4182	37.4485
		Bachelor	28.0667	3.94492	.000	17.5515	38.5818
		Doctorate	-10.2000	3.94492	.060	-20.7152	.3152
	Doctorate	High School	37.1333*	3.94492	.000	26.6182	47.6485
		Bachelor	38.2667*	3.94492	.000	27.7515	48.7818
		Master	10.2000	3.94492	.060	-.3152	20.7152
Job_ Satisfaction	High School	Bachelor	1.2000*	.38853	.017	.1644	2.2356
		Master	.9333	.38853	.091	-.1023	1.9689
		Doctorate	.4000	.38853	.733	-.6356	1.4356
	Bachelor	High School	-1.2000*	.38853	.017	-2.2356	-.1644
		Master	-.2667	.38853	.902	-1.3023	.7689
		Doctorate	-.8000	.38853	.182	-1.8356	.2356
	Master	High School	-.9333	.38853	.091	-1.9689	.1023
		Bachelor	.2667	.38853	.902	-.7689	1.3023
		Doctorate	-.5333	.38853	.523	-1.5689	.5023
	Doctorate	High School	-.4000	.38853	.733	-1.4356	.6356
		Bachelor	.8000	.38853	.182	-.2356	1.8356
		Master	.5333	.38853	.523	-.5023	1.5689

Based on observed means.
The error term is Mean Square(Error) = 1.132.

*. The mean difference is significant at the .05 level.

Figure 5m

Multiple Comparisons

Tukey HSD

Dependent Variable	(I) Position	(J) Position	Mean Difference (I-J)	Std. Error	Sig.	95% Confidence Interval	
						Lower Bound	Upper Bound
Income	worker	skilled worker	-13.3654*	4.03400	.009	-24.1180	-2.6128
		manager	-31.0598*	3.93225	.000	-41.5412	-20.5784
		director	-68.3077*	4.23752	.000	-79.6028	-57.0126
	skilled worker	worker	13.3654*	4.03400	.009	2.6128	24.1180
		manager	-17.6944*	3.71203	.000	-27.5889	-7.8000
		director	-54.9423*	4.03400	.000	-65.6949	-44.1897
	manager	worker	31.0598*	3.93225	.000	20.5784	41.5412
		skilled worker	17.6944*	3.71203	.000	7.8000	27.5889
		director	-37.2479*	3.93225	.000	-47.7293	-26.7665
	director	worker	68.3077*	4.23752	.000	57.0126	79.6028
		skilled worker	54.9423*	4.03400	.000	44.1897	65.6949
		manager	37.2479*	3.93225	.000	26.7665	47.7293
Job_ Satisfaction	worker	skilled worker	-1.1394*	.39730	.031	-2.1984	-.0804
		manager	-.7991	.38728	.180	-1.8314	.2331
		director	-1.3846*	.41734	.009	-2.4970	-.2722
	skilled worker	worker	1.1394*	.39730	.031	.0804	2.1984
		manager	.3403	.36559	.789	-.6342	1.3148
		director	-.2452	.39730	.926	-1.3042	.8138
	manager	worker	.7991	.38728	.180	-.2331	1.8314
		skilled worker	-.3403	.36559	.789	-1.3148	.6342
		director	-.5855	.38728	.439	-1.6178	.4468
	director	worker	1.3846*	.41734	.009	.2722	2.4970
		skilled worker	.2452	.39730	.926	-.8138	1.3042
		manager	.5855	.38728	.439	-.4468	1.6178

Based on observed means.
The error term is Mean Square(Error) = 1.132.

*. The mean difference is significant at the .05 level.

Figure 5m

Step 6. We need to find three critical *F* scores. The critical *F*-score for Level_Education and Position are both 2.20 (Put .95 in a new variable and call it *a* for area, then use Transform Compute Variable, call the target variable cf, click on Inverse DF and double click Idf.F. Use *a*, 6 and 92 as the three parameters.) The critical *F*-score for the interaction between Level_Education and Position is 1.80. (Use *a*, 14 and 92 as the parameters.)

Step 7. For Level_Education the obtained *F*-value 3.011 is greater than the critical *F*-value 2.20, alternatively the *p*-value .010 < .05. For Position the obtained *F*-value 11.094 is greater than the critical F-value 2.20, alternatively the p-value $0 < .05$. For the interaction between factors the obtained F-value is 1.590 is less than1.80, alternatively $.097 > .05$.

Step 8. We reject the null hypothesis that the population mean vectors of the dependent variables are equal in Level of Education. We also reject the null hypothesis the population mean vectors of the dependent variables are equal in Position. We accept the null hypothesis the factors are independent of each other.

Step 9. The Post Hoc Tests for the factor Level of Education shows that Bachelor and High School have the same income and they are different than Master and Doctorate. In Job Satisfaction the only difference lies between High School and Bachelor. The Post Hoc Tests for the factor Position shows all groups are different in Income, but the biggest difference in Job Satisfaction is between worker and director.

Let's review what we've learned in this chapter.

1. Two-Way ANOVA is used when we have two factors of several groups.

2. We have three hypotheses for both the null and alternative hypothesis.
   i.) Null and alternative hypothesis for all the means in factor 1.
   ii.) Null and alternative hypothesis for all the means in factor 2.
   iii.) Null and alternative hypothesis for the interaction between the two factors.

3. Degrees of Freedom, $k_i, i = 1,2$ is the number of groups in factor *i*, *n* is the total number of data values
   i.) df for factor 1 is $k_1 - 1$
   ii.) df for factor 2 is $k_2 - 1$
   iii.) df for interaction is $(k_1 - 1)\cdot(k_2 - 1)$
   iv.) df for the denominator is $n - 1 - (k_1 - 1) - (k_2 - 1) - (k_1 - 1)\cdot(k_2 - 1) = n - k_1 k_2$

4. The boxplots can illustrate if there is a difference in interaction between the two factors.

5. Two-Way MANOVA is used when we have two factors and two dependent variables.

6. We have three hypotheses for both the null and alternative hypothesis.
   i.) The different dependent variable vectors will not differ in factor 1.
   ii.) The different dependent variable vectors will not differ in factor 2.
   iii.) Factor 1 and factor 2 will not interact in the effect on the dependent variables.

7. The degrees of freedom for MANOVA need to be read from the output since their computation is very complex.

Class Work 15

A company wishes to test the effectiveness of its advertising. A product is selected and two types of ads are written; one is serious and one is humorous. Also, one ad is used on television, and one is used on radio. Sixteen participants are selected and assigned randomly to one of four groups. They listen or watch the ad and rate it on a scale, 1 - 20, for effectiveness. At $\alpha = .05$, analyze the data using two way ANOVA. Chpt15classwork.sav

	Radio	Television
Humorous	6, 10, 11, 9	15, 18, 14, 16
Serious	8, 13, 12, 10	19, 20, 13, 17

Homework 15

1. Using the data file BP.sav test if heart rate means while immersing a hand in ice water (hrcp) are no different with regard to a persons sex, or parental hypertension or a combination of these two factors.

2. Using the data file BP.sav test if heart rate means while doing mental arithmetic (hrma) are no different with regard to a persons sex, or taking prescription medicine or a combination of these two factors.

3. Using the data file BP.sav test if height (height) means are no different with regard to a persons sex, or persons race or a combination of these two factors.

4. Using the data file BP.sav test if weight (weight) means are no different with regard to a persons sex, or persons race or a combination of these two factors.

5. Using the data file Pheresis.sav test if platelet yield (platelet) means are no different with regard to type of machine (machine) or high white blood count (highwbc) or a combination of these two factors.

6. Using the data file Pheresis.sav test if platelet yield (platelet) means are no different with regard to low pH (lowph) or Left-Right Split (lr) or a combination of these two factors.

7. Using the data file Pheresis.sav test if platelet yield (platelet) means are no different with regard to type of machine (machine) or low yield (lowyield) or a combination of these two factors.

8. Using the data file Pheresis.sav test if platelet yield (platelet) means are no different with regard to type of machine (machine) or high pH (highph) or a combination of these two factors.

9. Using the data set Student.sav test if the means of GPA are different over the factor gender and over the factor major or a combination of these two factors.

10. Using the data set Student.sav test if the means of GPA are different over the factor eyes and over the factor dog or a combination of these two factors.

11. Using the data set Student.sav test if the means of GPA are different over the factor cigs and over the factor belt or a combination of these two factors.

12. Using the data set Student.sav test if the means of GPA are different over the factor resident and over the factor work category or a combination of these two factors.

13. Explain the differences between two-way ANOVA and one-way ANOVA.

14. Using the data file chpt9example1.sav test if redcomsc means are no different with regard to a person's age, or radio station listening preference or a combination of these two factors.
    Alpha =.10

15. Using the data file chpt15problem15.sav test if the mean population vectors of the dependent variables General Happiness and Respondents Income are the same or different across the factors degree and jobinc.

16. Using the data file chpt15problem15.sav test if the mean population vectors of the dependent variables General Happiness and Respondents Income are the same or different across the factors degree and tvhours.

Just for fun. Negate the following statement attributed to Abraham Lincoln: You can fool some of the people all of the time, and all of the people some of the time, but you can not fool all of the people all of the time.

Project 1

Class Survey

Use the class survey that was filled out in Chapter 1, classsurvey.sav, to answer the following questions. For each question list the eight-step process for hypothesis testing. Use $\alpha = .05$ (5%) for the significance level in each question. To find the class percentages of any of the variables use Analyze → Descriptive Statistics → Frequencies → Charts and click on pie charts. Double-click on the pie chart in the output and in the data editor click on Elements → Show Data Labels. Close the pop up box and the Data Editor. The percent will be listed on the pie chart. Round to the nearest tenth of a percent. You don't need to include any graphics in your report. Our working assumption is this class is a random sample (which it isn't) of the population of all students enrolled in the particular majors represented by this class and all conclusions will be based on that assumption. (Note to instructor: the null hypothesis will change according to the school and the class response. Feel free to adjust the null hypothesis to any question suitable to the data.)

In (fill in year here), the percent of women enrolled in the university/college was (fill in percent)%. Test if the percentage of females represented by this class is greater/less than (fill in percent)%. This is a test of proportions where $t = \dfrac{\hat{p} - p}{\sqrt{\dfrac{pq}{n}}}$, $\sigma = \sqrt{\dfrac{pq}{n}}$.

Find the mean, median, and mode for the student ages in this class. In (fill in year here) the mean age of all students in the university/college was (fill in years) years. Test the hypothesis that average age of students in this class is different from the average age of students in the university/college. Is the distribution for students in this class skewed or normal? Why? Make a histogram and check the skewness value in the frequency function. A measure greater than 1 or less than negative 1 indicates a strong skewness.

The percentage of Asian American students in the university/college in (fill in year here) is (fill in percent)%. Test if the percentage of Asian American students represented by this class is different from (fill in percent)%. This test is similar to the test in problem 1, except this is a two-tailed test.

Find a 95% confidence interval for the mean cost of attending the university/college for one quarter/semester. Based on this estimate, would you say a university/college education is cost-effective given your potential earnings?

Test if there is a correlation between the age of a student and the cost of attending the university/college for one quarter/semester. This is a two-tailed test because we have no idea if the cost goes up with age or down with age. Use Analyze-Correlate-Bivariate, move the two variables into the Variables window, and make sure Pearson Correlation Coefficient is checked. Hit OK. The null hypothesis is there is no correlation; the alternative hypothesis is there is a correlation. Observe the p-value and make a decision. Always use the eight-step process.

Project 2

Activities for Students Using Statistics to Check on Elvis the Dog

Elvis, a Welsh corgi, has captured the attention of the mathematics community because of his owner's claim that this cute dog knows calculus. In the May 2003 issue of the *College Mathematics Journal*, Tim Pennings writes that his dog Elvis chooses a path that is "remarkably" close to the optimal path (in terms of time) when a tennis ball is thrown from the water's edge at a point A to a point B in the water. Elvis runs along the shore from point A to D and then dives into the water and swims to the ball at point B (see Figure 1). Since Elvis can run faster than he can swim, the quickest route to the ball is different from the shortest route to the ball (swimming from A to B). The quickest route to fetch the ball involves a combination of running along the shore and swimming in the water.

The use of calculus allows mathematicians to determine optimal paths. Thus Pennings asks if Elvis knows calculus, or at least has an intuitive sense for the path that generates the shortest time. Since Pennings did not use statistics to back up his claim that Elvis did indeed choose the path with least time, then his claim is open to interpretation by what one may mean by "remarkably close." This project will test the hypothesis that Elvis chose the path with least time against the alternative hypothesis that he did not choose the path with least time.

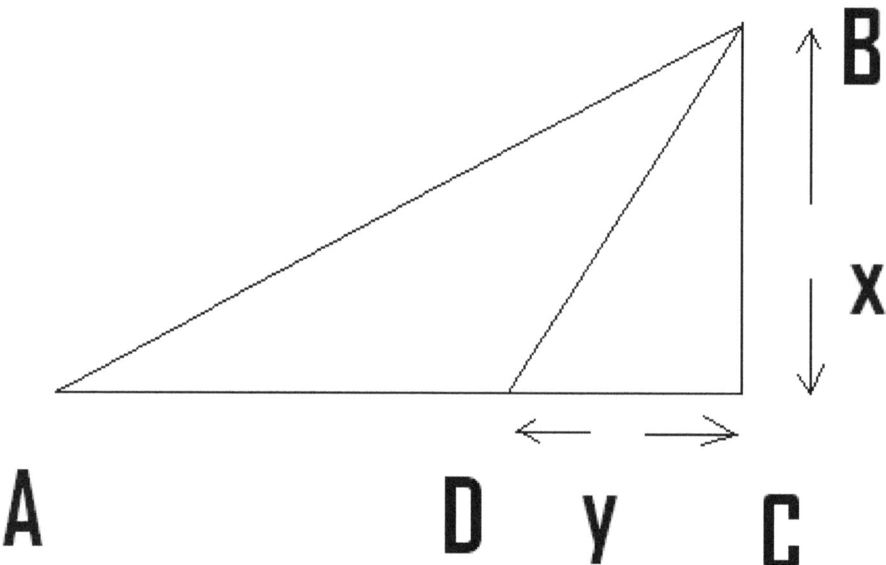

With the help of a friend, Pennings determined how fast Elvis can run along the sand and how fast he can swim. Next, Pennings gathered data to test his claim that Elvis was choosing the optimal path to the ball. Elvis and Pennings spent three hours gathering 35 pieces of data (see Table 1). Based on Elvis's speed and some problem solving in calculus, Pennings determined that if Elvis truly were choosing the optimal path to the ball, the data should fit the line $y = 0.144x$. (All units are in meters.) Pennings did not do any statistics to test his claim that Elvis was able to choose the optimal path. He writes, "It seems clear that in most cases Elvis chose a path that agreed remarkably close with the optimal path." The goal of this project is to do the statistical analysis of the data to determine whether or not the dog Elvis chose the optimal path from the shore to the tennis ball in the water.

Table 1
35 Throw-and-Fetch Trials

x	*y*	x	y	x	y	x	y	x	y
10.5	2.0	17.0	2.1	4.7	0.9	10.9	2.2	15.3	2.3
7.2	1.0	15.6	3.9	11.6	2.2	11.2	1.3	11.8	2.2
10.3	1.8	6.6	1.0	11.5	1.8	15.0	3.8	7.5	1.4
11.7	1.5	14.0	2.6	9.2	1.7	14.5	1.9	11.5	2.1
12.2	2.3	13.4	1.5	13.5	1.8	6.0	0.9	12.7	2.3
19.2	4.2	6.5	1.0	14.2	1.9	14.5	2.0	6.6	0.8
11.4	1.3	11.8	2.4	14.2	2.5	12.5	1.5	15.3	3.3

Sheet 1 Activities

1. Using Table 1, enter the x- and y-values into an SPSS data set. Make a scatter plot of the data and print it. Describe the shape of the scatter plot. What does the shape of the scatter plot tell us about the relationship between x and y?

2. Calculate the least squares regression line and graph it on your scatter plot. Click on the graph and a chart editor will appear. Go to Elements and click on Fit Line at Total. Close the chart editor and the line will be drawn on the graph. Go to Analyze → Regression → Linear to find the coefficients for the regression line; be sure to use the x-variable for the independent variable. State the least squares regression line equation in your project.

3. From the chart editor add the line $y = 0.144x$ (use a different color). Compare this line with the regression line. Does it appear that the two lines are very close? Do you think Elvis found the fastest route to the tennis ball? Explain.

4. Pennings claims that four of the data points in the upper right-hand corner of the scatter plot are outliers. Take these four data points out of the data set and recalculate the least squares line.
   a.) State the four outliers:
   b.) Calculate the new least squares regression line equation:

5. After the four outliers are removed, redraw the best-fit line and the reference line (use different colors). Now reconsider whether you think Elvis did a good job or not in finding the fastest route. Explain.

Sheet 2 Activities

1. State the null and alternative hypotheses regarding Elvis and the best-fit line. Calculate the Pearson Correlation Coefficient and check if it passes a two-tailed test at the 5% level. Check for both cases above; that is, check for the line with the outliers and the line without the outliers. Write your conclusions.

Sheet 3 Activities

According to Penning's calculations, if Elvis chose the optimal path from the shore to the tennis ball in the water, the data should follow the line $y = 0.144x$. So $\frac{y}{x} = 0.144$. In this activity you will calculate all the y/x values, find the mean, and test to see if the difference between the mean and the expected value is statistically significant.

1. Use transform to create a new variable and give it the value y/x, then find the mean.

2. State the null and alternative hypotheses to test if Elvis found the optimal path from the shore to the tennis ball in the water. Explain whether you should use a one-sided or a two-sided test.

3. Use a *t*-test to compare this mean with 0.144 to check if it's statistically significant at the 5% level of significance. Go to Analyze → Compare Means → One Sample *t*-test and input 0.144 in the test value. List the *t*-value and the *p*-value. Are your results statistically significant? Explain your answer.

4. Remove the four outliers referred to in sheet 1 and repeat #3. With the outliers removed, check again to see if the *t*-test is statistically significant at the 1% level. Can we conclude Elvis the dog understands calculus?

This is the end of the project. I hope you found this project educational and entertaining and also that you have learned many useful techniques you can now apply in doing research analysis for a company that employs statisticians. It's been a pleasure being your instructor for the quarter/semester.

INDEX

E

F

G

H

I

K

L

M

N

O

P

Q

R

S

T

V

W

Printed by Libri Plureos GmbH in Hamburg,
Germany